1001 THINGS EVERYONE SHOULD
KNOW ABOUT THE UNIVERSE

Produced by The Philip Lief Group, Inc.
Sketches by Brian Sullivan

*D*OUBLEDAY

NEW YORK ✦ TORONTO ✦ LONDON ✦ SYDNEY ✦ AUCKLAND

1001 THINGS

EVERYONE SHOULD KNOW

ABOUT THE UNIVERSE

WILLIAM A. GUTSCH, JR., Ph.D.

Former Chairman of the American Museum—Hayden Planetarium, New York

PUBLISHED BY DOUBLEDAY

a division of Bantam Doubleday Dell Publishing Group, Inc.
1540 Broadway, New York, New York 10036

DOUBLEDAY and the portrayal of an anchor with a dolphin
are trademarks of Doubleday, a division of Bantam
Doubleday Dell Publishing Group, Inc.

Book design by Maria Carella

Library of Congress Cataloging-in-Publication Data

Gutsch, William A.
1001 things everyone should know about the universe /
William A. Gutsch; produced by The Philip Lief Group,
Inc.
p.   cm.
1. Astronomy—Miscellanea.   I. Philip Lief Group.
II. Title.
QB52.G98   1998
520—dc21   96-52024
CIP
ISBN 0-385-48223-X

January 1998

First Edition

1   3   5   7   9   10   8   6   4   2

TO MY MOM AND DAD

who gave me my first telescope and never stopped
encouraging me to dream of the stars

ACKNOWLEDGMENTS

I'd like to thank Jim Trefil for writing *1001 Things Everyone Should Know About Science* (without whom this book probably never would have been written) and Judy Linden, Executive Editor at The Philip Lief Group, for asking me to be the one who would try to follow in Jim's footsteps. Thanks for believing in me, encouraging me, and gently prodding me along. Special thanks also to Brian Sullivan for creating the pen-and-ink sketches that will go a long way toward helping to make my words make sense and to editorial assistants Alexia Meyers and Diane Rhodes for their numerous letters and phone calls that helped secure some of the best images of the universe available today.

CONTENTS

INTRODUCTION   XI

1 ✦ ASTRONOMY, ASTRONOMERS, AND THE TOOLS
OF THE TRADE   1

2 ✦ A BRIEF TRIP BACK IN TIME:
ON THE SHOULDERS OF GIANTS   35

3 ✦ THE MOON: OUR NEAREST NEIGHBOR   43

4 ✦ THE SUN: LIGHT OF OUR LIVES   63

5 ✦ THE INNER SOLAR SYSTEM:
LAND OF THE ROCKY MIDGETS   87

6 ✦ THE OUTER SOLAR SYSTEM:
REALM OF THE GIANTS   109

7 ✦ ASTEROIDS, COMETS,
METEOROIDS, AND SPACE DUST   143

8 ✦ STARRY, STARRY NIGHTS:
BEYOND THE SOLAR SYSTEM   163

9 ✧ STAR SECRETS   177

10 ✧ THE LIVES OF THE STARS   193

11 ✧ STELLAR GERIATRICS: SUPERNOVAE,
BLACK HOLES, AND MORE   223

12 ✧ OUR HOME STAR CITY:
THE MILKY WAY   255

13 ✧ TO GALAXIES BEYOND:
ISLANDS IN TIME AND SPACE   275

14 ✧ COSMOLOGY: FROM THE END OF
THE UNIVERSE TO THE DAWN OF TIME   293

15 ✧ THE SEARCH FOR EXTRATERRESTRIAL
LIFE: ARE WE ALONE?   325

INDEX ✧ 341

I was honored and delighted to have been asked to write this complementary work based on my old friend Jim Trefil's wonderful book *1001 Things Everyone Should Know About Science*. I hope my effort measures up. Like Jim's book, this one is designed to be informative and fun. Some entries are simply "amazing facts" like . . . there's a canyon on Mars that would stretch clear across the United States or black holes bend light. Other entries are intended to explain how astronomers know such seemingly unknowable things as what stars are made of or the distance to a remote galaxy. Astronomy is marvelous detective work.

In general, the book is arranged in such a way that it starts close to home (in our own solar system) and then moves out to the stars and beyond. But like Jim's book, you should feel free to browse and ramble about this one, reading whatever strikes your fancy. Entry number 1 isn't necessarily more important than entry 500 or 1001. From the perspective of professional astronomy, some items are more profound or significant than others, for they go beyond individual facts and aim at a greater understanding of a general principle. Others are somewhat isolated bits of information but were chosen based on what years of experience have taught me people ask the most about.

To help add "thousands of words" to my words, we have also included over 150 photos and sketches. Some images are from the best observatories on Earth—the result of the great talents of astronomers like the Anglo-Australian Observatory's David Malin. Others are from top NASA space probes and the magnificent Hubble Space Telescope. All are designed to better enable you to visualize this amazing universe in which we live and the instruments we use to find out the things we do.

Why were exactly 1001 items chosen? If the truth be known, this is a publishing thing, not a science thing. Are there really 1001 things everyone should know about the cosmos? For some, there might be 10,000 things, all of which would be guaranteed to spark interest or be judged among the essential pieces of information no such book should be without. For others, a salient 50 or 60 items might well suffice. How many things about the cosmos do you really need to know to get into heaven? I'm not sure. Introductions are usually places where authors wax heavy and philosophic. I will simply say that in an age when more people are better informed about the top ten songs of the week or the latest rumors about the sexual exploits of Hollywood celebrities than they are about the incredible mind-boggling universe in which we live, I'm glad you decided to use some of your gray matter to learn about the latter.

Enjoy.

—BILL GUTSCH
Kinnelon, New Jersey
July 1996

1001 THINGS EVERYONE SHOULD
KNOW ABOUT THE UNIVERSE

CHAPTER 1

ASTRONOMY, ASTRONOMERS, AND THE TOOLS OF THE TRADE

1 **Astronomy is . . .** the branch of science dedicated to the study of everything in the universe (except the Earth). Astronomers, however, do study interactions between the Sun and the upper reaches of the Earth's atmosphere, including a celestial phenomenon known as the *aurora*.

2 **Most modern astronomers are really astrophysicists.** Before the late 1800s, astronomy was mostly descriptive or mathematical. Astronomers sketched or took pictures of objects in their telescopes and calculated such things as eclipses, the positions of the planets, or the positions of stars and the distances to them. Astronomers, however, lacked a real understanding of the physical properties of stars or the processes that governed why they shine and how they evolve. Since then, breakthroughs in our understanding of the atom and how matter interacts with energy have allowed astronomers to discover the intricate inner workings of the universe through the application of a wide range of the laws of physics. Thus, today most astronomers really do astrophysics and are astrophysicists. And the title usually impresses more people at cocktail parties.

3 **Astronomers are generally divided into observational astronomers and theoreticians.** While some astronomers work both sides of the fence, most fit into either one camp or the other. Although observational astronomers don't necessarily spend all or most of their time in observatories looking through telescopes, they are involved in the use and frequently the design of telescopes and instrumentation (such as cameras, photometers, and spectrographs) used to obtain and analyze data from objects in space. In contrast, theoreticians typically use high-speed computers to build mathematical models of phenomena in space.

4 **The work of observational astronomers and theoreticians frequently complements each other.** At times, observational astronomers may discover some new and unexplained phenomenon in space and theoreticians will attempt to use mathematics and the known laws of physics to come up with an explanation of what has been observed. At other times, theoreticians will develop a theory that predicts the existence of certain phenomena or physical conditions in space and observational astronomers will try to make observations to see whether or not the theory is correct or the object or condition, in fact, exists. An example of the former is the discovery of the first pulsars and the subsequent development of the theory of neutron stars. An example of the latter is the postulation of the theoretical existence of black holes followed by their actual discovery.

5 **For the most part, studying the universe is a frustratingly passive activity.** The physicist, the chemist, and the biologist all have one thing in common: they can all go into the laboratory or the field and effectively "muck about" with the phenomenon they are studying. They can touch it, manipulate it, work in direct contact with it. Ask a physicist how much a substance weighs and he can put it on a scale and immediately find out. Ask a chemist how much heat is given off in a particular chemical reaction and she can place a thermometer in a sample to find out. Ask a biologist whether or not a blood sample carries a particular genetic characteristic and he can immediately subject the blood to a series of carefully controlled tests. For the astronomer, the universe itself is the laboratory, but the universe, by definition, is "out there," far from our direct contact. The astronomer may wish to know the distance to a star, but she cannot run a tape measure out to that star to

find out the answer. An astronomer might want to know the temperature of the surface of the Sun, yet she cannot travel to the Sun and stick in a thermometer. An astronomer may wish to know what a distant galaxy is made of, but he cannot go there to gather samples and return them to Earth for analyses. And yet we do know the distances to the stars, the temperature of the Sun, and the chemical composition of distant galaxies. And that is precisely why astronomy is such a fascinating piece of detective work and such a tribute to the creativity and ingenuity of the human mind.

6 **Astronomers learn about the universe by collecting and analyzing light and other forms of radiation that come to us from objects in space.** Astronomers can't go to most of the universe to study planets, stars, and galaxies. Instead, they sit here and learn about the universe through information sent by *it* to *us*. The messengers that carry the information our way are beams of light and other forms of radiation. Thus, astronomers primarily study objects in space (which are made of matter), but do so from a distance by means of the radiation that these objects give off. We'll talk about radiation as we go along, and you'll also find a section on *matter* at the end of this chapter (see page 29).

7 **An optical telescope is simply a device that focuses light and thereby allows us to see and study fainter things than we can with our eyes or instruments alone.** The principle behind all telescopes is essentially the same. Light entering the telescope is concentrated, or focused, into a narrower and narrower beam by means of a system of lenses or mirrors or both. Since light and other forms of radiation are the means by which observational astronomers learn about the universe, the more radiation that can be gathered, the more that can be learned.

8 **There are two basic types of optical telescopes.** Most telescopes are either *refracting* telescopes (also simply called *refractors*) or *reflecting* telescopes (also called *reflectors*).

The 36-inch refracting telescope at the Lick Observatory. The telescope's main lens is on the far end of the tube. The astronomer is standing next to the "business end" of the telescope where cameras or other instruments would be attached. (UCO/Lick Observatory photo/image)

9 **Refractors use a system of lenses to focus light.** When most of us were kids, we had the experience of burning a leaf or a piece of paper with a magnifying glass on a sunny day. The "experiment" worked because the lens of the magnifying glass took all the heat that fell on its surface and focused it down, concentrating it into a point. In the process, the temperature at this point increased until it was higher than the kindling temperature of the leaf or paper. At the same time, it was even more obvious that the lens was also focusing the sunlight into a point. A refracting telescope uses lenses to accomplish the same thing. At the large end of a refracting telescope there are typically two lenses of equal size made of different types of glass. When light passes through them, they work together to bring the light to a focus at the other end of the telescope tube. At this point, an image is formed of whatever the telescope is pointed toward.

10 **Reflecting telescopes accomplish the same thing by using one or more mirrors.** In a simple reflecting telescope, light from a distant object is allowed to fall on a mirror. This mirror is not flat, however, but "bowl-shaped" or concave. The result is that the light that reflects off the mirror is also brought to a focus. A specific shape known as a *paraboloid,* commonly used in many reflecting telescopes, results in all the light conveniently coming to a focus at the

A reflecting telescope. An astronomer can be seen in the observer's cage high above the main mirror.
(UCO/Lick Observatory photo/image)

same point. As with the refracting telescope, an image of the distant object is formed at this *focal point*.

11 **A simple and common reflecting telescope design enjoyed by many amateur astronomers was invented by Sir Isaac Newton.** The design, known to this day as the *Newtonian reflector,* uses a concave parabolic mirror at one end of an open tube to bring light from a distant object to a focus. For the convenience of the observer, a small flat mirror is suspended inside the tube near the other end to catch the focusing light and redirect it to a hole in the side of the telescope tube where an eyepiece is inserted.

Many amateur astronomers have telescopes of this design, which can be easily made or purchased from catalogs.

12 **Inch for inch, refracting telescopes are a lot more expensive than reflecting telescopes.** An average 6-inch reflecting telescope, for example, can cost several hundred dollars, while a 6-inch refracting telescope will cost several thousand dollars. The reason is that, inch for inch, it is much cheaper to grind a mirror of astronomical quality than a system of lenses.

13 **For amateur astronomers who need portability, both refractors and Newtonian reflectors can get pretty cumbersome.** A typical 10-inch Newtonian reflector can be 6 to 7 feet long and weigh well over 100 pounds, while a 6-inch refractor can have a tube just as long. Clearly, unless you have a permanent place to house such an instrument, you may be faced with significant transportation problems.

14 **An alternative telescope design known as the *Schmidt-Cassegrain system* offers an interesting advantage.** It uses a combination of both lenses and mirrors. Inch for inch, Schmidt-Cassegrain telescopes are more expensive than Newtonian reflectors but are still significantly cheaper than pure refractors and have the advantage of producing similar powers to a Newtonian refractor, while having a tube that is only about a third as long. Thus, Schmidt-Cassegrain telescopes are more portable and can be accommodated in a smaller and, therefore, cheaper backyard observatory. Because they are shorter, they also vibrate less if there is a wind. This is important, since a telescope magnifies so much that even relatively gentle breezes can result in small vibrations of the telescope tube that, in turn, can be seen as large shakes in the telescope's image.

15 **We see things down to a certain level of faintness based on how much light can get into our eyes to be focused.** We are able to see because light enters our eyes through our pupils and is focused by the lenses in our eyes onto our retinas, from which signals are sent to the brain. The more light that enters our eyes, the more light falls on our retinas, the more signals are sent to the brain, and the brighter the object appears to be. When first entering a darkened room or the outdoors at night after having been in a brightly lit environment, we tend to be able to see very little. But "as our eyes adjust," we are able to see better in the dark up to a point. What, of course, is happening as our eyes adjust is that the pupils of our eyes are gradually growing larger and thus allowing more light to get through. There is a limit, however, to how well we can see in the dark because there is a limit to how large our pupils can grow and hence how much light can enter our eyes.

16 **Telescopes allow us to see fainter things than we can with our eyes alone because they allow more light to get into our eyes.** Even under the darkest conditions, the pupils of the average human being cannot expand to more than about a quarter of an inch across. So we can only see things down to a certain level of faintness in proportion to how much light can get through those quarter-inch-wide openings. But telescopes allow us to "cheat nature" by focusing down a lot of light into a beam small enough to fit through the pupil of our eye. Look at a starry sky with your unaided eyes and you are looking at the sky with pupils that are a quarter of an inch across. But look at a starry sky through a telescope with a lens or mirror 100 inches across and you are looking at the sky with the equivalent of pupils that are 100 inches across! No wonder telescopes allow us to see things in space that are far too faint to see with our eyes alone. Understand this basic principle and you understand the "magical power" of telescopes to reveal a hitherto unknown universe to us. As we'll see, professional astronomers don't typically look through their telescopes but instead use instruments that are far better and more objective than the human eye. But the point is the same.

17 **Astronomers tend to speak of a telescope in terms of the *diameter* of its main lens or mirror.** Astronomers tend to refer to a particular telescope as "the 36-inch" or "the 2.4-meter" telescope. When doing so, they are referring to the diameter of the telescope's main lens or mirror in inches or meters. The main lens or mirror of a telescope is called its *objective.*

---

18 The ability of a telescope to show us things that are fainter and farther away than we can see with our eyes alone depends on the *area* of its main lens or mirror. While astronomers refer to telescopes in terms of the diameter of their objective lens or mirror, the capacity of a telescope to focus light is proportional to the objective's area and not its diameter. Since the area of a circle is equal to 3.1416 times the radius of the circle squared, this means that a 10-inch telescope, for example, actually focuses *four* times as much light as a 5-inch telescope. A telescope's capacity for gathering light is sometimes referred to as its light-gathering *power,* but this has nothing to do with a telescope's magnification.

19 To magnify an image in a telescope, you need an eyepiece. Most telescopes bought by amateur astronomers and hobbyists come with an assortment of eyepieces. Each eyepiece is typically a small cylinder that contains a system of lenses. It is an eyepiece that you place your eye to and look through in viewing something through the telescope. Different eyepieces yield different powers or magnifications.

20 To figure out how much power a particular telescope gets with a particular eyepiece, you have to understand *focal length.* Every telescope objective and eyepiece has what is called a focal length. This is simply a distance, usually measured in millimeters (mm). (There are 25.4 millimeters in 1 inch.) If you ever burned a leaf or a piece of paper with a magnifying glass, the distance between the lens of the magnifying glass and the burning object was the focal length of the lens. In other words, it's the distance between the lens and the point where all the light and heat from a distant object (here it's the Sun) come to a focus. The focal length of each eyepiece is typically written on the side or end of the eyepiece barrel and the focal length of the telescope's main lens or mirror is always contained in the literature that comes with the telescope.

21 To calculate power, all you need to do is a simple division. To calculate the power you will get when you insert a particular eyepiece into a particular telescope, all you have to do is divide the focal length of the objective of the telescope by the focal length of the eyepiece. For example, if the focal length of a particular telescope is 2,540 mm (100 inches) and you insert into the telescope an eyepiece that has a focal length of 25.4 mm (1 inch), you will get 100 power. This, in turn, means that when you look through the telescope, you will see things 100 times bigger or closer than with the naked eye.

22 Theoretically, you can get any power with any telescope. To get higher and higher powers with a particular telescope, all you have to do is use eyepieces of shorter and shorter focal length. Thus, if an eyepiece with a focal length of 25.4 mm yields 100 power, then an eyepiece of half that focal length, namely 12.7 mm, will give 200 power on the same telescope; an eyepiece of 6.35 mm focal length will yield 400 power; and so forth. Theoretically, you could continue in this way up to a million power or more. But there's a problem with this argument, which is . . .

23 **There's power and then there's *useful* power in telescopes.** It must be remembered that all an eyepiece is really doing is magnifying an image of a distant object created in the telescope by the focusing of light from the telescope's main lens or mirror. All the eyepiece has to work with is the image created by the focusing of this light and there is a limit to how much it can effectively do with a certain amount of light. In short, the more light that the eyepiece has to work with, the more it can magnify it and still produce a reasonably bright and clear image on the retina of your eye. In other words, with any particular telescope there is a practical limit to how much you can magnify images and still see something that's reasonably clear and bright. Go beyond this limit and you come up against a law of diminishing returns. At higher and higher powers, the images you get will indeed be larger and larger, but they will also become increasingly fuzzy and dark and so, in reality, you will see less detail rather than more. So a far more important

question than "How much power can you get with this telescope?" is "What's the highest *useful* power you can get with this telescope?"

24 **The amount of useful power you can get with a particular telescope depends on the size of its main lens or mirror.** While the maximum amount of useful power you can get with a particular telescope will depend on a number of things, including the quality of the telescope's optics and the steadiness of the Earth's atmosphere on a particular night, a good general rule of thumb can be loosely applied. To get the approximately maximum useful power you should try to use with a particular telescope, take the diameter of the telescope's main lens or mirror in inches and multiply it by 40. Thus, with a 3-inch telescope, the maximum useful power you can hope to use on most nights will be about $3 \times 40 = 120$ power (also written 120 X), whereas, with a 6-inch telescope on the same night, you would get an equally clear and bright image at $6 \times 40 = 240$ power. For this reason, it usually is worth getting the telescope with the largest objective lens or mirror that you can.

25 **Sometimes it's wise to use a lower power rather than the highest power that you can.** Lower-power eyepieces will yield smaller images but will also give you an image that is typically sharper and brighter. In many cases, this can actually be more appealing to the eye. Also, for certain objects that are quite large, such as some star clusters, comets, and the Moon, a wide-angle, low-power eyepiece can give a much more pleasing view.

---

26 **Binoculars can be a very satisfying tool for enjoying the simple pleasures of the sky.** In keeping with the "low power can be more" theme for a moment, binoculars can be an affordable alternative to looking at the sky through a telescope. While most binoculars will not be able to show you the detail on the Moon or planets that a telescope can, they are wonderful for just lying back and sweeping the sky. Indeed, equipped with binoculars, you can enjoy many pleasurable moments, "cruising" along the Milky Way looking for some of the nebulae and star clusters you will read about elsewhere in this book, as well as looking at double stars, lunar eclipses, and occasional comets.

---

27 **The numbers on a pair of binoculars tell how big they are and what power they deliver.** Binoculars are typically described by a pair of numbers with the symbol × in between, such as 7 × 35 or 10 × 50. In each case, the first of the two numbers indicates the power or magnification of the binoculars, while the second number gives the diameter of the binocular's main lenses in millimeters (mm). Since there are approximately 25 millimeters in 1 inch, a pair of 10 × 50 binoculars have objective lenses that are

each 50 mm or about 2 inches across and produce a magnification of 10.

28 **7 × 50 binoculars are a good choice for nighttime use.** Many people feel that 7 × 50 binoculars offer more light-gathering power than a pair of 7 × 35 binoculars (which are fine for daytime sporting events) but aren't as heavy or cumbersome as higher-power binoculars. Higher-power, larger-aperture binoculars can give spectacular views of the Milky Way but are best used with a tripod to help steady them and support their weight.

29 **Most better-quality refracting telescopes and binoculars come with coated lenses.** These chemical coatings generally give the lenses a bluish cast and help reduce internal reflections that can produce "ghost images" within the instrument.

30 **While amateur astronomers can usually tell you the power they are currently using on their telescopes, professional astronomers usually don't think in these terms.** Magnifying power is something that professional astronomers usually are not very concerned about. This is because professional astronomers usually eliminate the eyepiece from their telescopes and use the rest of the telescope's optics to focus the light of whatever they wish to study onto a CCD, or charge-coupled device, that is being used as a camera or as part of a photometer or spectrograph. In this way, a professional astronomer is interested in things like the size of the image or the degree of detail that can be seen, as well as the wavelengths or colors of light that are reaching the CCD.

31 Instead of *magnifying* power, professional astronomers are typically more interested in a telescope's *resolving* power. The resolving power of a telescope is a bit of a misnomer, for the term doesn't really refer to power in the same sense as magnifying power does. Instead, resolving power might better be termed resolving *capability.* As such, it refers to the *fineness of detail* a particular telescope can theoretically allow you to see. This fineness of detail can be thought of, in effect, as how small an object can be seen or, more commonly, how close two objects (such as stars) can be to each other and still be seen, or resolved, into separate objects. The resolving power of a telescope is typically given in seconds of arc. (A circle is divided into 60 equal parts called *degrees.* Each degree is further divided into 60 *minutes* of arc. And, in turn, each minute of arc is divided into 60 *seconds* of arc.)

32 The theoretical resolving power of a telescope is an easy thing to figure out. The theoretical resolving power of an optical telescope in seconds of arc can be simply calculated by dividing the number 13 by the diameter of the telescope's main lens or mirror in centimeters (cm). (There are 2.54 centimeters in 1 inch). Thus, a 100-inch (254-cm) telescope has a theoretical resolving power of about 0.05 seconds of arc, while a

200-inch telescope has theoretical resolving power of about 0.025 seconds of arc. (That's only one thirty-six-thousandths the diameter of the full Moon.) In other words, under ideal conditions, the latter telescope could tell that a pair of stars only 0.025 seconds of arc apart in the sky were actually two stars, while the 100-inch telescope would see them as only a single star. Sharper images are better images, so astronomers typically want the best resolving power they can get. This is another reason why astronomers typically covet a telescope with as large a main lens or mirror as possible.

33 Hello, AAA? Please send me a map of the universe. Just as there are maps of Texas and maps of Afghanistan, so too are there maps of the sky. In the old days, they were painstakingly drawn by hand, but today astronomers rely more on actual photographs or computer images of the sky. One of the most extensive atlases of such photos and images consists of a combination of the *Palomar Observatory Sky Survey* taken in California and the *Southern Sky Survey* taken at the European Southern Observatory in Chile. Hundreds of images, each more than a foot on a side, show the entire sky down to at least magnitude + 20°. Another extensive map of the heavens is the *Guide Star Catalogue* compiled for the Hubble Space Telescope. It contains over 15 million stars down to magnitude + 15° and is only available on an extensive set of CD-ROMs. Before traveling to an observatory, an astronomer might typically take some Polaroid snapshots of the stars around

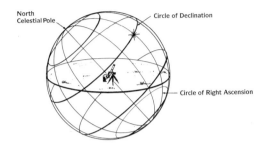

North Celestial Pole

Circle of Declination

Circle of Right Ascension

The astronomer uses a coordinate system similar to longitude and latitude. The celestial equivalents are called right ascension and declination.

his or her appropriate targets to serve as mini road maps to guide the astronomer to where he or she needs to look.

34 **Astronomers pinpoint objects in the sky by using a system similar to latitude and longitude.** Just as the location of any place or object on Earth can be designated by its latitude and longitude, so the location of any object in the sky can be designated by means of a similar coordinate system in which the term *declination* takes the place of latitude and *right ascension* takes the place of longitude.

35 **Declination is measured in degrees.** In the celestial coordinate system, a great circle in the heavens called the *celestial equator* lies directly above the equator on Earth. Just as with latitude, if an object in the sky is located north of the celestial equator, it is said to have a positive or north declination.

Similarly, objects found south of the celestial equator in the sky are given a negative or south declination. The distance north or south is given in degrees, minutes, and seconds of arc (again, just like with latitude). (In the photo on page 5, you can see a great circle on the telescope used to mark declination in degrees.)

36 **Right ascension is measured in units of time.** The right ascension coordinate is measured eastward around the sky, like longitude, and has a kind of zero point, just like the prime meridian that runs through Greenwich, England. The prime meridian in the sky is known as the *vernal equinox* and the right ascension of any celestial object is measured in how long it takes the sky to rotate around from when this celestial prime meridian is due south to when the object in question is due south. Thus, the right ascension of a celestial object is measured in hours, minutes, and seconds of time.

37 **Charts of the heavens frequently include the coordinates of the celestial objects they contain.** Just as an atlas of the Earth typically shows the longitude and latitude range of the map along its edges, so a chart of the sky frequently shows the right ascension and declination of the region it depicts. Tables and catalogs of celestial objects also usually list each object's coordinates. Right ascension is usually abbreviated *R.A.;* declination is written *Dec.* Thus, for example, the brilliant winter star Sirius may be found in the sky at R.A. 6 hours, 14 minutes, Dec. –16 degrees, 35 minutes, while the bright summer star Vega lies at R.A. 18 hours, 34 minutes, Dec. +38 degrees, 41 minutes. These coordinates define the positions of these stars in the sky as easily and accurately as the appropriate longitudes and latitudes would pinpoint a city like Los Angeles or a ship at sea on Earth.

38 **The celestial coordinates of objects that move around relative to the stars change continuously.** Because the Sun, Moon, and planets continuously move around in the sky relative to the stars, their right ascensions and declinations are continuously changing as well. Thus, tables that list their locations must do so for every night. In the case of objects whose apparent motion is quite large, such as the Moon, a different set of coordinates is sometimes listed for each hour.

39 **Why do astronomers need such a coordinate system? Can't they just point their telescopes where they want to look, like you point a pair of binoculars?** There are many good reasons why such a coordinate system is necessary. First, most large professional telescopes weigh many tons and are not easy to swing around. Second, telescopes are typically housed in observatories that only allow visual access to a rather narrow strip of the sky, so astronomers are usually flying pretty blind. Third, most objects that astronomers target for study are far too faint to be seen with the naked eye. And fourth, if

an astronomer in Germany wants to tell an astronomer in Chile where to point his telescope to find an object of interest, she can't just say, "Point your telescope *over there.*" It isn't very meaningful.

40 **Many telescopes are run with the assistance of computers, into which can be entered the right ascension and declination of the object the astronomer wants to study.** All professional telescopes and even some scopes used by amateurs are run with computers that automatically move and point the instruments to the appropriate celestial coordinates. In recent years, some amateur telescopes even come equipped with computers whose prepackaged memory contains the coordinates of the planets as well as all of the brighter stars and lots of other goodies, including the best-looking star clusters, nebulae, and galaxies. Just punch in the name of what you want to look at, press a button, and the telescope finds it for you.

41 **Astronomers hate twinkling stars.** A sky full of twinkling stars can be a very romantic sight, but, ironically, it's one that astronomers usually dread. That's because when stars twinkle they are telling us something bad about the state of the Earth's atmosphere. A telescope can produce nice clear images of objects in space only when the Earth's atmosphere is nice and steady. Sometimes, however, the Earth's atmosphere can be very unstable, meaning that it contains numerous currents of air moving up and down. Looking through the atmosphere at such times is like trying to see something clearly at the bottom of a rapidly moving stream. Objects on the streambed appear to rapidly ripple and distort in the turbulent water. In the same way, turbulent air bends and distorts rays of light from objects in space that pass through it. To the naked eye, such unstable air makes stars appear to twinkle. Telescopes compound the problem because in magnifying images of objects in space, they also magnify any distortions introduced by the atmosphere, smearing the images of stars into large blobs of light that can shimmer wildly and change shape from moment to moment. Astronomers refer to nights when the atmosphere is unstable as nights of "bad seeing." Thus, the actual resolving power of a particular telescope on a particular night may depend far more on the state of the atmosphere than the size of the telescope.

42 **Astronomers understandably try to build observatories in places where there is "good seeing" as much of the time as possible.** One of the major considerations in the search for the site for a new observatory is how consistently a particular location has steady atmospheric conditions or "good seeing." Such locations have typically been found on the tops of reasonably tall mountains where the prevailing winds approach the mountain over level terrain or miles and miles of ocean. Such flat topography produces a flow of air that remains smooth and level with as little vertical motion as possible.

Thus, for example, Kitt Peak National Observatory is located on a mountaintop overlooking miles of relatively flat Arizona desert. Some of the best "observatory real estate" in the world is located on the top of the extinct volcano known as Mauna Kea in Hawaii and on a variety of peaks in the Chilean Andes, all of which are bounded on their windward sides by thousands of miles of ocean. Yet even in such ideal locations, the actual resolving power or "seeing" achieved by the big telescopes rarely exceeds about 1 second of arc.

43 **In seeking out places to build their observatories, astronomers also try to find places that are as clear as possible.** Understandably, astronomers not only want steady atmospheric conditions for their work, they also want skies that are as clear as possible. This, of course, means as many cloudless nights as possible in the course of the year.

Some of the telescope facilities of the Cerro Tololo Inter-American Observatory in the Andes Mountains of Chile. High desert locations, such as this one, are ideal for doing modern optical astronomy. (NOAO)

Parts of Hawaii are covered by rain forests, but at over 13,000 feet, the summit of Mauna Kea is so high that, except for an occasional snowstorm, it lies *above* the weather. The Chilean observatories look out on a high-altitude desert so dry that years can go by without a single drop of rain.

44 **Another important factor in picking a spot for an observatory is an absence of pollution.** This again may seem pretty obvious, but when it comes to pollution, optical astronomers are concerned with more than just unwanted chemicals in the air. They're also concerned about another form of pollution most people don't think about at all: *light* pollution. Stray light from cities and cars can shine up into the sky and wash out the light of faint stars and galaxies, making many kinds of astronomical research virtually impossible from all except extremely rural lo-

A satellite image of North America at night. The challenge of placing optical observatories far from city lights is becoming greater and greater. (USAF and NOAO)

cations. Places such as Mount Wilson and Mount Palomar, which were major astronomical research sites in California earlier in this century, have become increasingly unusable because of the spread of the light pollution from the Los Angeles and San Diego metropolitan areas and even Kitt Peak is increasingly threatened by the ever-swelling population of Tucson. Astronomers have moved on to more remote peaks in places like Hawaii and Chile.

45 **Communities can help reduce light pollution.** Without reducing the amount of lighting necessary to maintain safe illumination levels on streets and highways at night, states and communities can take simple and inexpensive measures to significantly reduce the amount of light pollution they create. Just placing shields over streetlights and using different lights to illuminate our highways can help bring back beautiful dark starry skies that some see as not only critical for astronomical research but also an increasingly rare "natural resource." To learn about what your community can do, contact:

> Dr. David Crawford
> Dark Sky Association
> 3545 Stewart Street
> Tucson, Arizona 857161

46 **When it comes to studying the universe, sometimes there's more to "clear" than meets the eye.** Sometimes the sky can look perfectly clear to the eye and yet not be acceptable for certain types of astronomical research. This can be particularly true for the branch of astronomy known as *photometry,* which seeks to very accurately measure the brightness of stars. Very high thin clouds, for example, that are virtually invisible to the naked eye can cause wild fluctuations in such equipment and render the data useless.

47 **There are technical limitations to just how big you can build individual telescopes.** The bigger a telescope's main lens or mirror, the brighter and shaper the images it can produce. So why not simply build telescopes with gigantic lenses and mirrors? The problem is one of simple limits in the strengths of the substances out of which the lenses and mirrors are made. In order for a telescope's lens or curved mirror to accurately focus light into a clear image, the surfaces of these lenses and mirrors have to have very specific shapes—shapes that are accurate to a fraction of the wavelength of light itself—shapes that are accurate to a few millionths of an inch. Modern mirror- and lens-grinding procedures are capable of such accuracy, but beyond a certain size, a lens or a mirror can simply become so heavy that it actually sags under its own weight. Not enough to be visible to the eye but enough to distort beyond the point where it can accurately focus light.

48 **The largest refracting telescope in the world is in Wisconsin and the largest single reflector is in Russia.** The largest refracting telescope in the world has lenses that are 40 inches across. It is located at the Yerkes Observatory run by the University of Chicago in Williams Bay, Wisconsin. In 1948, work was completed on the 200-inch reflecting telescope on Mount Palomar in California. For several decades, it was the largest reflecting telescope in the world. In the 1970s, a telescope with a 237-inch (6-meter) mirror was completed in the Caucasus Mountains, but unfortunately, its optics were never very good.

49 **In time, new materials and technologies have led to the creation of even larger telescopes.** In the 1980s, an exciting new advance in telescope design was pioneered that allowed astronomers to defy what were previously thought to be limits on the size of optical telescopes. The concept involved incorporating several individual mirrors into a single telescope tube and combing the light that fell on all the individual mirrors to create one integrated image. In

The twin observatory domes of the Keck telescopes. (Keck Observatories, University of California)

and send signals to computer-controlled motors that support the telescope's main mirror to prod the mirror moment by moment and actually change its shape in just the right ways to counteract the turbulent effects of the Earth's atmosphere. If successful, such telescopes could achieve images of unprecedented clarity.

51 **Another telescope design technique combines the light from several telescopes to yield images of great clarity.** In a recent experiment at the University of Cambridge in England, astronomers electronically combined the light from three separate telescopes all aimed at the same object to create an image. The principle is known as *interferometry* because the image is obtained after a computer analyzes how the light beams from each of the telescopes interfere with each other. From such an analysis, the computer can ascertain a great deal of information about the target object and ultimately produce an image equivalent to what could be obtained with a single telescope whose main lens or mirror was as big as the separation between the individual telescopes used. In the initial experiment, the three telescopes were about 20 feet apart and thus simulated or synthesized a single telescope with a mirror 20 feet across. The result was an image of the stars that make up the Capella star system with a clarity equivalent to that which would allow you to read a license plate at a distance of 600 miles! In time, the telescopes will be moved even farther apart to create even better resolution. Using a different interferome-

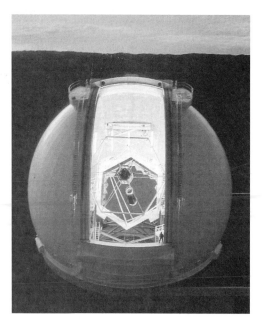

The Keck Telescope segmented mirror. Light from the different mirrors is combined and thus acts as one. (Keck Observatories, University of California)

this way, all the individual mirrors act as if they are a single mirror whose area equals the combined areas of all the separate mirrors. The Keck Telescope on Mauna Kea in Hawaii uses the combined light that falls on 36 mirrors, each of which is 1.8 meters in diameter. It was first tested in 1990 and, in 1996, was joined by a twin telescope (Keck II) placed next to it. Even larger "multiple-mirror telescope" designs are currently being pursued.

50 **Other telescope designs use lasers and computers to fool Mother Nature.** In an area of research known as *adaptive optics,* scientists are investigating the use of lasers to continuously probe the state of the atmosphere above a telescope ter, a team of American astronomers recently did ten times better, resolving a binary star pair only 0.0032 seconds of arc apart in the sky—the equivalent apparent size of a car on the Moon!

52 Similar telescope arrays are planned for elsewhere. Other optical interferometers are planned or under construction from Chile and Australia to the United States. Indeed, the Keck and Keck II telescopes can and will be linked in this way. As computers become faster and able to handle more and more data, such systems will undoubtedly play a significant role in astronomy as we enter the twenty-first century. There are drawbacks to such a system, however, since in science as well as in the rest of life, there's really no such thing as a free lunch. For one, the system can require a lot of computer power. Second, the creation of the final image can take hours or even days of telescope time.

53 Astronomers spend almost none of their time actually looking through their telescopes. This might seem very strange, but it's true. Time on large telescopes is just too valuable a commodity and the human eye is too insensitive and subjective a device. Instead, modern optical astronomers spend most of their time at observatories looking at computer screens. One might typically show an image of the planet, star, galaxy, or other object that the astronomer is studying. But just as often, the image might be of a nearby unrelated object. And the im-

Years ago, optical astronomers usually had to spend chilly nights at their telescopes inside the unheated observatory dome. Today, they can work in the warmth and comfort of control rooms nearby, such as this one at the 4-meter telescope at Kitt Peak National Observatory in Arizona. Instead of peering through the telescope, astronomers spend their time monitoring images and data on computer screens. (NOAO)

age might not even be coming from the main telescope in the observatory but rather from a smaller telescope attached to the main one. With this smaller telescope and the corresponding image on-screen, the astronomer tracks the main telescope across the sky. On other monitors, he or she keeps track of incoming data from scientific instruments that are far more trusted than the eye alone and analyzes radiation coming through the main telescope from the object under study.

54 In some cases, astronomers don't even have to travel to the observatory. The sophistication of remote control devices has reached a point where an astronomer can sometimes direct a telescope's operation via a computer link from his or her home or office with only a night assistant present at the observatory to open and close the facility and correct any problems that might arise with the equipment.

---

55 **In some cases, astronomers can't travel to the observatory at all.** Of course, astronomers using the Hubble Space Telescope and other orbiting spacecraft to make observations must rely completely on remote control from the ground. (Only astronauts occasionally visit the HST to make repairs or install new equipment. Astronomers are wisely not allowed near it.) In such cases, specially trained engineers and technicians translate each astronomer's desire to look at specific celestial objects with specific equipment into computer commands that are radioed up to the spacecraft. The astronomers

The Hubble Space Telescope seen above the cargo bay of the Space Shuttle *Endeavour* during its first servicing mission in 1993. This mission successfully installed new optics that corrected for mistakes in the shape of Hubble's main mirror. (NASA)

can be present at the ground station at the time their observations are being made (as long as they promise not to touch anything) or just stay home and receive their data in the mail or via computer link for later analysis.

56 **There are certain basic instruments in the astronomer's "tool kit." One of the most common is a camera.** Photography was first introduced into astronomy in the mid-1800s. The development was exciting because, for the first time, it allowed astronomers to objectively record what their telescopes were pointed at rather than relying on subjective sketches done by hand, which could vary significantly from one astronomer to another. For many years, the main limitation of film's use in astronomy was its relative insensitivity to light, especially given the faintness of most astronomical objects. As time went by, film increased in sensitivity and a variety of techniques were developed by astronomers to help it along, from baking the film in an oven before its use to chilling it down. While some astronomical images exist in color, most images taken for the purposes of doing astronomical research are recorded in black and white.

57 **In recent years, an electronic alternative to film has swept astronomy.** It is known as the CCD, or charge-coupled device, and you can find one in your home camcorder. The device consists of a square or rectangular array of tens of thousands to millions of very tiny electronic light-sensitive cells known as *pixels* that build up an

A CCD, or charge-coupled device, is the astronomer's modern alternative to photographic film. (NOAO)

electric charge when exposed to light. By electronically reading the charge on each pixel, a computer can reconstruct the pattern of light and dark (the image) that fell on the CCD to make a picture that can then be displayed on a monitor or printed out as

A close-up of a portion of a CCD showing the individual photon receptors, or pixels. The object cutting across the top of the picture that looks like a cable is, in reality, a human hair. (European Southern Observatory)

a hard copy. The advantage of the CCD over photographic film is that CCDs are typically much more sensitive to light and, whereas film can only be used once, the same CCD can be used over and over again. In addition, once the CCD's image is stored in a computer, it can be electronically manipulated, just like any other digital image, to change its contrast, bring out hidden detail, and so forth. CCDs and other clever technical advances now allow today's astronomers to wring hundreds of times more data out of every hour of telescope time than their predecessors could only a few decades ago.

58 **CCDs are typically used to take images onboard spacecraft.** If a conventional photograph is taken in an observatory, it can be easily developed in a convenient darkroom. But when it comes to spacecraft, it's not as easy to run up and change the film. So modern spacecraft use CCDs and similar cameras that take images electronically. The images are stored onboard the spacecraft in a computer or on magnetic tape in the form of a series of numbers that are then transmitted back to Earth via radio and reconstructed back into an image by a computer on the ground.

59 **Another common tool of the astronomer's trade is known as a photometer.** A photometer is an electronic device that is used to very accurately measure how bright an object is. The object might be a planet, a star, a galaxy, or any other astronomical target. Photometers used by astronomers are basically the equivalent of a very, very sensitive light meter like you would find in a typical 35-mm camera. The heart of the photometer is a substance that emits an electric current when light falls on it. The brighter the light, the stronger the electric current. The level of the current is then recorded in a computer. Usually, a series of colored filters are placed one at a time between the source of light and the photometer. In this way, the relative brightness of the star, planet, galaxy, or whatever can be measured in different colors. Sometimes a polarizing filter is placed into the light beam and rotated to see if the light from the target object is polarized itself.

60 **Perhaps the most versatile tool of the modern astronomer is the spectrograph.** A spectrograph is a device that uses a prism or a polished surface with many fine, parallel lines scratched onto it called a *diffraction grating* to split up the light from a celestial object into its spectrum of colors. This spectrum is then recorded on a piece of film or, if a CCD is used, data about the spectrum is collected and stored in a computer for later display and analysis. An incredible number of things can be determined about an object from its spectrum, such as its temperature, chemical makeup, size, rate of spin, speed toward or away from us, presence and strength of magnetic fields, and more. Again, in all cases, astronomers are gathering and studying light or other forms of radiation.

61 **There are three basic types of spectra.** These are generally known as *continuous, absorption,* and *emission* spectra.

62 **A hot solid or a hot gas under high pressure produces a continuous spectrum.** A continuous spectrum is simply one where the colors are spread out continuously from, for example, red through violet. A hot iron poker, the incandescent filament inside an electric lightbulb, or the interior of a star all produce continuous spectra.

63 **Most stars have absorption, or dark line, spectra.** An absorption, or dark line, spectrum, as the name suggests, is a continuous spectrum that has dark lines cutting across it. While the interior of a star produces a continuous spectrum, this radiation must pass through the star's atmosphere before it can travel across space to us here on Earth. The cooler gases in the star's atmosphere absorb certain discrete wavelengths of radiation from the continuous spectrum and redirect, or scatter, these discrete colors in all directions. Thus, very little of the light from these particular wavelengths winds up traveling in our direction and these wavelengths show up as dark lines (missing radiation) in the spectrum of the star. As discussed later, each form of each element in the star's atmosphere removes only certain specific wavelengths and so the identification of specific lines tell us that specific substances are in the atmo-

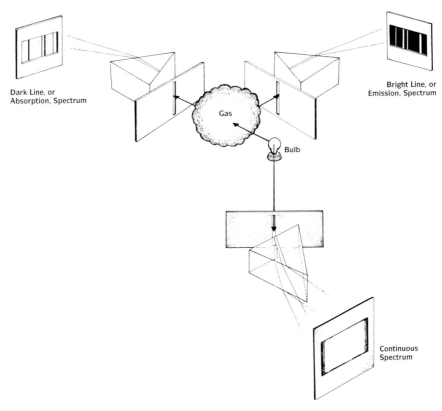

Three different kinds of spectra. A hot solid or a gas under relatively high pressure (such as the filament of a lightbulb or the interior of a star) produces a continuous spectrum. A cooler gas intervening between such an object and the observer produces a dark line, or absorption, spectrum. The gas by itself against a dark background (such as interstellar space) produces a spectrum of bright lines that correspond exactly to the dark lines in the absorption spectrum of that gas.

sphere of a particular star as well as many other things about the star.

64 **The planets also exhibit absorption spectra.** Planets don't give off light of their own but merely reflect the light of the Sun back into space. As a result, the spectra of the planets is really the spectrum of the Sun with the addition of some extra dark lines that correspond to more wavelengths that were subtracted out as the Sun's light traveled down and back out through the planet's atmosphere.

65 **Gas under very low pressure frequently produces an emission, or bright line, spectrum.** In space, these conditions frequently exist in the hot rarefied atmospheres of stars (such as a region of the Sun's atmosphere known as the *chromosphere*) and in shells of gas blown off stars known as *planetary nebulae*. As the name implies, emission spectra consist of a series of bright lines superimposed on a continuous or black background.

66 **What is light?** It's quite ironic. Light is all around us and we see because of it, but it's not easy to describe exactly what it is. Light may be thought of as behaving at times like a wave traveling through space and at times like a stream of particles. It's almost as if light has a dual personality. If you want to picture it as a wave, imagine rows of waves in the ocean or a lake. Of course, light waves aren't made of water but rather waves of electrical energy and magnetic energy traveling through space together. We call these *electromagnetic waves* or *electromagnetic radiation.* In the vacuum of space, these light waves travel at, well, the speed of light in a vacuum or 186,000 miles per second. The distance from the top (or crest) of one wave to the next is called the wave's *wavelength.* The number of wave crests that pass a stationary point in a second is called the *frequency* of the wave.

67 **Light waves have very short wavelengths.** Whereas you may be used to seeing waves on an ocean or lake that have wavelengths of several feet or yards, light waves have wavelengths that range from about 16 millionths of an inch to about 27 millionths of an inch.

68 **These differences are what we refer to as** *color.* When electromagnetic waves with wavelengths around 27 millionths of an inch long hit you in the eye, you see *red.* Not because you're angry but because electromagnetic waves of this length stimulate the retinas of people with normal color vision to see the color red. If electromagnetic radiation with wavelengths around 16 millionths of an inch strike your eye, you see *violet.* Radiation with wavelengths falling in between stimulate our eyes to see the other colors from red to orange and then through yellow, green, and blue to violet. Different *colors* just come down to a matter of *wavelength* and nothing more. This range of colors, or wavelengths, of radiation to which our eyes are sensitive is referred to as the *visible* spectrum.

69 **Beyond the visible spectrum lies much, much more.** Just because our eyes can't see electromagnetic waves with shorter wavelengths than violet light doesn't mean nature doesn't create them. Indeed it does. These are the energetic rays that can cause us to get a sunburn and make certain minerals fluoresce. Since these rays have wavelengths that lie beyond the violet, we call them *ultraviolet rays.* At still shorter wavelengths, we find radiation with high enough energy to pass right through human flesh. We call them *X rays.* And at still shorter wavelengths and higher energies we find *gamma rays.* In the other direction, beyond red light we find radiation that stimulates our skin to feel warm instead of our eyes to see. We call it *infrared.* At longer wavelengths, we encounter radiation that can get your dinner cooked in a hurry called *microwaves.* And at even longer wavelengths (now in the range of

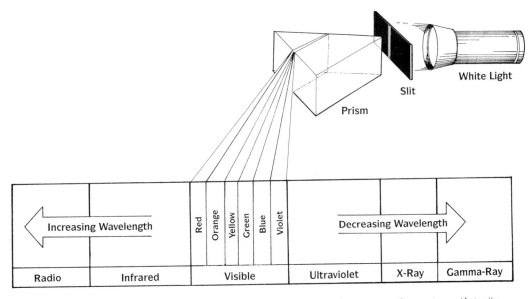

The spectrum of visible light occupies only a tiny portion of the entire electromagnetic spectrum. (Actually, much less than shown here.)

inches, feet, and yards), we have waves that are used to carry music, news, and information around the world—*radio waves.*

70 **All these different forms of electromagnetic radiation have different names because we discovered them at different times.** The important point is that they are really all the same. They are all electromagnetic waves. They just vary in wavelength. Together, this full range of waves from radio to gamma rays comprises what is known as the *electromagnetic* spectrum.

71 **The human eye is sensitive to only a tiny part of the entire electromagnetic spectrum.** The visible spectrum comprises only a minuscule slice of the entire electro-

magnetic spectrum. For this reason, we actually only see a tiny slice of what is all around us. Imagine, by way of comparison, only being able to hear a single key on a piano or a few notes around middle C being played by an orchestra. That's an indication of how much of the universe we see with our eyes or optical telescopes alone.

72 **Most objects in space give off radiation across much more of the spectrum than we can see with our eyes.** Our Sun normally gives off more radiation at visible wavelengths than anywhere else (and perhaps it is not coincidence that this is the very wavelength range where our eyes are sensitive), but the Sun actually radiates all up and down the spectrum. Our Sun naturally gives off radio waves, infrared, and ultraviolet, as well as X rays and gamma rays. And virtually all other stars and galaxies do as well. Given the right equipment, continuous, absorption, or emission spectra or direct images of celestial objects in other parts of the spectrum can be obtained and studied.

73 **Objects in space frequently look bizarrely different when "seen" at different wavelengths.** Using the right instruments, astronomers can create images of objects in space as they would look to our eyes, if we could actually see in wavelengths other than those of visible light. (Night vision glasses that convert infrared radiation to visible light to allow us to see things in the dark or an X ray we might get taken at the hospital are simple examples from outside of astronomy.) Astronomical images beyond the

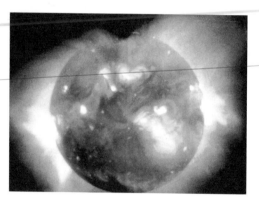

The Sun takes on a very unusual appearance in X rays as seen by the Japanese, American, and British satellite *Yokhoh* (Japanese for "sunbeam"). The photosphere is quite dark, but active regions around sunspots and flares are bright with powerful X-radiation that shines from the photosphere up into the solar corona. (Yokhoh Soft X-Ray Telescope, Lockheed Missiles, and Space Company, Inc.)

visible spectrum can be striking. In X rays, for example, the bright disk of our Sun is almost black, but magnetic storms that appear dark in visible light are brilliantly aglow with X-radiation that can change explosively from day to day and even hour to hour. Indeed, while the night sky can look tranquil and unchanging to our eyes, in X rays and gamma rays, the universe is seen as a place of chaotic and sudden violence.

74 **The more wavelengths of electromagnetic radiation astronomers can gather and study from an object in space, the more they can learn about that object.** Since objects in space can appear radically different when viewed in different parts of the spectrum, the more wavelengths of radiation we can gather and analyze, the more we learn about an object. Indeed, making these formerly invisible aspects of the universe visible for the first time has been one of the major developments and triumphs of twentieth-century astronomy. What used to be called just astronomy is now more correctly referred to as *optical* astronomy and the past half-century has seen the rise of radio, microwave, infrared, ultraviolet, X-ray, and gamma ray astronomy. While images of the same celestial object at different wavelengths typically yield very different views, these images all complement each other and together provide us with a much fuller understanding of that object and the universe at large. In effect, these images add up to a new image that is greater than the sum of its parts.

75 **Only some of the electromagnetic radiation from objects in space can be gathered from the ground.** Only visible light, a bit of the infrared and ultraviolet, and much of the radio portion of the spectrum passes quite easily through the Earth's atmosphere. (Some radio waves even penetrate clouds and so can reach the ground from objects in space when it's cloudy out.) For this reason, optical and radio astronomy is mostly done from the surface of the Earth.

76 **Some radiation from space can only get part of the way down in the atmosphere.** Infrared rays have trouble penetrating water vapor. Since the lower portions of the Earth's atmosphere typically have lots of water vapor in them, infrared telescopes are usually located in dry climates and on the peaks of tall mountains as well as in balloons and high-flying jet aircraft.

77 **Some radiation can't get through the atmosphere at all.** X- and gamma radiation cannot penetrate the Earth's atmosphere (which is a good thing for us) and, with the unfortunate exception of the ozone hole, most of the ultraviolet spectrum has trouble getting through as well. Thus, astronomers who want to do ultraviolet (UV), X-ray, and gamma ray astronomy have no alternative but to send their instruments above the atmosphere and the development of these branches of observational astronomy had to wait until the dawn of the space age. Since infrared (IR) astronomy is also hampered by the atmosphere, IR telescopes also have flown with increasing frequency onboard satellites in Earth orbit.

78 **Telescopes for different parts of the spectrum can look very different.** Telescopes for doing infrared and ultraviolet astronomy usually look very much like reflecting telescopes used for optical astronomy. Radio telescopes, however, can take on the appearance of giant satellite or radar dishes. X-ray telescopes can't use conventional mirrors to focus the X-radiation because these rays are so powerful they pass right through the mirrors without being reflected! Instead, the insides of many X-ray telescopes look like a stack of polished metal bowls with their bottoms knocked out. The incoming X rays scatter off the polished metal surfaces and are brought to a focus. And gamma ray "telescopes" that record this most powerful of all electromagnetic radiation are more like Geiger counters.

79 **At first glance, radio telescopes may seem to look very different from optical reflecting telescopes, but they're not.** Radio telescopes may look like satellite dishes on steroids, but they really work very much the same as optical reflecting telescopes. A large bowl-shaped dish takes the place of the optical reflector's main mirror and brings the radio waves from distant objects in space to a focus where the radio astronomer places instruments. Since this is *radio* astronomy, the instrument will not be a camera or photometer but instead a very sensitive radio receiver. The analogy to your TV satellite dish (or even your boom box or car radio with a whip antenna) is completely valid. All are devices for receiving radio waves. But radio telescopes are millions of times more sensitive to incoming radio waves than your personal electronics equipment.

80 **Just as with optical telescopes, the bigger a radio telescope is, the more radiation it can collect. But radio telescopes typically need to be big for another reason as well.** As we noted earlier, the resolving power of a telescope depends on the size of its main lens or mirror. However, the simple formula given at the time was a bit too simple, for it only works for visible light. In reality, the resolving power of a telescope also depends on the wavelength of radiation you are focusing. The longer the wavelength, the fuzzier the image you will get for the same size telescope. Since radio waves are thousands to millions of times longer than light waves, a radio telescope's collecting dish would have to be thousands to millions of times bigger than an optical telescope's mirror to yield images of the same clarity. While such engineering feats are not yet possible, radio telescopes do exist with dishes that are hundreds of feet across. The largest single-dish radio telescope is strung on posts over a mountain valley near Arecibo, Puerto Rico, and measures 1,000 feet across. Still this telescope can't "see" nearly as clearly as most optical telescopes can see in the visible part of the

spectrum. Indeed, a 300-foot radio telescope doesn't see the sky as clearly as a person with 20/20 vision can see with the naked eye!

81 **As with optical telescopes, radio telescopes can be connected together to create interferometers.** Astronomers can overcome the natural handicap that the sheer lengths of radio waves present by combining the signal from two or more radio telescopes and effectively synthesizing a telescope with a dish as large as the separation between the individual dishes. An example is the VLA, or Very Large Array, consisting of 27 radio telescopes each about 80 feet across arranged in a giant **Y** shape on railroad tracks in southern New Mexico. The maximum separation of the dish is 26 miles. The result is similar to having a single radio telescope as big as the Washington Beltway. The VLA can see details in objects in the radio spectrum with a

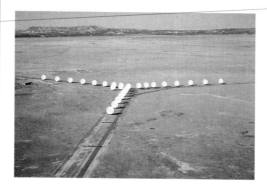

The VLA, or Very Large Array, is a system of 27 radio telescopes in New Mexico. Astronomers can change the spacing between these different telescopes and combine their signals to create images of celestial objects of very high resolution in radio waves. (NRAO)

resolution of 0.1 seconds of arc—better than any single optical telescope on Earth.

82 Beyond the VLA.

Radio telescopes in various parts of the world have even had their signals combined to simulate an antenna virtually as large as our entire planet. Such an array is referred to as the VLBI, or Very Long Baseline Interferometer. One such network stretches from Hawaii in the mid-Pacific to St. Croix in the Caribbean. The farther apart the individual radio telescopes are, however, the longer it takes computers to piece together the data.

83 Other telescopes search for things called *cosmic rays.*

Cosmic rays are not electromagnetic waves, as their name might suggest. Instead, they are tiny subatomic particles (mostly protons and the nuclei of helium atoms) that stream down through our atmosphere from space at close to the speed of light. Their origin is still a matter of some debate, but it seems likely that most are created in supernova explosions or in interactions between pairs of stars that contain tiny compact members called *neutron stars.* The particles are then accelerated by our galaxy's magnetic field and come streaking at us from every direction imaginable. When cosmic rays enter the Earth's atmosphere, they can collide with the air high above our heads and give off faint bursts of light that can be detected by very sensitive equipment. Cosmic rays can also be directly studied by instruments akin to Geiger counters flown in balloons or spacecraft.

84 Some telescopes are buried underground.

And to make matters even stranger, sometimes they are filled with dry-cleaning fluid! These "telescopes" are more correctly known as *detectors* and consist of enormous tanks holding tens to hundreds of thousands of gallons and are used to record and study tiny subatomic particles called *neutrinos* that are produced by the Sun, other stars, and exploding stars called *supernovae.* When a neutrino passes through such a tank, it has a very small but finite chance of colliding with one of the atoms inside and changing it into a different atom. By periodically flushing and examining the contents of the tank, scientists can determine how many neutrinos have passed through the detector.

A tank filled with dry-cleaning fluid buried in an old gold mine near Lead, South Dakota, served as an early detector of solar neutrinos. (Brookhaven National Laboratory)

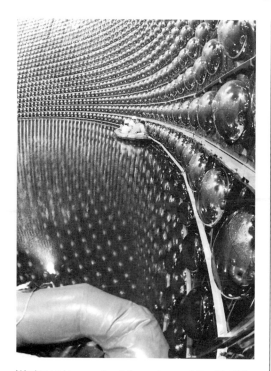

Workmen in a small raft inspect one of the 11,000 phototubes inside the Super Kamiokande solar neutrino detector in Japan. When operational, the entire space is filled with 50,000 tons of highly purified water. Minute flashes of light within reveal the passages of neutrinos through the detector. The detector is located 0.6 miles below ground in a zinc mine and is 120 feet in diameter. (Yoji Totsuka, Institute for Cosmic Bay Research, University of Tokyo)

Other neutrino detectors are filled with very pure water. When neutrinos pass through and interact with the water, minute flashes of light are given off which are detected by very sensitive light meters lining the walls of the enclosure. The tanks are all buried far underground (in places like an abandoned gold mine in South Dakota and an old salt mine under Lake Erie) to screen out all other particles, such as cosmic rays, and only let the neutrinos through.

85 Other detectors seek out *gravity waves* from deep space. According to Einstein's General Theory of Relativity, objects that are undergoing rapid changes in motion should generate gravity waves, which are actual distortions in space and time. The more massive the object and the more rapid its acceleration, the greater the amplitude of the gravity waves. As gravity waves pass through objects on Earth, they should create minute momentary distortions in these objects that might be recorded if these objects are sufficiently isolated from vibrations and have sensitive enough measuring devices at-

Joseph Weber makes adjustments to an early device designed to detect gravity waves. The rectangular plates attached to the cylinder are extremely sensitive sensors capable of detecting minute amounts of deformation in the cylinder that would be a sign of gravity waves passing through it. Far more sensitive gravity wave detectors are now under construction. (University of Maryland)

tached. An early gravity wave detector built by Joseph Weber at the University of Maryland proved to not be sufficiently sensitive. New detectors now under construction in various parts of the United States should be able to detect gravity waves released by such catastrophic events as the collision of neutron stars out to a distance of 70 million light-years from Earth.

*O*n Matters of Matter

86 **The most fundamental forms of substances are called *atoms*.** There are billions of natural and man-made substances in the world from water to Teflon, but all of these can be broken down in chemistry labs to simpler substances. For example, using an electric current, water can be broken down into two gases, namely hydrogen and oxygen, while, in other ways, ordinary table salt (sodium chloride) can be broken down into sodium, which is a metal, and a poisonous gas called chlorine. Each of these four new substances—hydrogen, oxygen, sodium, and chlorine—has unique properties. And none of them can be further broken down without losing those properties and no longer being hydrogen, oxygen, sodium, or chlorine. They are as fundamental as substances get and so they are called *elements.* The smallest unit of any element that has the properties of that el-

ement is known as an atom. However, atoms may be thought of as consisting of still smaller particles called *protons, neutrons,* and *electrons.* Normally, the protons and neutrons are huddled close together in the atom's nucleus, while the electrons orbit the nucleus at a distance. There are actually an entire "zoo" of additional exotic subatomic particles, but, with rare exception, we will not encounter them in this book.

87 **When atoms stick together, they make *molecules*.** Two or more atoms attached to each other, or bonded, form a molecule. For example, 1 atom of carbon bonded to 1 atom of oxygen creates 1 molecule of carbon monoxide. One atom of carbon bonded to 2 atoms of oxygen makes 1 molecule of carbon dioxide. Molecules containing only a *few* atoms are typically called *simple* molecules, while molecules with *many* atoms are called *complex* molecules. How many atoms you need to go from simple to complex depends on who you talk to. When radio astronomers started finding molecules in interstellar space that had 6 or 8 atoms, they referred to them as complex because no one expected such things in the hostile environment of space. But a biochemist would typically call such a molecule very simple indeed.

88 **In the whole universe, there are only 92 naturally occurring elements.** The single thing that makes a particular element that element and no other is the number of protons in its atoms' nuclei. For example, every atom in the universe that has 1 proton in its nucleus is hydrogen and only hydrogen. Every atom with 2 protons in its nucleus is a helium atom and nothing else. Carbon atoms have 6 protons, oxygen atoms have 8, and so on, all the way up to an atom that has 92 protons in its nucleus and that's uranium. Many elements with similar numbers of protons and electrons have similar chemical properties and, for convenience, scientists group all the elements in order of the number of protons they have (known as their *atomic* number) in a nice little table called the *Periodic Table of Elements*. There's usually one plastered in the front of virtually every chemistry classroom in the world. It is literally the alphabet soup or atomic blueprint of the entire universe because from these 92 basic ingredients all other things are made.

In a wonderful science fiction short story written many years ago by Armand Deutsch, a group of future archaeologists are excavating the remains of an ancient Martian civilization and have uncovered what is clearly a university campus. They remain frustrated at not being able to crack the Martian language until they break into the chemistry building. There hanging on the wall is the Periodic Table of Elements—an object that is immediately recognizable to them because it represents something that is universal, transcending culture or even species. Thus, the Periodic Table becomes the Rosetta stone used to crack the Martian language.

Elements with relatively few protons in their nuclei are sometimes referred to as *light* or *simple* elements, while atoms with more protons are called *heavy* or *complex* elements.

89 **What are ions?** Two other words often heard when conversations turn to "atomic matters" at cocktail parties are *ion* and *isotope*. In talking about *ions,* we have to concern ourselves with those little things called electrons that orbit the atomic nucleus. Normally, atoms are said to be *neutral* (as a whole) because the number of positively charged protons in their nuclei are matched by the number of negatively charged electrons whirling about each nucleus. But because some of the electrons can hang out rather far from the nucleus, they are frequently rather easy to knock off. This leaves the atom with more total positive charge than negative charge. Similarly, electrons can also frequently be added to atoms, giving the atoms a net negative charge. Simply put, atoms that carry a net positive or negative charge are called *ions.*

90 **What are isotopes?** In any discussion about isotopes, we must concern ourselves with those other little particles in the nucleus, the neutrons. *Isotopes* are forms of the same element that have different numbers of neutrons in their nuclei. There are, for example, three isotopes of carbon known as carbon 12, carbon 13, and carbon 14. The numbers in this case refer to the number of *neutrons* in the nucleus of each atom. Since it is the number of protons in a nucleus that make something a particular element, all atoms with 12 protons in their nuclei are all atoms of carbon, regardless of whether they have 12, 13, or 14 neutrons. Each isotope simply has a slightly different amount of mass. Anything made of carbon from graphite to diamonds will have a mixture of these different carbon isotopes.

91 **Some isotopes of some elements are radioactive.** Radioactive atoms spontaneously change into other atoms, in effect, doing a fast-change act. Sometimes certain atoms will lose a neutron in their nucleus and thus become a different isotope of the same element. Sometimes certain atoms will lose a proton in their nucleus and thus become a different element. Such processes are known as *radioactive decay.* Uranium, for example, can undergo a series of radioactive decays and ultimately wind up being lead. Some isotopes of some elements are very radioactive, meaning they decay into other things very fast, while others decay very slowly or virtually not at all. Such atoms are said to be very *radioactively stable.*

92 **Radioactive decays can occur at different rates.** In any given sample of radioactive stuff, you can't predict in advance which specific atoms are going to spontaneously decay. The atoms just don't tell you what they're going to do in advance. But by watching and carefully measuring, scientists have found that the rate at which whole samples of the same isotope decay is consistent from sample to sample. The amount of time it takes half the isotopes in any given sample

to decay is called that isotope's *half-life.* Highly radioactive isotopes have short half-lives, while very stable isotopes have very long half-lives.

93 **Radioactive decay can be an important scientific tool.** All of this knowledge turns out to be an important tool in trying to figure out how old something is. For example, if you compare the amount of a particular radioactive isotope there is in something (from a dinosaur bone to the Shroud of Turin to a Moon rock) to the amount of a very stable isotope of the same element in the sample, then compare those numbers to the amounts of those isotopes you find in something that you know is not very old, and you know the half-life of the radioactive isotope in question, you can figure out how old the object under study is. Anthropologists, archaeologists, and paleontologists use this technique a lot, especially using isotopes of carbon, a procedure known as *carbon dating.* Astronomers use the technique on occasion as well, using carbon or isotopes of other elements when convenient.

94 **Matter typically is said to exist in three *states.*** These states are known as *solids, liquids,* and *gases.* Just which state something is in at a particular time and place depends on things like the chemical nature of the substance itself as well as the temperature and pressure of its environment. Here on Earth, we find an example of something which we can see existing in all three states.

It consists of 2 atoms of hydrogen bonded to 1 atom of oxygen: H_2O. Under normal conditions, we call this stuff ice when its below 32°F, water at temperatures between 32°F and and 212°F, and steam above 212°F. (At very high temperatures, the bonds between atoms of hydrogen and oxygen are torn apart and the substance is no longer steam, i.e., H_2O, it's just hydrogen and oxygen as separate gases.)

95 **In the extraordinary ranges of temperature and pressure we find across the universe, matter can take forms and behave in ways that might seem strange.** On Mars, for example, a barometer would hardly budge because there is very little atmosphere and thus very little air pressure even on the surface of the planet. Under these conditions, H_2O goes directly from the solid state (ice) to the gaseous state (water vapor) without passing through the liquid state. Hence there are no rivers or lakes on Mars today. We call this process *sublimation.* The stuff mothballs are made of does this on Earth (they just don't leave little puddles in your closet). In short, what's normal and what you have to learn to expect depends on where you are. When it comes to astronomers, hey, the whole universe is our beat.

96 **When ions exist in a gaseous state, they are said to form a *plasma.*** Some refer to this as a fourth state of matter because plasmas are charged and regular gases

are neutral. It's a bit of semantics, as long as you understand what's what. Stars are typically made of gas, but most of the gas is so hot, it's in plasma form. This becomes significant because plasmas respond to magnetic fields, whereas neutral gases don't. Some magnetic fields, which most stars have, can seriously affect how the star and its atmosphere and material move around and through the star.

97 *Fluids:* **What you see is what you put them in.** To help confuse the tourists in scienceland, liquids, gases, and plasmas are all frequently referred to as fluids because they typically assume the shape of their containers. (Put a quart of water in a pie plate and it takes the shape of a pie plate; pour it into a goldfish bowl and it takes the shape of a goldfish bowl. Similarly, the gas inside a "neon sign" takes on the shape of the sign.) When you rub two solids together (like your hands on a cold day), the action encounters a resistive force called *friction,* which in turn generates heat. We normally don't think of fluids as having friction, but they usually do have some. However, under certain conditions of temperature and pressure, this resistance can effectively become zero. Such substances, under those conditions, are called *superfluids.* Most stars are thus made of fluids, but neutron stars are actually made of a superfluid of neutrons.

98 **And then there are *solids.*** Some solids have a crystal structure, meaning their atoms are arranged in a regular geometrical pattern. Salt and carbon, in diamond form, are examples. Other solids, like modeling clay, are called *amorphous* because their atoms are not rigidly arrayed. The insides of white dwarf stars resemble crystals, in that the nuclei of their atoms are arrayed in a regular pattern, although the electrons "swim free" in a vast "sea" around the nuclei.

99 **"Dark matter" is a different matter.** Based on studies of the motions of stars in galaxies and groups of galaxies, astronomers know that much of the universe does not exist in the form of *any* of the matter we have described above. Instead, the majority of matter consists of other types of particles. So far this matter has eluded direct observation because it seems to interact little, if at all, with ordinary matter or radiation in any part of the electromagnetic spectrum. For this reason, astronomers simply refer to it as "dark matter." Its exact nature remains one of the great unknowns of latter-twentieth-century astrophysics.

100 **And finally, there's antimatter, Captain.** On *Star Trek,* Scotty and Geordi are always concerned about antimatter. Antimatter is made of particles, the same as "ordinary" matter. Indeed, particle for particle, they are identical, except that their charge is opposite. Thus, the antiparticle twin to the electron, called the *positron,* is equal in mass to the electron but has a positive charge. The antimatter twin to the proton, simply called the *antiproton,* is identical in mass to the proton but has a negative charge. If a particle of matter comes into contact with one of its antimatter twins, they annihilate each other in a burst of pure energy. (That's why Scotty and Geordi like them so much.) Antimatter does exist in the universe, but because there is so much more ordinary matter around, it typically faces annihilation whenever it happens to come into existence. Large pieces of antimatter or even antimatter atoms are not likely to exist in our universe. Other universes composed primarily of antimatter are theoretically possible.

CHAPTER 2

A Brief Trip Back in Time: On the Shoulders of Giants

101 Many early views of the universe placed the Earth at the center of everything. From the ancient Greeks to the Hindus and the Chinese, most cultures first developed a *geocentric,* or Earth-centered, view of the universe. The illusion, after all, was strong. The Earth felt very much like it was stationary and the lights of heaven appeared to turn around it each day and night.

102 Influenced primarily by Aristotle, many ancient Greeks differentiated between the realm of the sky above and the Earth below. To Aristotle, all things on Earth or immediately above the Earth were made of four *elements,* or essences: Earth, air, fire, and water. Above, the Sun, Moon, and the five known planets were each fixed to a crystalline sphere. These spheres were encased in a celestial sphere that contained all the stars. All revolved around the Earth in circles. They had to travel in circles, Aristotle said, because a circle was the "perfect form" and nothing in the heavens could travel with anything except perfect motion in a perfect

form. These heavenly objects and their crystalline spheres were made of a fifth element, or *quintessence.* In the realm of Earth below, there was constant change in the form of birth, death, and decay. But in the heavenly realms above, all was pure, unblemished, and unchanging. The heavens were forever serene and constant in appearance. To be anything less would not be perfect.

103 Aristotle's picture of the universe was very neat but not very accurate. Ancient Chinese sky watchers, unaware of and therefore uninfluenced by the writings of Aristotle, saw and recorded changes in the sky. These included dark spots that came and went on the supposedly unblemished Sun, comets that swept across the sky like "brooms," and "guest stars" that suddenly blazed brightly enough to be seen in full daylight. (Westerners undoubtedly saw such things as well, but it was wise not to speak too loudly against the philosophies of great men.) One observation, however, was so obvious and persistent that it couldn't be ignored.

---

104 **Some of the planets just wouldn't behave themselves.** It was well known to anyone who regularly watched the sky that certain planets, namely Mars, Jupiter, and Saturn, each in its own way and in its own time, would stop its usual eastward motion, do a U-turn, travel westward for a while, do another U-turn, and finally resume its eastward trek among the stars. To make matters worse, these backward, or retrograde, loops and zigzags were rarely the same shape or size. In order to "save the phenomenon" as it was called (or, more correctly, Aristotle's image), numerous astronomers, philosophers, and mathematicians tried to come up with clever ways of explaining all these complex motions while still keeping Aristotle's "sacred" notion that everything in the heavens had to travel on spheres and in perfect circles.

105 **The complex celestial machine of Ptolemy.** In the second century A.D., a Greek mathematician and astronomer

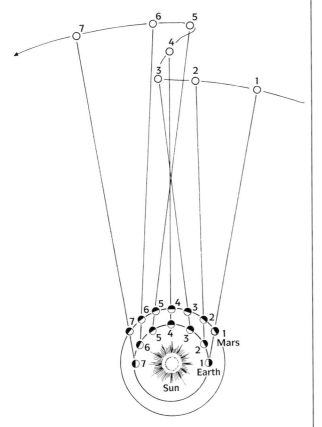

Since the Earth goes around the Sun faster than Mars, we get the illusion that Mars is going through a retrograde loop as Earth passes Mars by. The same phenomenon is also seen in the motion of all the other planets that are farther from the Sun than the Earth.

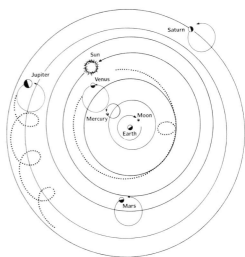

The model of the solar system developed by Ptolemy used a complex system of epicycles to explain the retrograde motion of the outer planets. The planets were envisioned as moving on smaller circles, which in turn moved on the larger circles.

named Ptolemy took the Aristotelian system about as far as it would go by adding little spheres to the big spheres of the outer planets. In effect, Ptolemy set each of the outer planets traveling about on one of these little spheres (called *epicycles)* whose centers, in turn, traveled on the main crystalline spheres around the Earth. In some variations on the original model, additional spheres were added (to a total of *80)* to try to duplicate all the variations in observed retrograde patterns. In this clever way, Ptolemy and his followers were able to get the outer planets to perform their retrograde loops and still keep everything traveling in circles. The system was to remain the accepted model of the universe in Western civilization for another fourteen centuries.

106 **In the sixteenth century, a shy Polish cleric started a revolution and changed the universe.** Down through the centuries there were those who remained unsatisfied with the utter complexity of the Ptolemaic system of circles upon circles. Nicholas Copernicus, however, had the mathematical skill and insight to try to do something about it. He realized that he could eliminate all of Ptolemy's epicycles and greatly reduce the messy complexity of the accepted view of the universe if he made one little change. And that was to take the Earth out of the center of the picture, put the Sun there instead, and have the Earth revolve around the Sun as one of the planets. The solution was simple and, therefore, mathematically had great appeal. This is referred to as a *heliocentric,* or Sun-centered, model of the universe.

107 **The Ptolemaic system, however, had the weight of the ages and, for that matter, the Holy Mother Church behind it.** Copernicus was careful not to come out and say that he felt that his new vision of the cosmos was correct, for that could only prove unfavorable to people in high places and even dangerous to one's health. Indeed, he simply offered it to the world and the papal authorities as a "mathematical exercise." Deciding to risk nothing, Copernicus also held off publishing it until he was on his deathbed.

108 **The Italian astronomer Galileo found proof of the Copernican model of the universe in a place few had thought to look.** To Aristotle and those who followed him, science was based as much, if not more, on pure reasoning as on scientific experiment. To Galileo, the proof was in the pudding and, if you wanted to know how the heavens worked, your pudding was in the sky. Hearing of the invention of a device that used lenses to make things that were far away look closer (a telescope), Galileo built several of his own design and turned them on the heavens. He noted that the Moon was not perfect, "as a great number of philosophers believe it to be," but "instead varies everywhere with lofty mountains and deep valleys." Galileo also recorded dark spots on the Sun and discovered that the planet Jupiter had four moons orbiting around it (in a universe where everything was supposed to be orbiting around the Earth). Finally, he observed the planet Venus and saw that, like the Earth's Moon, this world also went through phases. The discovery sounded a death knell for the Aristotelian/Ptolemaic vision of the universe because Venus could only go through phases in the manner observed if it orbited the Sun, not the Earth. Galileo's findings, however, were hardly applauded in his day. Instead, the Church, which favored the views of Aristotle and Ptolemy, forced Galileo to rescind his views and placed him under house arrest for the rest of his life.

109 **Two of Galileo's contemporaries also helped to smash the crystalline spheres of Aristotle.** Galileo effectively disproved Aristotle's vision of the heavens and proved the Copernican vision to be correct. But even Copernicus had not discarded the notion that all motion in the heavens was circular. A contemporary of Galileo's named Tycho Brahe did not use a telescope for his work but nevertheless made the most accurate measurements of the changing positions of the planets up to that time. An associate, Johannes Kepler, who was more than a bit of a mystic but also a very able mathematician, took the observations and used them to examine these planetary motions better than anyone had been able to do before.

110 **Kepler first suggested that planets made elliptical orbits around the Sun.** When he examined Tycho's data, Kepler realized that the planets could not be traveling around the Sun in circles as everyone had assumed. Instead, they traveled in paths called *ellipses.* (Stick two thumbtacks in a tabletop several inches apart, tie a loop of string around the thumbtacks somewhat larger than the distance between the tacks, stick a pencil in the loop, hold it taut as you run the pencil around the tacks, and you'll have an ellipse.) Kepler also deduced *three laws of planetary motion* that all the planets seemed to obey.

These are known as Kepler's three laws of planetary motion.

Kepler's first law states that the orbit of every planet around the Sun is an *ellipse* and the Sun lies at one of the *foci* of that ellipse. (In the ellipse you made above, the tacks mark the ellipse's two foci). Kepler's second law states that planets don't travel around the Sun at the same speed all the time. The closer a planet is to the Sun in its orbit, the faster it travels at that time. Finally, Kepler's third law states that the closer a planet is to the Sun, the less time it takes that planet to travel around the Sun; in other words, the shorter is its year.

111 **A genius named Isaac Newton took Kepler's work to the next step.** In the year that Galileo died (1642), Isaac Newton was born. Kepler had shown that the planets did not travel in circles but rather in ellipses and seemed to follow complex and exact rules of motion, but he didn't understand why. Newton coinvented a new branch of mathematics called *calculus* and used it to explain the motion of objects (falling on Earth

Newton showed that placing a satellite in orbit around the Earth was a natural extension of throwing an object off a tower. Objects in space are "weightless" not because there is no gravity at the height of these satellites, but rather for the same reason a person in a falling elevator would feel "weightless"—because both are falling toward the Earth. For the satellite, however, the Earth conveniently curves out of the way, so the satellite never "hits bottom."

and traveling through space) in terms of a force we now call *gravity*.

112 **Newton probably never got hit on the head with an apple as the legend suggests.** But he probably did see an apple fall from a tree, which inspired him to think about the force of gravity. If this invisible force could reach up and pull an apple from a tree toward the Earth, could it not also reach up and hold the Moon in its orbit around the Earth? Using a mathematical description of how gravity behaved, Newton was able to prove that the same force did indeed control the motion of the apple and the Moon and also *all other moving bodies in the universe.*

With one powerful stroke of insight, Newton showed that gravity was a universal force and described in precise mathematical form the single law by which all motion in the universe was governed. He not only showed that the physics we experience on Earth is the same physics that is experienced across the universe, he showed that the human mind was capable of comprehending it.

113 **In addition to his Law of Universal Gravitation, Newton also described three *laws of motion.*** These basically say that:

✧   An object will remain stationary or in motion in a straight line at constant speed unless something pushes or pulls it.

✧   If something pushes or pulls on something else, it will cause that object to change its speed and/or its direction.

✧   If an object pushes on another object, the second object pushes back by an equal amount.

These laws of motion control everything from hockey pucks to race cars to rocket ships to the planets orbiting around the Sun and the stars orbiting each other.

114 **In the early twentieth century, Einstein went one better than Newton.** In 1913, Albert Einstein published his Special Theory of Relativity. In it, he showed that while Newton's laws worked well at speeds experienced in everyday life, they broke down at very high speeds, that is, speeds that approach the speed of light. A fundamental postulate of the theory was that the speed of light is a constant (about 186,000 miles per second) and is independent of the speed of the source of the light or the speed of the person observing the light. While this may seem extremely paradoxical, it has been proven true by numerous independent experiments and led to three other equally curious consequences. These are that *mass, length,* and *time* are all dependent on the speed of the observer. Thus, for example, a spaceship flying past you at close to the speed of light will appear more massive, shorter in the direction of travel, and its onboard clocks will appear to be running slower than when the same spaceship is parked next to you. Again, while strange, it has all been proved to be true and has to be taken into account when observing objects in space that are moving at very high speed relative to us.

115 **A few years later, Einstein published his General Theory of Relativity.** The General Theory deals with Newton's force of gravity and suggests that one object influences the motion of other nearby objects not because its mass exerts a force but because its mass actually *warps the space around it.* Furthermore, the masses of objects not only affect space but also time, in effect causing time to slow down. Again, while sounding bizarre, numerous scientific observations have confirmed the validity of the theory (as we shall see at various points in this book).

116 **The progress of astronomy has been the result of the efforts of numerous people.** Of his accomplishments, Newton said, "If I have seen further, it is because I have stood on the shoulders of giants." There have been many more giants both before and after Newton's time and you shall read about some of them and a little of what they have discovered about our incredible universe in the chapters that follow.

CHAPTER 3

THE MOON:
OUR NEAREST NEIGHBOR

117 The Moon, a small desolate world, is our nearest neighbor in space. The Moon is 2,160 miles across, making it small enough to fit between the east and west coasts of the United States. It is an airless and waterless place (except for small amounts of ice near its south pole) where no life has ever existed. The average distance from the Earth to the Moon is about 238,000 miles.

The full Moon. Dark areas are the maria, or lunar lowlands. The bright areas mark mountain chains and the cratered regions of the lunar highlands.
(UCO/Lick Observatory photo/image)

118 During the day it gets very hot on the Moon, but at night things really chill off. The denser a planet or satellite's atmosphere, the less difference there is between its daytime and nighttime temperatures. Since the Moon has no atmosphere, the local thermometer really gets a workout. At noon on the lunar equator, for example, temperatures hover around 210°F, while the same spot at midnight could come in with a bone-chilling reading of −250°F.

119 The most common lunar features are *craters.* There are millions of craters on the Moon, mostly created when asteroids, meteoroids, and comets crashed into the lunar surface. Most of these impacts occurred long ago. Others still happen today. The Moon has no atmosphere to protect it, so

The rugged cratered terrain near the Moon's south pole. Old craters have younger craters within them. (UCO/Lick Observatory photo/image)

such objects plunge unimpeded into the lunar surface at tens of miles per second. Many lunar craters can be seen in even a small telescope. Craters range in size from microscopic pits to vast depressions over 100 miles across. Some craters have rim walls that are over 20,000 feet high.

120 **And little lunar craters have littler craters and so on . . . ad infinitum.** If we count up how many craters of different sizes there are on the Moon, we find that there are relatively few really big ones but progressively more smaller and smaller ones. One reason why is that there are far fewer large objects sailing around space that can crash into the Moon than small ones. (This is because, over time, many of the large objects

have already collided with each other and fragmented.) Another reason for different numbers of different-sized craters deals with how craters are made. In short, it's not typically a one-step process. First comes the formation of what is called the *primary* crater when a meteoroid or other object from space strikes the Moon. As the impact occurs, smaller chunks of lunar rock are blasted out of the primary crater in all directions. These pieces of material, in turn, crash down onto the lunar surface and create surrounding *secondary* craters. Debris thrown skyward in the formation of secondary craters creates still smaller *tertiary* craters. And so on.

121 **Some lunar craters have bright systems of *rays*.** Lunar craters like Tycho and Copernicus display bright rays up to hundreds of miles long that look like the spokes of a wheel. The adornments are the result of the light-colored material that was blasted out of the crater when it formed. Ray craters are relatively young (perhaps only a few hundred million years old). As such craters age, their rays gradually fade from view as the extreme temperature differences between lunar day and night create minute surface expansions and contractions that eventually erode them away.

122 **Studying lunar features can help us to determine their relative ages.** Some craters have sharp, clearly defined rims. Others appear broken or crumbled. The former are newer, while the latter show signs of age from meteoroid bombardment or a type of lunar

The crater Copernicus. A system of bright rays can be seen radiating from the crater like the spokes of a wheel. (UCO/Lick Observatory photo/image)

"erosion" that results from constant expansion and contraction of the lunar soil due to the big differences in day and night temperatures. Sometimes we see one crater superimposed on another. A crater that crosses another's rim or lies embedded within another crater is clearly the younger of the two.

123 **If you have 20/20 vision, you can actually make out a lunar crater with the naked eye.** The crater in question is, as you might imagine, one of the largest on the lunar surface. It is called Grimaldi and has a dark floor that helps it stand out from its lighter surroundings. If you think of the full Moon as the face of a clock, look for Grimaldi near the left-hand edge of the Moon at just about

nine o'clock. It appears as a tiny dark oval, but it is over 100 miles across.

124 **The Earth has been struck by many more objects from space than the Moon, yet it has far fewer craters.** Because of its greater size and mass, the Earth has attracted far more meteoroids over its lifetime than the Moon. Yet the Moon looks like a truly cratered world, while the Earth does not. Our planet's weather and crustal movements perpetually wear it smooth, while the Moon's lack of such forces allows it to preserve its scars from earliest times.

125 **The Moon also has impressive mountain ranges.** The Moon has several mountain ranges. One of the most prominent is known as the Apennine Mountains. Some of its peaks are higher than Mount Everest. Unlike the Earth, the Moon does not have plate tectonics, nor does it have erosion due to wind and rain. Hence, once mountains form, except for crumbling due to impacts from objects in space, there is little wearing away as with mountain ranges on Earth.

126 **The Moon also has features known as "seas," cliffs, and rills.** The lunar "seas" are not actually bodies of water but rather vast relatively smooth plains of dark solidified lava hundreds of miles across. Elsewhere, we find *cliffs* that run for scores of

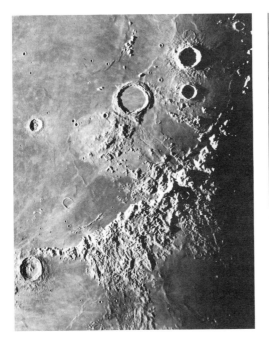

The mighty Apennine Mountains contain peaks as high as the Himalayas. (UCO/Lick Observatory photo/image)

maria, or "seas." The name dates back to earlier days when the smooth dark appearance of these areas led some astronomers to speculate that they might actually be bodies of water. Today, we realize that they are really large plains of solidified lava that welled up from deep inside the Moon during its early evolution and flooded lowland regions. The term "seas" is still applied, however, and many of these features still carry poetic names like the Sea of Serenity and the Sea of Clouds.

miles and sinuous valleys called *rills* that may be places where subterranean tubes of lava collapsed just below the lunar surface.

127 **The Moon's face displays a variety of light areas.** Even a casual look at the Moon with binoculars or the naked eye reveals that its surface is not uniformly bright. Instead, the Moon has a somewhat mottled appearance. The lighter areas are generally regions of higher elevation known as the lunar *highlands.* Much of this is mountainous or heavily cratered terrain.

128 **The Moon also has dark areas.** The darker places on the Moon are roughly circular in shape and are known as the lunar

129 **What creates the illusion known as "the man in the Moon"?** The inter-

The large crater Plato has a smooth dark floor because it was flooded by lunar maria material after it was formed. (UCO/Lick Observatory photo/image)

play between the light highland regions and the darker lava plains, or "seas," creates what is commonly called "the man in the Moon." Two lunar plains, the Sea of Serenity and the Sea of Tranquillity (where astronauts first set foot on the Moon), make up the left eye, while the Sea of Rains is the right eye. A mountain range known as the Apennine Mountains forms the bridge of the man's nose and a conglomerate of other plains, including the Sea of Vapors, creates the man's puckered little mouth. Various other cultures interpreted these light and dark areas in other ways, seeing instead a "lady in the Moon," a "rabbit in the Moon,"

and even a "frog in the Moon"—all of which makes Moon gazing kind of a cosmic Rorschach test.

---

130 **The Moon goes through a regular cycle of phases.** The Moon goes through a regular cycle of phases that repeats every 29.53 days. The major "points" in this cycle are the new Moon, the waxing crescent, the first quarter, the waxing gibbous, the full Moon, the waning gibbous, the third or last quarter, and the waning crescent. The cycle begins with the new Moon.

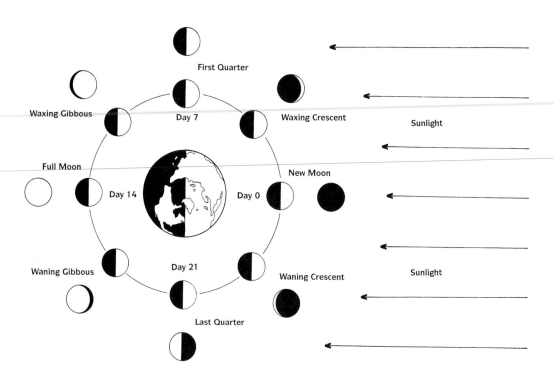

As the Moon orbits the Earth, it goes through a cycle of phases. The inner circle of the Moon images shows the light and dark hemispheres of the Moon as the Moon orbits the Earth. The outer circle of the Moon images shows the resulting phases of the Moon as seen from the Earth.

131 What's a *waxing* Moon vs. a *waning* Moon? When the Moon is waxing, it is growing fuller from night to night. When the Moon is waning, it is growing less full from night to night. Between the new Moon and the full Moon, the Moon is waxing. Between the full Moon and the new Moon, it is waning.

132 You can watch the cycle of lunar phases progress across the sky from night to night. Except at times of a solar eclipse, we cannot see the *new Moon* because, during this phase, the Moon is located between the Earth and the Sun, so the Sun is illuminating the side of the Moon that is pointed away from the Earth.

Within a few days of the new Moon, we see a thin *waxing crescent* Moon appear low in the west after sunset. From night to night, this crescent grows thicker until, a little over a week after the new Moon, we see the *right* half of the Moon illuminated in the south at sunset. Ironically, this phase is called the *first quarter* because the Moon has now completed a quarter of its cycle of phases.

Over the coming nights, the Moon grows fuller and takes on the appearance of an egg. This is the *waxing gibbous* phase.

A little more than two weeks after the new Moon, the Moon is positioned opposite the Sun as seen from the Earth. Now the Sun illuminates the entire hemisphere of the Moon seen from the Earth and so we see a *full Moon*.

During the following two weeks, the Moon changes from being a *waxing* Moon, that is, one that grows a bit fuller from night to night, into a *waning* Moon. Now, from night to night, we see less of the hemisphere illuminated by the Sun as the Moon gradually becomes slimmer and slimmer.

First comes the *waning gibbous* Moon, which grows increasingly flatter on its left side. A little more than three weeks after the new Moon, we see the *left* half of the Moon lit up by the Sun. But because we are now three-quarters of the way through the lunar cycle of phases, we refer to this as the *third-quarter* or *last-quarter* Moon.

Finally, the Moon becomes a thinner and thinner *waning crescent* over the last days of its cycle in the predawn sky and ultimately rejoins the Sun in the sky as the next new Moon.

133 You can understand how and why the Moon goes through phases by using a ball and a lamp. Just how and why the Moon goes through its cycle of phases can be a pretty tricky thing to understand, but in a darkened room with a ball and a lamp you can demonstrate the whole thing to yourself in moments and you'll understand it for the rest of your life. All you need is a ball (any size will do) and a lamp without a shade, that is, one that casts good sharp shadows.

Turn the lamp on and place it on something that raises its lightbulb to about the same height from the floor as your shoulder. Turn off all the other lights in the room. Next stand several feet away from the lamp with the ball in your hand. The lamp represents the Sun, the ball is the Moon, and your head is the Earth. Now hold the ball straight out in front of you at arm's length and at shoulder height. Slowly turn yourself around while continuing to hold the ball straight in front of you. As you turn, you will see the ball go through phases, just like the Moon. Indeed, the lamp, the ball, and your head will be in the same orientation as the Sun, Moon, and Earth for any particular lunar phase at any time.

As you bring the ball between your head and the lamp, you will see the ball cut off the light from the lamp and thus create an "eclipse of the Sun." As your head passes between the ball and the lamp at full Moon, the shadow of your head will fall on the ball and thus create an "eclipse of the Moon." You may want to try this little exercise with the shades drawn (so the neighbors don't start talking about you) . . . but do try it.

134 The *terminator* **marks the line along which sunrise or sunset is occurring on the Moon.** When the Moon is waxing, that is, growing fuller in the sky, the terminator marks the line along which sunrise is occurring along the lunar surface. From night to night in a telescope, you can see this line slowly advance to the west, just as it does hour after hour on Earth. When the Moon is waning, the terminator marks the line along which sunset is taking place. From night to night, the terminator progresses westward, engulfing more and more of the Moon's face in the shadow of night as the Moon grows slimmer and slimmer in our skies.

135 *Earthshine* **is a beautiful phenomenon to look for when the Moon is a slender crescent.** A few days before or after the Moon is new, it appears as a delicate crescent in the sky. But look carefully with a pair of binoculars and you will see the rest of the Moon faintly visible as well. This phenomenon, known as Earthshine, is the result of sunlight reflecting off the Earth and then spilling on the darkened portion of the Moon.

When applied to the waxing crescent seen in evening twilight, the phenomenon is sometimes poetically referred to as "the old Moon in the new Moon's arms."

136 **Quite often you can see the Moon in the daytime sky if you know where to look.** While the Moon is waxing, it rises before sunset and so can be seen at least during the afternoon hours. When the Moon is a waxing crescent, look for it in the southwestern sky during the late afternoon. At first quarter, the Moon can be spotted in the southeastern and southern sky during the afternoon, while the waxing gibbous is quite conspicuous in the east and southeast late in the day. After full, the Moon rises after the Sun sets but also sets after the Sun rises the following morning. So look for the waning gibbous Moon in the west early in the morning, the last-quarter Moon in the south early in the day, and the waning crescent to the right of the Sun all day long.

137 **We always see the same side of the Moon from Earth.** The Moon always keeps the same hemisphere pointed toward the Earth. This means that we always see the same face of the Moon. Indeed, if you could shine a giant searchlight on the new Moon, you would see the same features that you see during a full Moon.

---

138 **There's no such thing as "the dark side of the Moon."** It's a great rock album but, for the record, there is no "dark side of the Moon." It's true that half of the Moon is in darkness at any particular time, but as the Moon orbits the Earth, the specific part of the Moon that's lit by the Sun constantly changes. The side of the Moon that is dark at the moment will be the brightly lit side in two weeks and vice versa.

139 **Over time, more than half of the Moon's surface is actually revealed to our eyes.** While the Moon always keeps one side turned toward the Earth, our satellite actually wobbles a bit on its axis. This motion, called *libration,* allows us to alternately peek around the Moon's eastern edge and then its western edge. Also, because the Moon's orbit is tilted a bit relative to the plane of the Earth's orbit around the Sun, we also periodically get to glimpse a bit over the top and bottom of the Moon as well. In all, this allows us to see a total of about 59 percent of the Moon's total surface instead of just half.

140 **The side of the Moon that faces the Earth looks quite different from the side that points the other way.** While the near side of the Moon displays both light and dark regions, the far side of the Moon has far fewer maria and instead more cratered and highland regions. No one really knows why, but the huge impacts that created the deep basins that ultimately filled with lava to form the lunar "seas" primarily occurred in only one hemisphere for some reason. It might be tempting to suggest that the Earth somehow exercised some influence on the situation, but the bombardment that led to these great lunar depressions likely occurred at a time before the Moon's rotation slowed to the point of keeping only one side of the Moon facing the Earth.

141 **The Moon typically looks larger when it is near the horizon than when it is higher in the sky.** Technically speaking, the Moon is actually about 4,000 miles farther away from us when it rises than when it is high overhead. But a rising full Moon can really look huge. The effect, referred to as the "Moon illusion," is psychological rather than physical. Various explanations for the "Moon illusion" have been put forth, the most notable of which is that the brain interprets the Moon as being closer when it is near the horizon because the human mind is somehow influenced by the presence of foreground objects. No explanation of this curious apparition is universally accepted, however. So the next time you see a big beautiful full Moon on the rise, revel in its romantic appearance, but remember . . . its size is just an illusion.

142 **The full Moon always rises in the east as the Sun sets in the west.** When full, the Moon always lies opposite the Sun in our sky. (This positions the Moon in such a way that we see the full hemisphere lit up by the Sun.) For this reason, the full Moon always pops up above the eastern horizon as the Sun goes down behind the western horizon. As the full Moon sets in the west the next morning, the Sun rises again in the east.

143 **The full Moon may appear very bright, but . . .** The Moon can shine brightly in our skies, but it is actually a very dark object. Most of its rocks and soil are dark gray and, on average, the Moon only reflects about 7 percent of the light that falls on it from the Sun. Some of the moons of the outer planets reflect over 80 percent of the Sun's light because they are made largely of ice. Imagine how bright a moonlit night would be on Earth if we had such an "ice Moon" in orbit around us.

144 **In talking about the brightness of the Moon or other planets or their moons, astronomers frequently use the term** *albedo.* The albedo of an object is simply the fraction of sunlight that the object reflects or scatters back into space. Thus, we would say that the albedo of the Earth's Moon is about 7 percent, while that of some of the moons of the outer planets is over 80 percent.

145 **Different full Moons take very different paths across the sky.** Because the axis of the Earth is tilted to the plane of its orbit and the orbit of the Moon is also tilted relative to the Earth's orbit, the Moon can take very different paths across our sky. On a chilly winter night in midlatitudes, for example, the full Moon can climb almost to the top of the sky, while at the beginning of summer, the full Moon rides low across the southern sky. In some areas, this June Moon is seen through a lot of water vapor during what is a humid time of year. The humidity scatters the blues and violets out of the Moon's light, giving the Moon a yellow or orange cast. No wonder this Honey Moon occurs at a time of year traditionally associated with weddings.

146 **Our word *month* is derived from the word *moon.*** Not surprisingly, since a cycle of lunar phases is just about a month long, it became a convenient way of measuring time and was used as such by many cultures.

147 **Many Native American cultures traditionally give different names to the different full Moons of the year.** Among the more common names are:

January	Old Moon
February	Snow, Hunger, or Wolf Moon
March	Sap or Crow Moon
April	Grass or Egg Moon
May	Planting or Milk Moon
June	Rose, Flower, or Strawberry Moon
July	Thunder or Hay Moon
August	Green Corn or Hay Moon

September	Fruit or Harvest Moon
October	Hunter's Moon
November	Frosty or Beaver Moon
December	Long Night Moon

148 **Because of its location in the heavens, the Harvest Moon really did help farmers harvest their crops.** Because the Moon's orbit is tilted relative to the orbit of the Earth, the time of moonrise can vary considerably throughout the year. Around the first of spring, the nearly full Moon rises well over an hour later from one night to the next. Around the first of fall, however, the situation is reversed and the nearly full Moon rises at almost the same time for several nights in a row. This Harvest Moon thus provided farmers with a lingering source of light in the east after the Sun had set in the west at just the time of year when they needed to put in long hours to harvest their crops.

149 **The date of the Harvest Moon each year is determined by another celestial event.** The Harvest Moon is the full Moon that occurs closest to the *autumnal equinox* or, in other words, the full Moon that occurs closest to the first day of fall. Because the first day of fall occurs around September 22 or 23 and a full Moon can occur as much as about two weeks before or after this date, the Harvest Moon can take place anytime between about September 7 and October 7. Stevie Wonder's song *I Just Called to Say I Love You* has a lyric that goes "No Harvest Moon to light one tender August night." It's a great song and has sold millions of copies, but for the record, a Harvest Moon can never occur in August.

150 **Many cultures today still use a lunar calendar to fix the time of religious festivals or solemn occurrences.** The Chinese, Hindus, Jews, Muslims, and others still use a lunar calendar in this way. The Muslim holy month of Ramadan, for example, begins at the first sighting of a particular waxing crescent Moon and ends at the next, while the Jewish feast of Passover and even the modern date for Easter are also still determined by the Moon. The date for Easter, for example, floats from year to year but always occurs on the first Sunday after the first full Moon after the *vernal equinox* (the first day of spring). Since the date for Passover is set by the same full Moon, Easter and Passover usually take place around the same time each year.

151 **The worst time to look at the Moon through a telescope is when the Moon is full.** Full Moons can be very romantic, but they can be very disappointing when viewed through a telescope. When the Moon is full, the places at the center of the Moon's face are experiencing "noon" (when the shadows cast by the Sun are the shortest). With no lunar shadows to help topography stand out, the face of the Moon appears almost featureless. The best time to look at the Moon through a telescope is when the Moon is near first or last quarter. Then craters and mountains stand out in stark relief, especially along the Moon's flattened edge, and cast impressive long shadows.

152 **The notion that the Moon is up in the sky all night is true only about one night each month.** That night is the night of the full Moon. Most of the time the Moon is only visible for part of the night and for three or four days each month, around the time of the new Moon, there is essentially no Moon to be found in the sky at all. If you went out at random times on random nights, you would find the Moon in the sky about half the time.

153 **There are two kinds of "blue Moon."** The length of time between one full Moon and the next is 29.53 days. This means that, except during February, if a full Moon occurs at the very beginning of the month, a second full Moon will just sneak in before the month is over. This second full Moon in the same month is sometimes given the nickname the "blue Moon." On average, a "blue Moon" occurs every two and a half to three years. So if you have always wondered just how often something happens that happens only "once in a blue Moon," there's your answer. Of course, the "blue Moon" doesn't actually look blue; it just has that name. Can the Moon actually ever have a bluish cast? The answer is yes. Smoke from a fire or ash from a volcanic eruption can scatter the reds and oranges from the Moon's light and the Moon can take on a bluish color when seen through such a veil.

154 **The Moon's orbit around the Earth is not a perfect circle.** As with virtually all other astronomical bodies, the Moon's orbit is an *ellipse*. During the course of a month, its distance from Earth varies from about 221,460 to 252,700 miles. The nearest point to Earth (known as *perigee*) and the farthest point (called *apogee*) slowly move around the orbit, however, so no single phase always corresponds to a particular distance. Thus, at times, the Moon is closest when it's full and, at other times, the Moon is closest when it's new. A set of photographs carefully taken throughout a lunar cycle clearly shows the Moon's apparent change in size as its distance varies.

155 **What causes an *eclipse* of the Moon?** An eclipse of the Moon takes place when the Earth, Moon, and Sun line up in space with the Earth in the middle. When this happens, the Moon passes through the shadow that the Earth casts out into space and we see that shadow slowly crawl across the face of the Moon. Therefore, an eclipse of the Moon can only take place when the Moon is full. But a lunar eclipse does not take place every time the Moon is full. This is because the orbit of the Moon around the Earth is tilted a bit relative to the orbit of the Earth around the Sun. This results in the full Moon usually passing a little above or below the shadow of the Earth.

156 **An eclipse of the Moon clearly shows that the Earth is round (spherical).** People at different latitudes on the Earth can see the same lunar eclipse and can also see that the shadow that the Earth casts onto the

Moon is round. Yet people at different latitudes see the Moon at different places in the sky during the eclipse. These circumstances would only be possible if the Earth were round and not flat. Indeed, this argument was used as early as 350 B.C. by the Greek philosopher Aristotle as proof that our planet was a sphere. So much for the myth that everyone except Columbus in Columbus's time thought the Earth was flat.

157 **Eclipses of the Moon can be either** *total, partial,* **or** *penumbral.* The Earth's shadow consists of a dark inner region called the *umbra* and a fainter outer region that surrounds the umbra (like a doughnut) known as the *penumbra.* When the Moon passes completely through the umbral portion of the shadow, a total eclipse of the Moon takes place. If the Moon passes through only a portion of the umbra, we see a partial eclipse of the Moon. Should the Moon only pass through the Earth's penumbral shadow, a penumbral eclipse results. Because the penumbra is so faint, penumbral eclipses go by virtually unnoticed.

158 **When a lunar eclipse takes place, everyone on the same side of the Earth gets to see it.** Because half of the Earth is in darkness (nighttime) at any given time, the entire hemisphere that is facing the full Moon when it passes through the Earth's shadow is turned toward the eclipse. Thus, weather permitting, millions of people can witness the same lunar eclipse at the same time.

159 **You can easily get the impression that there are fewer eclipses than there really are.** Unless you purposely globe-trot to the right place at the right time, you will see fewer lunar eclipses than actually take place. This is because you have to be on the nighttime side of the Earth during the time the Moon passes through the Earth's shadow to see the eclipse. So sometimes when an eclipse of the Moon occurs, we are on the right side of the Earth (the nighttime side) and sometimes we have the "cheap seats" on the daytime side, which means folks on the other side of the Earth are treated to the eclipse instead of us.

160 **Some lunar eclipses are very dark, while others are light or even quite colorful.** In some instances, a totally eclipsed Moon can be so dark that it seems to virtually disappear from the sky. At other times, the eclipsed Moon can remain very visible, even when passing through the very heart of the umbra, and can take on a reddish or coppery hue. The reason for these differences has nothing to do with the Moon but rather the state of the Earth's atmosphere at the time of the eclipse. Typically, some sunlight passes through the ring of atmosphere surrounding the Earth as seen from the Moon. The Earth's atmosphere scatters the blues and violets in the Sun's light but allows most of the reds and oranges to pass through. Some of these, in turn, spill into the Earth's shadow and onto the face of the Moon. Hence, the totally eclipsed Moon can take on the appearance of a softly glowing lantern. During times of large volcanic activity, however, large amounts of ash and dust can be present in the Earth's atmosphere. These particles absorb all the colors of sunlight and so can leave the shadow and the eclipsed Moon very dark.

161 **A lunar eclipse can be a rather long affair.** When only a portion of the Moon skirts the umbra, the resulting eclipse can be quite short. But because the diameter of the Earth's umbral shadow at the distance of the Moon is more than twice the diameter of the Moon and because the penumbral shadow, in turn, surrounds that, a total eclipse of the Moon from beginning to end can last about five hours (with the total phase almost two hours long). I refer to short and long lunar eclipses as "one and two six-pack eclipses," respectively.

162 **Lunar eclipses are perfectly safe to look at.** Unlike an eclipse of the Sun, a lunar eclipse will not harm your eyes in any way. After all, the full Moon itself is safe to look at and, indeed, has been gazed at by lovers from time immemorial. During an eclipse, the Moon is passing through the Earth's shadow, so less light, rather than more, is falling on your eyes. Binoculars and telescopes enhance the view of the eclipsed Moon and help to bring out sometimes subtle colors in the Earth's shadow.

163 **The tides are caused by both the Moon and the Sun.** It is a common notion that the Moon causes the tides, but actually the Sun also plays a role, albeit a smaller one. Although the Sun is farther from the Earth than the Moon is, the Sun is so much more massive that it still exerts a significant gravitational influence.

164 **Two tides occur each day.** Two tides typically occur at the same spot on Earth every day. This is because the Moon raises two bulges of water on the Earth—one that faces the Moon and another on the side of the Earth that lies opposite the Moon. The Earth then rotates under these tidal bulges, creating two high tides per rotation. Midway between these "watery mountains" or regions of high tide are two corresponding troughs or areas of low tide. Thus, two low tides, midway between the high tides, also occur each day. The times of high and low tide change a bit each day because the Moon is constantly advancing in its orbit around the Earth, so the time when the Moon is high in the sky advances a bit from day to day.

165 **The tidal bulge in the direction of the Moon is pretty easy to understand, but the bulge on the side of the Earth that points away from the Moon is a bit harder for most folks to fathom.** The secret to understanding the second bulge lies in understanding what really causes the tides. Most people think the Moon's gravity causes the tides, but this isn't exactly correct. It isn't the Moon's gravity per se but rather the *difference* in the *amount* of *gravitational pull* the Moon exerts on the *near* side of the Earth vs. the *far* side of the

Earth. The amount of gravitational force exerted by the Moon on the Earth depends on the *distance* between the Earth and the Moon. But the side of the Earth facing the Moon is almost 8,000 miles closer to the Moon than the opposite side. This means the Moon pulls with more force on the side of the Earth closest to it, less force on the side of the Earth farthest away, and an intermediate amount of force on the center of the Earth. The Moon pulls the water on the near side of the Earth away from the rocky core of the Earth, thereby creating the tidal bulge facing the Moon. But—and here's the tricky part—the Moon pulls *harder* on the Earth's *core* than on the *opposite side* of the Earth. Thus, the core of the Earth is *pulled away from* the water on the *far* side of the Earth in the direction of the Moon. This creates the second tidal bulge.

166 *Spring* tides and *neap* tides are signs of a cosmic tug-of-war. When the Earth, Moon, and Sun all line up in space (at the times of a full and new Moon), the Moon and Sun pull on the Earth along the same line. This produces enhanced tides known as spring tides. (The name, however, is misleading, for these tides occur every few weeks throughout the year and not just in the spring.) In contrast, when the Moon and the Sun are positioned at right angles to each other as seen from the Earth (when the Moon is in its first- and third-quarter phases), they pull at cross-purposes, so the resulting tides are minimized. These are known as neap (rhymes with *pep*) tides.

167 Although the Moon is the major influence on the tides, it is erroneous to think of the Moon generating tides in human beings and, in so doing, influencing human behavior. The key to understanding this frequently misunderstood notion is to again remember that tides are the result of how much *harder* the Moon pulls on *one* side of the Earth than the *other*. In the case of the Earth, this *difference* in gravitational pull can be substantial because the Earth is nearly 8,000 *miles* across. Thus, one side of the Earth is always nearly 8,000 miles *closer* to the Moon than the other. Human beings, on average, however, are less than 6 *feet* tall and only about 1 foot thick. Thus, the difference between the Moon's gravitational pull on your head and your feet or your front side and your back side is millions of times less than across the diameter of the Earth. The capillary forces within the body are millions of times greater. Thus, the human body is neither aware of nor influenced by the Moon.

168 But isn't there proof of the Moon's influence in the close correlation between the length of time from one full Moon to the next and the human menstrual cycle? Coincidence, yes. Influence, no. After all, of the many species that exhibit menstrual cycles, only some come reasonably close in length to the cycle of the Moon's phases. And the length and regularity of menstrual cycles within the human species varies significantly. Furthermore, the start of each woman's menstrual cycle is different from most other women, so if a few females seem synchronized to a particular phase of the Moon, the vast majority are not.

169 Yet many nurses and doctors who work in maternity wards claim that the frequency of births and multiple births goes up significantly around the time of the full Moon. This is an interesting point but one that can easily be tested. In a study some years ago, Dr. George Abell, an astronomer at UCLA, decided to find out. He simply looked at the birth records and the phases of the Moon for the past several years. The result? No correlation whatsoever. The human mind can be a funny thing.

170 Some suggest that a higher incidence of violent crimes or other abnormal behavior occurs around the time of the full Moon. Again, most scientific studies of such situations, when large enough data samples are used, show little or no correlation. In this case, however, we must differentiate between the actual influence of the Moon and the influence a person is willing to believe the Moon has over them. In short, if a person believes strongly enough that the full Moon will influence his or her behavior, then his or her behavior is indeed more likely to show some variation. It is clearly important, however, to differentiate between the power of the Moon and the power of the human mind. To paraphrase the great philosopher Pogo: "We have met the phenomenon and it is us."

171 The Moon was probably formed when the Earth was struck by a huge object from space. Before we knew about *plate tectonics* (that is, the movement of large sections of the Earth's crust), some people noticed that the Moon was about the size of the Pacific Ocean and suggested that it somehow spun off this part of the Earth. Others have suggested that the Moon was formed elsewhere in the early solar system and, upon passing close by, was captured by the Earth. The most commonly held theory today, however, suggests that the Moon formed early in our solar system's history when an object about the size of Mars underwent a grazing collision with the Earth. The matter that was dislodged initially formed a ring around the Earth but ultimately coalesced to form the Moon. However, the Earth was probably still molten at the time.

172 The Moon has had a violent past. The history of the Moon (and most other objects in our solar system, for that matter) is a violent one. The Moon formed out of a cloud of gas and dust about 4.6 billion years ago. Smaller pieces of space debris continued to be swept up by the Moon as it solidified. Between 4.2 and 3.9 billion years ago, this great period of bombardment created many of the lunar craters that we see today.

By 3.8 billion years ago, the radioactive decay of materials at the Moon's center caused the interior to heat up, become molten, and trigger *volcanism* on the lunar surface. Lava flowed onto the surface, flooding low-lying basins and creating the lunar maria, or "seas."

By 3.1 billion years ago, the period of volcanism was over, the maria solidified and, except for occasional meteorite impacts, the Moon took on the general appearance it has today.

173 **The interior of the Moon is lumpy.** When spacecraft were first sent into orbit around the Moon, scientists noticed that they unexpectedly slowed down and speeded up at various points along the way. In time, scientists surmised that the changes in speed were due to the presence of large dense meteoroids that had collided with the Moon shortly after it formed and sunk below the molten lunar surface. The denser lumps created stronger gravity in their vicinity, which increased the speed of the spacecraft. Thus, the careful measurement of the motion of lunar probes allowed scientists to map the invisible distribution of material inside the Moon.

174 **The interior of the Moon also can be studied via seismographs that were left behind by the Apollo astronauts.** Almost 10,000 moonquakes have been recorded by this equipment. Some were due to objects striking the Moon at high velocity, but the vast majority were the result of stresses inside the Moon induced by tidal forces exerted by the Earth. Most moonquakes seem to occur in a zone that ranges from about 400 to 750 miles below the lunar surface. Below this lies the Moon's core, which many scientists think is still molten. Above the zone of moonquakes lies the Moon's mantle and its surface crust. The crust, on average, is only about 45 miles thick.

175 **A total of twelve astronauts have walked on the surface of the Moon.** Only a dozen human beings have left footprints on another world. They are the six teams of two astronauts each that rode the lunar lander to the Moon's surface in Apollo missions 11 through 17. Originally, there were several more lunar landing flights planned, but they were canceled due to NASA cutbacks. The ill-fated *Apollo 13* orbited the Moon without being able to land, as

Apollo astronaut Gene Cernan, one of only 12 men who ever walked on the Moon, stands next to the lunar rover. (NASA)

portrayed in the famous movie of the same name. According to detailed plans that were kept secret during the Cold War, the Soviets also attempted to at least send men around the Moon, but all attempts were aborted before takeoff.

176 **In all, the astronauts brought back 381 kilograms (840 pounds) of material from the Moon.** The material ranged in size from rocks the size of your head to fine grains of dust and came from a wide range of locations, including the vast lunar plains and the mountainous highlands. The youngest rocks brought back were about 3.1 billion years old, while the oldest were about 4.42 billion years old—almost as old as the solar system itself.

177 **Lunar rocks are generally much older than terrestrial rocks.** The Moon became geologically inactive 3.1 billion years ago and many places there have been geologically dead for far longer. By comparison, the Earth remains active with volcanism and plate tectonic movements till this day. As a result, most rocks on Earth are far younger than 3 billion years old, while most lunar rocks date back 4 billion years or more. Thus, the study of Moon rocks gives us clues about the early history of our solar system that we cannot obtain by studying rocks on Earth.

178 **There are similarities and differences between Earth rocks and those brought back from the Moon.** The rocks brought back from the Moon were all types that are familiar to geologists. In the maria and highland regions explored by the Apollo astronauts, some of the rocks are *breccias,* that is, rocks that are a mixture of different types of rock that have been "welded" together under pressure. In the maria, however, most of the rocks are *basalts,* which are fine-grained rocks containing metals and silicates.

The lunar rocks showed that the Earth and Moon appear to be similar chemically, at least near the surface, although the Moon rocks recovered contain no traces of water and some elements that are rare on Earth, including uranium and thorium, are found in greater abundance on the Moon. Perhaps someday mining the Moon will be financially feasible.

179 **The Moon has only about one-sixth the gravity of the Earth.** Because of its much smaller mass, the Moon has significantly less gravitational pull than the Earth. A person weighing 100 pounds on Earth would weigh less than 17 pounds on the Moon because the Moon pulls down on the person with only one-sixth the force that the Earth does. Astronauts quickly developed a kind of combination "saunter and hop" maneuver that allowed them to get around quite well on the lunar surface. If it weren't for their cumbersome space suits, however, they would have been able to jump six times higher and farther than they can on Earth. Given more flexible suits, the first Lunar Olympics someday will be one for the record books.

180 **If you drop a hammer and a feather at the same time on the Moon, they hit the ground at the same time.** Try this little experiment in your backyard and the hammer clearly hits the ground first. This is because the feather has a lot more surface area for its weight than the hammer and so is buoyed up by more air resistance as it falls. On the Moon, there is no air, so there's no air resistance. The Moon pulls down on the hammer with more force than the feather, but the hammer, because of its *inertia,* requires more force to fall as fast as the feather. The *inertia* of an object depends on its mass and is a measure of how easy or difficult it is to get the object moving or stop it once it is moving. A Cadillac, for example, has more mass than a hockey puck and so has more inertia. It takes a lot more force to get a Cadillac going from 0 to 60 than a hockey puck, but a Cadillac is also a lot harder to stop once it's going that fast. So the Moon pulls harder on the hammer than the feather but in so doing, it only causes the hammer to achieve the same acceleration, thus causing both it and the feather to hit the ground at the same instant.

181 **As seen from the Moon, the Earth goes through phases.** For the same reason the Moon goes through phases as seen from Earth, the Earth goes through phases as seen from the Moon. These phases, however, are always the exact complement of each other. In other words, when we see a full Moon on Earth, astronauts on the Moon would experience a new Earth. When a first-quarter Moon is seen in Earth's skies, a third-quarter Earth hangs over the lunar landscape. Also, since it's almost four times larger, the Earth looks almost four times bigger in the Moon's sky than the Moon looks to us on Earth.

182 **The Moon is gradually moving away from the Earth.** The Moon is gradually spiraling away from the Earth. Each year it is about an inch farther away than it was the year before.

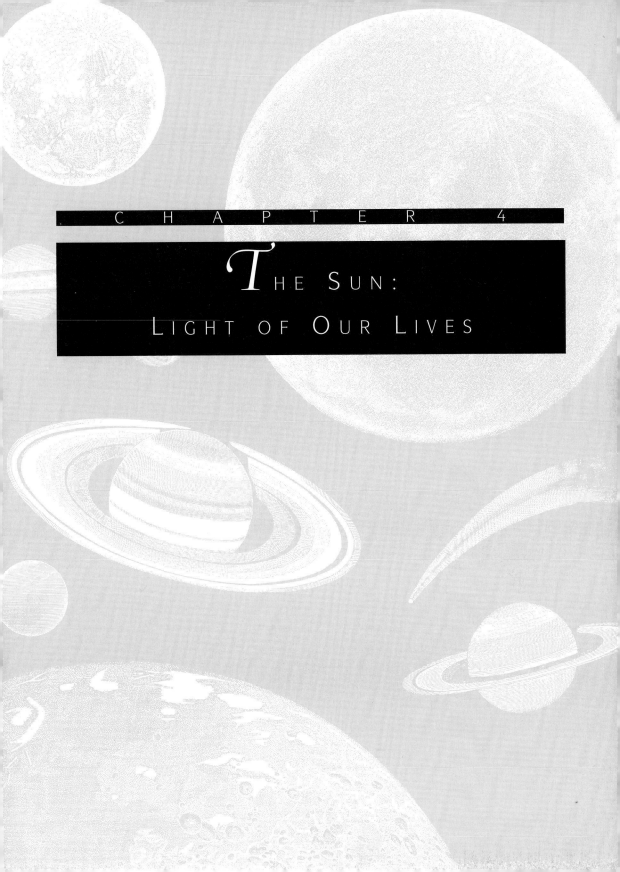

CHAPTER 4

THE SUN:
LIGHT OF OUR LIVES

183   **The Sun is an average garden-variety star.** The Sun is but one of trillions upon trillions of stars in the universe. While many stars are very similar to our Sun, there are many stars that are smaller, larger, hotter, and cooler. All in all, the Sun is about as average a star as you can find.

184   **The Sun rises in the east and sets in the west—but only two days a year.** Ask the average person where the Sun comes up in the morning and he or she will answer that it rises in the east. Similarly, he or she will probably say that the Sun sets each night in the west. In reality, however, the Sun only does this two days a year: the first day of spring and the first day of fall. Between the first day of spring and the first day of fall, people living in the northern hemisphere see the Sun rise to the *north* of east each day and set to the *north* of west. The situation reaches an extreme on the first day of summer when the Sun typically rises well to the north of east and sets well to the north of west. Between the first day of fall and the first day of

spring, the Sun rises to the *south* of east and sets to the *south* of west, reaching its farthest points south on the first day of winter. For people living south of the equator, the opposite is true.

185   **The Sun's path across the sky also changes throughout the course of the year.** Just as the rising and setting points of the Sun change during the year, so too does its path across the sky. Indeed, the two phenomena are connected. On the first day of summer, when the Sun rises as far to the north of east as it is going to get and sets an equal distance to the north of west, it also takes its longest and highest path across the sky. Thus, the first day of summer is also the longest day of the year. Conversely, on the first day of winter, when the Sun rises well to the south of east and sets well to the south of west, it also takes its lowest and shortest path across the sky. The first day of winter is therefore also the shortest day of the year. On the first of spring and fall, the Sun takes a middle path across the sky and indeed, the length of day and night are the same.

186 **The first days of spring and fall are known as the *equinoxes.*** The word *equinox* comes from Latin and literally means "equal night." While the term is a bit off the mark, it is still used and refers to the fact that on these two days each year, the lengths of day and night are equal.

187 **The first days of summer and winter are known as the *solstices.*** The term *solstice* comes from Latin and literally means "sun halt." This, however, requires a bit of explanation, for everyone knows that the Sun does not stop its motion across the sky. Instead, the term refers to the fact that, each year, between the first of winter and the first of summer, the Sun takes a longer and longer and higher and higher path across the sky each day. Finally, on the first day of summer, the Sun takes its longest and highest path across the sky, thus its *northward progression* comes to a *halt.* The *summer solstice* has arrived. Similarly, between the first of summer and the first of winter, the Sun reverses its course and takes a slightly lower and shorter path across the sky each day. And this trend continues until the first day of winter when again the Sun ceases its migration and gets ready to reverse its course once more. Hence another day when the Sun "halts": the *winter solstice.*

188 **Many cultures have been acutely aware of the changing positions and paths of the Sun across the sky.** On England's Salisbury Plain, a neolithic culture erected a circle of standing stones over three thousand years ago. Today, Stonehenge marks the rising and setting points of the Sun on the equinoxes and solstices as accurately as it did then. A thousand years ago, a great Native American settlement we call Kohokia stood on the banks of the Mississippi near modern-day St. Louis. In the ground there, scientists have found where a large circle of timber poles once stood. And to this day, the Hopi of the southwestern United States and the indigenous peoples of the Andes carefully note the changing point of sunrise by using mesas and mountain peaks as guides. The reasons for this sky watching were and are practical as well as spiritual, for the changing

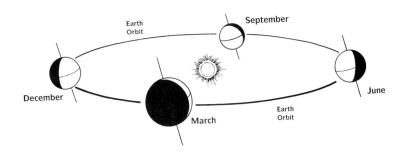

The tilt of the Earth's axis is the reason for the seasons.

positions of the Sun mark the turning pages of a celestial calendar and tell when to plant and harvest as well as when to observe key religious services.

189 **The cause of the Sun's changing paths across the sky, of course, is the tilt of the Earth's axis.** As the Earth revolves around the Sun, its axis of rotation remains tilted to the plane of its orbit. On the first day of summer in the northern hemisphere, this hemisphere is tilted most directly toward the Sun, so the Sun takes its highest path across the sky. Six months later, the Earth's northern hemisphere is tilted the most away from the Sun, so the Sun takes its lowest path across the sky. Midway between these points in its orbit, our planet's hemispheres are both pointed equally toward the Sun, so the Sun takes a midpath across the sky.

190 **By Earth standards, the Sun is a very big place.** The visible surface of the Sun is 864,000 miles in diameter. If the Sun were a large goldfish bowl, it would take more than 1 million marbles the size of the Earth to fill it up.

191 **The chemistry of the Sun is pretty simple.** While the Sun contains most of the elements we find across the universe, like the universe itself, most of the Sun is made of the simplest element of all: hydrogen. Indeed, hydrogen and the next most abundant element in the universe, helium,

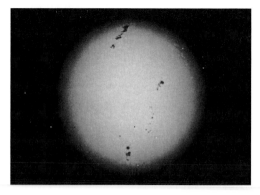

The Sun's visible surface, or photosphere, along with several sunspot groups. (NSO)

make up about 99.9 percent of the Sun. Other elements, including oxygen, carbon, nitrogen, and iron, comprise the other 0.1 percent.

192 **What we see as the surface of the Sun is not really a surface at all.** To our eyes, the Sun appears to have a solid-looking surface and the disk of the Sun has a distinctive sharp edge. In reality, however, the Sun is

The McMath-Pierce Solar Telescope at Kitt Peak National Observatory in Arizona. Sunlight is reflected off a large mirror at the top of the telescope (where the vertical and diagonal towers meet) and is directed down the diagonal tower to a room far below ground where it is studied. (NSO)

Two astronomers examine the image of the Sun deep inside the McMath-Pierce Solar Telescope. (NOAO)

depth when we look directly into the center of the Sun's disk than along its edges. Lower levels of the photosphere are hotter and therefore look brighter.

194 **The Sun's color tells us its surface temperature.** If you were to stick an iron poker into a very hot furnace and pull it out after a few minutes, you would probably find that it was glowing dull red. If you measured its temperature at this point, you would find that the glowing tip was radiating heat at about 5,000°F. If you were to put the poker back in the furnace and wait a little longer before pulling it out again, you might find it glowing bright yellow. Measuring its temperature, you would discover that it had climbed to about 11,000°F. The poker would now be the same color as the surface of the Sun and indeed would also have the same temperature. Similarly, the colors of other stars also indicate their surface temperatures, from red stars, which are the coolest, to blue-white stars, which are among the hottest.

195 **The surface of the Sun has a very mottled look.** Telescopic images of the Sun reveal that it looks a bit like a street of bright cobblestones embedded in darker cement. This appearance is due to the fact that we are actually looking down on the tops of countless cells of gas. The bright regions (each about the size of Texas) mark places where hot jets of gas are rising up from below, while the darker intervening areas show where slightly cooler gas is sinking to lower

a huge ball of gas and has no solid surface. What we see as a surface is merely the place where the density of the Sun's gases has fallen off to a level where they become transparent to light. Interior to this point the density is sufficiently high that the Sun turns opaque and we can see no deeper into its fiery layers. Although they really know better, modern astronomers still refer to this layer of the Sun as its surface, or *photosphere.* The name comes from the fact that the Sun releases the photons of light that ultimately travel to our eyes from this region.

193 **The Sun looks brighter in the center of its disk than along its edges.** This phenomenon is called *limb darkening* and is due to the fact that we are seeing to a greater levels in the Sun where it will be heated anew. Because this mottled appearance also looks a bit like rice soup, the effect is referred to as *granulation.*

196 **The Sun's granules flock together in giant clusters.** Studies of large-scale motion on the surface of the Sun show that granules are actually grouped into huge, roughly polygon-shaped regions. Matter seems to well up in the middle of these regions, moving out in all directions and sinking down along the edges. The polygons are about 20,000 miles across and are appropriately referred to as *supergranules.*

197 **The Sun's face has dark blemishes.** Chinese astronomers as early as the second century B.C. reported seeing occasional dark spots on the surface of the Sun. In Western tradition, they were not widely pub-

licized until nearly 1,800 years later when Galileo first examined the Sun through a telescope. Today, we know that these "sunspots" are places on the surface of the Sun where powerful magnetic fields constrain and slow the Sun's hot gases. The slowing down of the gas results in a lower temperature in these regions, which causes them to look darker than the surrounding areas. This, however, is somewhat of an illusion. If you could pluck a sunspot off the Sun and hold it against the night sky, it would shine more brightly than the brightest stars. The dark inner region of a sunspot is known as the *umbra,* while the lighter outer region is called the *penumbra.*

198 **Sunspots come and go in a cycle.** In the mid-1800s, an amateur astronomer named Samuel Heinrich Schwabe discovered that the number of spots on the Sun is not a constant. Instead, the number varies from relatively few to many and back to few again in a cycle that, on average, lasts a little over 11 years. At the beginning of a cycle, spots appear in each hemisphere at a latitude of about 30°. As the cycle continues, the number of spots increases and the spots in both hemispheres gradually migrate toward the equator. As the cycle comes to an end, the spots thin out and finally disappear at latitudes of about 5° while the spots of the next cycle begin to appear at higher latitudes once again. The most recent sunspot maximum occurred around 1990. The next sunspot maximum is due around the year 2001.

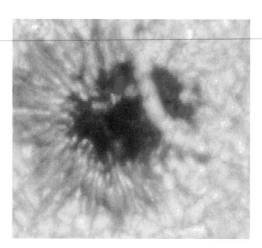

This close-up of a sunspot shows its dark inner region (umbra) and its lighter outer region (penumbra). Surrounding it, some granulation in the photosphere can be seen. (NSO)

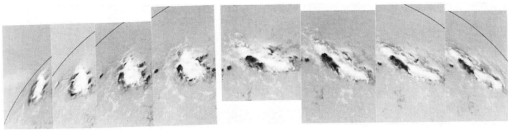

Images of a large sunspot group taken at one-day intervals show the rotation of the Sun. (NOAO)

199 **Sunspots come in pairs.** Because sunspots are magnetic in nature, they always come in pairs. One spot has a positive polarity, while the other has a negative polarity—just like the ends of a magnet. Indeed, you can kind of think of a sunspot pair as being like a horseshoe-shaped magnet with its ends poking up through the surface of the Sun.

200 **Frequently, sunspots occur in large groups and, as a whole, the group has magnetic properties.** Such groups can consist of two or more pairs of spots. If the leading portion of a sunspot group has one polarity, the trailing portion will have the opposite polarity. Furthermore, if the leading sunspots in the Sun's northern hemisphere have a positive polarity, the leading spots in the southern hemisphere will have negative polarity.

201 **The Sun's magnetic polarity reverses with every sunspot cycle.** Every 11 years or so, the polarities of the lead spots in the two hemispheres reverse; indeed, the magnetic polarity of the *entire Sun* reverses. Thus, magnetically speaking, the 11-year sunspot cycle can really be thought of as a 22-year cycle.

202 **Sunspots can be big.** Sunspots can easily be larger than the entire Earth and groups of sunspots have been known to stretch more than 100,000 miles across the solar surface.

203 **Some believe that major variations in the number of spots on the Sun have affected the Earth's climate.** Historical astronomical records show that during the period from 1645 to 1715 there were very few sunspots (a period referred to as the *Maunder Minimum*). Meteorological records from the same period also reveal an unusually prolonged cold spell for much of Europe. Could there be a link? Others believe they have found a correlation between periods of drought as evidenced in tree ring data and other sunspot cycle peculiarities.

204 **Recently, scientists may have found a physical effect link between sunspots and climate.** Dr. Sallie Baliunis of Stanford University has found evidence that the number of sunspots on the Sun is related directly to the total energy output of the Sun and this, in turn, may have an effect on the Earth's climate. One must be careful, however, because a statistical correlation alone can be coincidental or indicative of other variables at work and does not offer direct evidence of a physical cause and effect. There is, after all, a good statistical correlation in most American cities between the number of drunks and the number of churches, but common sense tells us there's more to the story. When it comes to phenomena as complex as the Sun and the Earth's climate, most scientists welcome additional studies.

205 **The Sun vibrates.** It has recently been discovered that the surface of the Sun vibrates a little like a ringing bell or a gong. These motions can be as rapid as hundreds of feet per second and seem to be combinations

A computer model illustrating one of many up-and-down vertical motion patterns the Sun's surface can undergo. The study of such motions in the Sun has led to the science of helioseismology and allows astronomers to gain insight into conditions in the Sun's interior, just as the science of seismography on Earth leads to a better understanding of conditions in the Earth's interior. (NOAO)

of millions of smaller motions of a few inches per second. From the study of these vibrations, astronomers are able to gain insight into the interior of the Sun, much as seismologists use vibrations from earthquakes to study the interior of our planet. To continue the study, the Global Oscillation Network Group (GONG) was established in 1995. It uses 6 specially equipped telescopes spread around the world so that the Sun can be kept under continuous surveillance.

206 **Powerful explosions frequently take place on the surface of the Sun.** These explosive eruptions are known as *solar flares* and they can unleash the energy of millions of nuclear bombs in only seconds. When a solar flare erupts, it blows a gigantic hole in the Sun's atmosphere and blasts light, radio waves, high-energy X rays, and trillions upon trillions of charged particles out into space. The event is usually referred to as a *solar storm.* Solar flares and storms occur more frequently and are fiercest around times of sunspot maximum.

207 **Solar flares can trigger some interesting reactions on Earth.** Charged particles of a solar flare aimed in the direction of Earth will arrive within a few days of the eruption. The particles get trapped in the Earth's magnetic field and come crashing down into the upper atmosphere at tens of thousands of miles per second. The result is a *geomagnetic storm,* a great disturbance in the Earth's magnetic field that can make compass needles swing off-course, create radio inter-

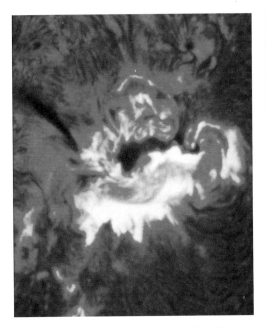

A solar flare explodes on the surface of the Sun, unleashing the energy of millions of nuclear bombs in only seconds. (NSO)

ference, make circuit breakers on power lines trip out, and . . . trigger displays of the *northern* and *southern lights.*

208 **The northern and southern lights, also known as the *aurora,* are a result of the atmosphere of the Sun touching the atmosphere of the Earth.** When charged particles from a solar flare reach the Earth, they interact with a huge doughnut-shaped magnetic field that surrounds our planet. This magnetic field redirects the incoming particles and channels them down toward the north and south magnetic poles. Streaming down into the Earth's upper atmosphere at a significant fraction of the speed of light, the solar particles crash into the oxygen and nitrogen atoms dozens of miles above our

heads. When these collisions occur, they cause the atoms of air to light up (like the gas in advertising signs at night). We see this glowing display from the ground as the northern and southern lights.

209 **The aurora can have many different forms and colors.** Sometimes, the aurora looks like amorphous patches of light that flash quickly across the sky. At other times, the lights can take the form of curtains or tapestries that appear to slowly ripple and flap in a silent breeze. Yet another form of the aurora resembles a burst of lines radiating from a point high overhead. Auroras can be white or shades of red, orange, green, or blue. The colors depend on the types of atoms in our atmosphere that are being struck by the solar particles as well as the energy of the particles. Sometimes the aurora will come and go in only minutes, while at other times it can last all night long. It depends on many things, including the length and severity of the triggering solar storm.

210 **Folklore offered many colorful explanations for the aurora.** Some Eskimo tribes believed the aurora was the spirits of departed ancestors running across the sky and playing a kind of soccer game with the skull of a walrus. To the ancient Chinese, the twisted streamers of light looked like dragons fighting in the heavens, while the Vikings held that the aurora was the light of torches held high on the "back side of the sky" and used to guide newly departed souls to Valhalla.

211 **There is no "best time of year" to see the aurora.** Auroras do not occur more frequently during any particular season of the year. Instead, because they are triggered by solar flares, the frequency and intensity of auroral displays comes and goes with the sunspot cycle. There are more frequent, brighter, and more active auroras during times around sunspot maximum and fewer during times of sunspot minimum. The next sunspot maximum will likely be around the year 2001, but occasional displays of the northern and southern lights might be seen at any time.

212 **Displays of the northern and southern lights do not occur most frequently directly over the north and south poles.** The charged particles from the Sun that produce the aurora are not channeled directly down toward the Earth's poles. Instead, they are spread out into rings or bands that surround the north and south *geomagnetic* poles. Thus, auroras most frequently occur within these bands. Typically, for the northern lights, this zone passes over central Alaska, dips southeastward across Canada, and curves northward again, exiting the North Atlantic over Labrador. In turn, it enters Europe over northern Scandinavia and continues across extreme northern Russia. From Earth-orbiting satellites, the aurora indeed looks like a continuous ribbon of light looping around the Earth. This auroral zone can expand and contract, however, and auroras over the continental United States are not uncommon. Auroras have even been seen in the Caribbean and within a few degrees of the equator, but these occurrences are rare. The southern lights occur most frequently in a band that cuts across portions of Antarctica and the surrounding Antarctic Ocean and hence are not commonly seen.

213 **Auroras usually occur simultaneously in both hemispheres.** The particles from the Sun that strike the Earth's upper atmosphere and cause the aurora are forced by our planet's magnetic field to bounce back and forth between the north and south geomagnetic poles. Frequently, the particles are accelerated to such high speeds that a particle that strikes an oxygen atom over Alaska one moment may be all the way "down under" striking an oxygen atom over Antarctica a fraction of a second later! Thus, auroras usually occur in both hemispheres at the same time and also tend to take on the same shape in both places.

214 **Auroras usually occur at altitudes of 50 to 100 miles above the surface of the Earth.** It is at these altitudes that the incoming particles from the Sun are best able to strike atoms of oxygen and nitrogen in our atmosphere. From space the vertical extent of the aurora is very evident. Dramatic pictures taken by the astronauts aboard the Space Shuttle show auroral bands of light hanging down like curtains through 50 miles or more of the Earth's atmosphere.

215 **Some other planets also have auroras.** Other planets with significant magnetic fields and atmospheres also have auroral displays. The Hubble Space Telescope has captured pictures of the aurora on Jupiter and Saturn and data from the *Voyager* spacecraft suggest Uranus and Neptune probably also exhibit the phenomenon.

216 **A day on the Sun can last different amounts of times.** Like the Earth, the Sun rotates on its axis. But unlike the Earth, the Sun is not solid. As a result, different latitudes on the Sun rotate at different speeds. At the equator, the Sun turns around once in about 25 Earth days. At higher and higher latitudes, however, the Sun rotates slower and slower. Near its poles, a day on the Sun lasts over 31 Earth days long. On Earth, a place that lies due south of you this week will also lie due south of you next week and a year or century from now as well. But on the Sun this isn't the case, as places closer to the equator continually speed ahead and slip off to the east. It's all a rather fluid situation.

217 **The Sun acts like a giant magnet in space.** Like the Earth, the Sun acts as if it has a gigantic bar magnet stuck inside of it. This magnet creates an enormous invisible magnetic field that stretches for hundreds of millions of miles out into space and controls the flow of all the hot gases in and around the Sun. Every 11 years or so, at the beginning of each sunspot cycle, this entire magnetic field flips over and reverses its polarity so that the north magnetic pole of the Sun becomes the south magnetic pole and vice versa. The direction of rotation of the Sun, of course, remains the same.

218 **The Sun has an atmosphere.** Above the visible "surface" of the Sun, or *photosphere,* lies an extensive atmosphere of very hot electrically charged gas. The lowest layer of this atmosphere is called the *chromosphere* (from the Greek *chromos,* meaning "color"). The name derives from the fact that

Jets of gas called spicules spout out of the Sun's photosphere. Weaving in the ever-changing magnetic field of the Sun, they make the Sun's surface look like a "burning prairie." (NOAO)

this region has a red or pinkish hue. The chromosphere is about 7,000 miles thick and is somewhat hotter than the photosphere, with temperatures ranging from about 11,000°F to about 30,000°F.

219 Through a telescope, the chromosphere looks like a "burning prairie." The chromosphere is peppered throughout with giant jets of hot gas called *spicules,* which are about 500 miles across and about 5,000 miles high. At any one time, half a million or more may be found shooting up like fountains of flame from the Sun's surface. The Sun's ever-changing magnetic field forces the charged particles that make up the spicules to weave and dance like shafts of wheat in the wind, giving this region of the Sun's atmosphere the appearance of a "burning prairie."

220 Above the chromosphere lies the Sun's outer atmosphere. Above the chromosphere lies the *corona.* This pearly white region of charged gases extends from the top of the chromosphere far out into space. Within it lie clouds of glowing gas called *prominences* that take on the appearance of jets, loops, or great arches that can span hundreds of thousands of miles.

SIZE OF EARTH

An enormous eruptive prominence rises over 350,000 miles into the solar atmosphere. It reached this altitude in only 90 minutes. A dot shows the size of Earth for comparison. (NASA/JPL)

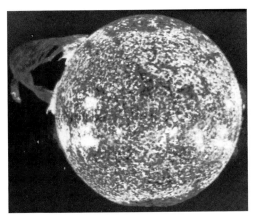

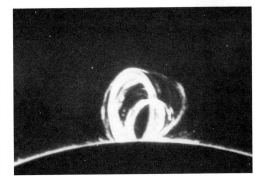

A graceful loop prominence forms in the lower atmosphere of the Sun. It is made of charged particles that move along the curved paths of the Sun's magnetic field. (NSO)

A spectacular prominence stretches over 367,000 miles of the Sun, arcing high into the solar atmosphere. (NASA/JPL)

221 **Astronomers can study the lower portions of the corona with a special instrument.** The special instrument is called a *chronograph* and it is essentially a telescope with an opaque disk located within that blocks the intense light of the photosphere. Such instruments must typically be located in high mountain observatories where the air is very clear and steady. Under such circumstances, scientists are able to see the Sun's chromosphere and the lowest regions of the corona.

222 **The outer corona can only be seen during a very special circumstance.** The outer corona is so faint that it can only be seen during the brief moments of a total eclipse of the Sun when the dark disk of the Moon blocks out the glare of the brilliant photosphere. This is one of the reasons that total solar eclipses are of interest to astronomers, for they give scientists a few precious moments to see and study our Sun's atmosphere.

223 **The Earth swims in the Sun's atmosphere.** During a total eclipse of the Sun, we can see the Sun's corona stretching out to several times the diameter of the Sun, but in reality the corona reaches much farther into space. Indeed, it actually engulfs the entire solar system, so the Earth and the other planets are literally moving through its atmosphere as they orbit the Sun. This "Earth-Sun connection" is most visible to us during displays of the northern and southern lights when particles shot from solar flares stream out through the corona and collide with the upper atmosphere of Earth.

224 **Portions of the corona have temperatures as high as 4 million°F, but if you were placed there you would freeze to death.** This sounds like a great paradox, but actually it's not. To understand the riddle, however, we must understand the difference between *temperature* and *heat.* Many people use these two words as if they are synonymous and, in everyday experience, something that is hot does indeed have a high temperature. But when it comes to the Sun's atmosphere, this is not the case. The reason why is because the gas that makes up the corona is *extremely* thin. It's so thin, in fact, that we would be hard-pressed to produce gas this tenuous in a very good vacuum chamber in a laboratory on Earth.

Now, when we refer to the *temperature* of an object (be it a solid, a liquid, or a gas), what we are really referring to is the *average speed* of the atoms and molecules that make up that object. (In a gas or liquid, the atoms or molecules actually bounce around. In a solid, they vibrate in place.) The higher this average *speed,* the higher the object's *temperature.* It's as straightforward and simple as that. But the *heat* content of an object is a measure of the *total energy* possessed by all that object's atoms and molecules. Since the Sun's atmosphere is so thin, there are very few atoms of gas in each cubic foot of the Sun, so the *total energy* present in each cubic foot is very *low,* even though the average *speed* of the atoms in the cubic foot is very *high.* Thus, the corona has a very *high temperature* (fast atoms) but holds very *little* total *heat energy,* so you would quickly freeze to death if you hung around there.

225 **The Sun is continuously leaking bits and pieces of itself out into space.** A constant flow of charged particles (mostly electrons) is forever streaming off the Sun in what is known as the *solar wind.* When flares erupt on the surface of the Sun, they greatly enhance the solar wind. The wind blows past the Earth and the other planets and probably extends halfway out to the nearest stars where the wind from the Sun encounters that of neighboring stars. Astronauts in space don't feel the pressure of the solar wind the way you feel the air brush past your face because the corona is far too tenuous.

226 **Someday spaceships may sail on the solar wind.** Futurists have envisioned fashioning great sheets of plastic far thinner than a dry-cleaning bag but miles on a side. Stretched on a lightweight frame and sprayed with a coating of aluminum only a molecule thick, such "solar sails" would be large and light enough to actually use the solar wind to coast around the solar system like a sailboat on the ocean. For a fun read, check out Arthur C. Clarke's short story "The Wind from the Sun." It's all about the first solar sail races from the Earth to the Moon.

227 **The Sun and its planets were all born at the same time.** The Sun and its planets formed from a giant cloud of gas and dust about 4.6 billion years ago. Inside the cloud, gravity gradually overcame turbulence. At the center of the cloud, as gas and dust fell in upon itself, temperatures steadily rose. In time, the region ignited and became our Sun. Smaller clumps of material also formed in orbit around the central star. These became the planets and their satellites as well as asteroids and comets.

228 **The Sun has undergone some changes over the last 4.6 billion years.** From studying other stars, astronomers deduce that billions of years ago our Sun was probably somewhat cooler than it is today and perhaps even appeared quite orange in the early skies of Earth. If this is true, it also means that the early Earth may have received a little less light and heat from the Sun than we do today. Within the last few billion years, however, the Sun has likely remained pretty much as we see it today—a stable yellow star.

229 **The Sun remains stable because it performs a perfect balancing act.** Each day, except for its changing spots, the Sun looks about the same in our sky, growing neither larger nor smaller. (As we will see in the section of this book on the lives of the stars [see page 163], many stars are not like this.) The Sun's stability is due to a perfect balancing of forces that is going on throughout the Sun. Gravity is the force that first formed the Sun, causing it to collapse out of a giant cloud of gas and dust. But at some point long ago, the Sun stopped collapsing (in other words, it stopped getting smaller). The reason is quite simple. As the Sun collapsed and its nuclear fires ignited, the heat these fires generated plus the action of so many atoms colliding furiously inside the Sun created enormous pressure. This pressure pushed outward and counterbalanced the force of gravity pulling inward. Today, the Sun performs a miraculous balancing act, for at each point inside the Sun, the force of the radiation and gas pressure pushing outward ex-actly matches the force of gravity pulling inward. So the Sun remains poised as a stable healthy star. And it's a good thing because life on Earth would be a far more difficult proposition without such stability.

230 **The center of the Sun is a really hot place.** While the temperature on the surface of the Sun isn't exactly frigid, its central temperatures are probably somewhere around 30 million°F.

231 **The Sun shines because it is a giant thermonuclear power plant in the sky.** The Sun creates its energy, including its light, by a process known as nuclear *fusion*. Nuclear power plants on Earth use a different process called nuclear *fission* in which complex elements, like uranium, break apart (or fission) into simpler elements and generate heat. The Sun's energy comes from combining (or fusing) simple elements into more complex ones. Scientists would like to be able to duplicate this process on Earth. Fusion is a lot "cleaner" than fission because it doesn't create dangerous radioactive waste as a by-product. So far, scientists have not been able to create high enough temperatures in their laboratories to be successful.

232 **The Sun creates its energy by performing a great disappearing act.** The Sun creates its energy by fusing the simplest and most abundant element in the universe, hydrogen, into the next simplest element, helium. Each second, inside the Sun, 600 million tons of hydrogen are converted into 596 million tons of helium. The missing 4 million tons a second are converted into pure energy, according to Einstein's famous equation $E = mc^2$. In this equation, m is the missing *mass, E* is the equivalent amount of *energy* that gets created when the mass disappears, and c^2 is the *speed of light times itself.* Now, the speed of light is a really big number, so the speed of light multiplied by itself is a really, really big number, which means that a small amount of mass is equivalent to a whole lot of energy. Since 4 million tons of missing mass each second amount to a lot of mass, the equivalent amount of energy that is created by the Sun each second is, well, astronomical. This energy ultimately works its way to the Sun's surface and out into space as heat and light and that, in essence, is what makes the Sun shine.

233 **While doing its great energy generating trick, the Sun still obeys a law of the universe called the Conservation of Matter and Energy.** This law states that the sum total of matter plus energy must remain constant in any physical process, whether it's burning a candle or making a star shine. The Sun (and other stars) can create enormous amounts of energy but only by losing an exactly equivalent amount of mass.

234 **The inside of the Sun is a very dark place.** Even though the Sun shines brilliantly with visible light, it doesn't make light in its interior. This is because its central temperatures are so high the energy given off by its nuclear reaction is in the form of high-energy gamma rays to which the human eye is insensitive. Thus, if you could somehow withstand the incredible temperatures and look around deep inside the Sun, it would be darker than the darkest night.

Gradually, these gamma rays work their way to the surface and, in the process, are gradually changed into visible light. At the photosphere, this visible light streams out into space and eventually into our eyes.

235 **Once free of the Sun, its radiation travels fast, but . . .** Once visible light reaches the Sun's photosphere from its interior, the light streaks the 93 million miles to Earth in only 8 minutes and 20 seconds. This is because outer space is pretty empty and does little to impede the radiation. But the interior of the Sun is so opaque that radiation suffers a very different fate. Indeed, a gamma ray created deep inside the Sun is passed from atom to atom for several hundred thousand years and gradually converted to visible light before finally making its way out to the photosphere. So while the final 93 million miles of the radiation's journey take place in little more than 8 minutes, the first few hundred thousand miles take many, many millennia to traverse.

236 **The Sun has better climate control than your home or office.** You wouldn't want to live inside the Sun, but in terms of temperature variations, the Sun does a very good job of climate control. While temperatures deep in the Sun's interior are millions of degrees higher than they are at the Sun's surface, this drop in temperature is spread out over a distance of nearly half a million miles. The result is less temperature variation across 25 feet of the Sun than over the same 25 feet of home or office space. This very gradual drop in temperature is enough to drive the radiation from the Sun's interior to its surface. But it is the fact that this temperature drop is *so* gradual that accounts for the long, long time it takes the radiation to travel to the surface.

237 **When the Earth was young, the Sun looked a bit different than it does today.** Around 3.5 billion years ago, when life was first developing on the Earth, the Sun looked a little different in the sky than it does today. In short, its surface was a few hundred degrees cooler (which means it was a tad yellower) and it was also about 8 percent to 10 percent smaller and only about 70 percent to 75 percent as bright. The Sun has been gradually growing larger, hotter, and brighter ever since, a span of some 3.5 billion years. However, this has nothing to do with the current concerns about "global warming," which appears to have occurred in only the last century or two.

238 **About 5 billion years from now, real estate values are really going to bottom out.** For the next 5 billion years or so, the Sun should remain quite stable, getting a little brighter, larger, and hotter as hydrogen continues to be converted into helium. Thereafter, however, the Earth is in for some big changes. In about 5 billion years, the Sun's core of helium will have grown so large and massive that it will collapse under its own weight. When this happens, temperatures inside the core will soar and trigger the burning of helium to create carbon (while hydrogen will continue being converted to helium in a shell surrounding this region). The extra energy released by these additional thermonuclear reactions will gradually push the photosphere of the Sun outward and the Sun will expand to become a *red giant star*. In time, the Sun will swell to engulf and incinerate first Mercury and then Venus. It may even reach out almost to the Earth's orbit. Its red surface color will mean that this future Sun will be cooler than the one we see today, but the closeness of its surface to Earth will cause our planet's polar caps to melt and its oceans to boil.

239 **After becoming a red giant star, more changes will be in store for our Sun.** Late in life, the Sun will gently blow off some of its photosphere. The material will engulf the Sun in one or more expanding luminescent shells of gas and dust known as *planetary nebulae*. Then, as its nuclear fires shut down, the Sun will settle down to become a *white dwarf star* about the size of the Earth. Thus, the Sun will have gone from being over 800,000 miles across (as it is today) to being a red giant star almost 200 million miles in diameter to finally become an object only 8,000 miles across. With its nuclear fuel spent, the Sun will gradually cool off, changing from white-hot to yellow, orange, and red before ultimately becoming a black lump in the cold darkness of space. How's that for a happy ending? We'll have more on the evolution of the Sun and other stars later.

---

240 **What causes an eclipse of the Sun?** An eclipse of the Sun occurs when the Earth, Moon, and Sun line up with the Moon in the middle or, in other words, at the time of the new Moon. When this happens, if we're in the right place, we see the dark disk of the Moon pass across the bright disk of the Sun.

---

241 **Although the Moon must be new at the time of a solar eclipse, an eclipse doesn't take place every time the Moon is new.** If this was the case, a solar eclipse would occur somewhere on Earth every month. Instead, the new Moon typically passes just a little above or below the Sun, so its shadow misses the Earth and no eclipse takes place. This is the result of the orbit of the Moon being tilted a bit relative to the orbit of the Earth around the Sun.

242 **Solar eclipses can be *partial, total,* or *annular.*** If the new Moon passes in front of only a portion of the sun's disk, we see a partial eclipse of the Sun. Should the Moon pass directly in front of the Sun, its dark disk can cover the entire solar disk, resulting in the awesome spectacle known as a total eclipse of the Sun. A third type of solar eclipse is called an annular eclipse. Because the Moon's orbit is an ellipse, the Moon's dis-

tance from the Earth—and, therefore, its apparent size—varies. Should an eclipse occur when the Moon is near the far point in its orbit, its disk is not quite big enough to completely cover the Sun. At mideclipse a ring, or *annulus,* of the Sun's bright photosphere remains visible, encircling the Moon like a brilliant gold band.

243 **While a total eclipse of the Moon can be seen simultaneously by everyone living on the same side of the Earth, a total eclipse of the Sun is witnessed by a lucky few.** The partial phases of a solar eclipse can be seen over a large area, but the total phase, where the Moon completely blocks the Sun, is confined to a narrow strip along the surface of the Earth. This strip, called the *path of totality,* marks the path of the Moon's shadow as it races across the surface of the Earth (as the Moon itself crosses between the Earth and the Sun). The Moon's shadow typically first strikes the Earth at its western limb where sunrise is taking place and then sweeps across land and sea at over 1,000 miles per hour before exiting the Earth's surface on its eastern limb at sunset. Along the way, the point where the shadow touches the Earth traces a narrow path. Only people who manage to find their way into this narrow path get to see a total eclipse. People outside the path of totality see only a partial eclipse or none at all.

244 **A total eclipse of the Sun is one of the most awesome spectacles in nature.** If you are ever fortunate enough to be in the path of a total eclipse, it will be an experience

The "diamond ring effect," which occurs just before and after totality in a solar eclipse, is caused by a tiny spot of the Sun's brilliant photosphere shining out from behind the dark disk of the Moon. Surrounding the rest of the new Moon, the lower portion of the Sun's atmosphere can be seen. This picture was taken in Hyderabad, India, on February 16, 1980. (NOAO)

that you will never forget. As the Moon closes in on the Sun, it produces an ever-thinning crescent of sunlight. The landscape grows noticeably darker and can even take on strangely colored hues. Moments before totality, light and dark bands are sometimes

A total eclipse of the Sun. Only during moments of totality can the Sun's pearly outer atmosphere, the corona, be seen. (NOAO)

seen racing across the ground and the last brilliant patch of photosphere still peeks out from the Moon, creating a dazzling "diamond ring effect." Finally, in an instant . . . *totality.* The pale pearly atmosphere of the Sun, the corona, frames the dark disk of the new Moon and makes it look like a black hole in the sky. The brighter stars and planets come out in a surrealistic false twilight and brilliant prominences ring the Moon's disk, looking like reddish tongues of fire. Then, as quickly as it began, totality is over. The precious seconds will go by quickly, but standing "in the shadow of the Moon" is something you will always remember.

245 **Totality can last for different amounts of time.** The length of totality depends on your position and that of the Moon at the time of the eclipse. The Moon's orbit is not a perfect circle but rather an ellipse, which means that at different times the Moon is different distances from the Earth and its apparent size also differs. If the Moon is at the near point in its orbit at the time of the eclipse (a point known as *perigee),* it will appear a bit larger than the Sun and totality can last almost seven and a half minutes. By contrast, if the Moon is near *apogee* (its farthest point from Earth) at the time of an eclipse, the Moon will not appear large enough to even cover the Sun, so it will only produce an *annular* eclipse. Also, if you are going to travel to see a total eclipse, try to position yourself as close to the center of the path of totality as possible, for the closer you are to the center line, the longer totality will last for that particular eclipse.

246 **Astronomers wishing to prolong their viewing time of eclipses have found a way to cheat the system.** Some astronomers have beat nature at its own game by racing across the sky in a jet full of astronomical instruments. By getting into the shadow of the Moon and flying along with the shadow, they have prolonged their view of totality for an hour or more.

---

247 **The Earth is the only planet other than Pluto where a total eclipse of the Sun can be seen.** The fact that we can witness a total eclipse of the Sun is the result of a rare coincidence. It just so happens that our Moon, while about 400 times smaller than the Sun, is also, at this point in history, about 400 times closer. Thus, the Moon and the Sun appear to be just about the same size in our sky, making total eclipses possible. In our solar system, a total eclipse can occur on no other planet with a solid surface except Pluto because the satellites of the other planets are either too small or too far away from their planets to completely cover the Sun. And so, here on a planet where there are beings to appreciate the awesome beauty of a total solar eclipse, nature has produced it.

---

248 **Among early peoples, a solar eclipse could be a pretty scary thing.** If you knew the Sun was responsible for growing your food and sustaining your life and then one day it began to disappear from the sky, you might understandably get a little concerned. One ancient Chinese belief was that, during an eclipse, a dragon was eating the Sun. Various other cultures saw evil spirits at work. Many solutions were sought, including the beating of drums, the shooting of arrows into the sky, and the sacrificing of something or someone to the gods. And unfortunately for the victims of these sacrifices, the solutions always worked.

249 **Eclipses can be predicted with great accuracy.** Because we understand the orbit of the Moon and the motions of the Earth, the Moon, and the Sun very well, eclipses can be predicted down to the minute for many years to come. Eclipses follow a regular cycle known as the *Saros cycle,* which is 6,585.32 days long (18 years, 11.32 days). About 71 solar eclipses of all types (total, annular, and partial) occur during each Saros cycle, after which the basic pattern repeats. Thus, every 6,585.32 days a similar eclipse takes place. It doesn't happen over the same place on Earth, however, because of the 0.32 days left over at the end of each Saros cycle. During this extra 0.32 days, the Earth has time to revolve an extra 117° on its axis, thus displacing the corresponding eclipse almost, but not exactly, a third away around the globe from where it was last time. For this reason, even though there is a cycle to solar eclipses, most people are unaware of it. To really see the pattern, you have to think *globally* and not in terms of how often an eclipse is seen over a particular city.

250 **According to legend, the failure to predict an eclipse once proved fatal.** The story goes that two royal astronomers who lived in China in the second century B.C. failed for some reason to predict an eclipse of the Sun. Since the ancient Chinese emperors believed themselves to be the "sons of heaven," it was of great importance that they be kept aware of all "signs" in the heavens that might be signals from the gods. For this reason, the court typically employed a number of astronomers to perpetually watch the skies for important celestial signs. Bright comets and shooting stars could not be predicted, so these were reported the next morning. But eclipses were predictable even then. The royal astronomers' failure to warn the Emperor of such an important celestial event reportedly led to such anger and embarrassment for the son of heaven that he had them beheaded! Needless to say, being an astronomer in those days was a bit more dangerous than it is today.

251 **The Sun doesn't give off any special radiation during an eclipse.** Safe viewing of a solar eclipse is a very misunderstood thing. The Sun does not give off any special radiation at the time of an eclipse (after all, the Sun doesn't know or care that an eclipse is taking place on its third little planet), so there is nothing harmful about being outside during an eclipse. Under no circumstances, however, should you stare at the Sun or even sneak a peek during the *partial* phases of a solar eclipse. This is because, while the Sun isn't any more harmful on an eclipse day than any other day, it also isn't any *less* harmful. Common sense normally dictates that you should not stare at the Sun because you can seriously and permanently burn your retina. But knowing that an eclipse is taking place can get the better of one's curiosity and it can become very tempting to sneak a peek to see what's going on. Of course, even the partially eclipsed Sun is still glaringly bright, so you squint and stare a little longer and, before you know it, you've burned your retina. In short, eclipses can cause eye damage not because of a difference in the Sun, but because people get curious and don't take the right precautions.

252 **With the right precautions, the *partial* phases of a solar eclipse can be viewed safely.** It's very unfortunate that misunderstandings lead to some people not enjoying one of nature's neatest sky shows. When it comes to *safely* viewing the *partial* phases of a solar eclipse, however, there are some very important do's and don'ts.

✦ DON'T look directly at the Sun even for a second without proper protection, whether an eclipse is taking place or not.

✦ DON'T use so-called "smoked glass" (that is, a piece of glass on which is deposited the soot from a burning candle).

✦ DON'T use mirrored sunglasses, even several layers of them stacked together.

✦ DON'T use either black-and-white or color film negatives, even stacked several layers thick.

✦ DON'T look at a reflection of the Sun in water or a mirror.

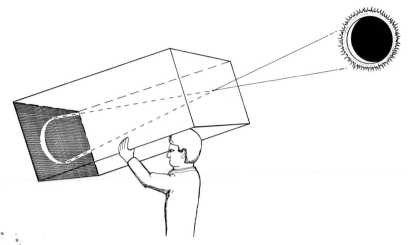

A box with a pinhole in one end creates a projection of the Sun on the opposite inside wall of the box. This projection is safe to look at.

❖ DO look at the Sun through a piece of No. 14 welder's glass (available from welding supply companies).

❖ DO use special coated Mylar or coated Mylar glasses (these are NOT mirrored sunglasses) that are sometimes available from reputable planetariums and science museums.

❖ DO construct a pin-hole projector.

With the right precautions, a partial eclipse or the partial phases of a total solar eclipse can be safely enjoyed.

253 The *total* phase of a solar eclipse is perfectly safe to look at. During, and only during, totality itself can you safely view a solar eclipse with the naked eye. During totality, all of the brilliant disk of the Sun's photosphere is covered by the Moon and only the corona is visible. The corona itself is only about as bright as the full Moon, so viewing the corona falls well within acceptable safety limits for the eye. But even a tiny sliver of the photosphere is bright enough to cause eye damage, so up to the very moment when totality begins, you must use one of the safe viewing methods outlined above. And the moment the "diamond ring effect" begins to burst out from behind the Moon, signaling the end of totality, you must immediately divert your eyes and again return to the safe viewing methods for the partial phases.

254 Interesting things can happen during a total solar eclipse. If you are near or within the path of totality, be aware of some curious consequences of the oncoming false twilight in the middle of the day. As more and more of the brilliant center of the Sun's disk is covered, the landscape will not only grow darker, but may take on an odd coloration. In addition, the temperature may drop several degrees and the wind may pick up. And birds may actually begin to roost, thinking that the Sun is about to set.

255 **During a total solar eclipse, the Moon wears a necklace.** Just before the first moment of totality, as the "diamond ring effect" fades from view, you can see tiny brilliant specks of the photosphere just peeking out through the valleys along the edge of the Moon while being blocked by intervening hills or mountains. The result makes the Moon look like it's wearing a necklace of jewels. These bright beads of light have come to be called *Bailey's beads.* At the very end of totality, the phenomenon is repeated along the opposite limb of the Moon as the Sun begins to emerge from behind the Moon.

CHAPTER 5

THE INNER SOLAR SYSTEM: LAND OF THE ROCKY MIDGETS

Mercury

256 **Mercury is the closest planet to the Sun.** Averaging only 36 million miles from the fiery solar surface, Mercury is our Sun's closest planetary neighbor.

---

257 **Mercury really runs hot and cold.** Under the searing rays of the Sun, daytime temperatures on Mercury can climb to 800°F—hot enough to melt tin and lead. At night, however, temperatures can plummet to nearly −350°F. The fact that Mercury is an airless world makes such great extremes possible.

---

258 **From Earth, Mercury looks like little more than a tiny blob.** Mercury is so small and always stays so close to the Sun that it is particularly difficult to observe from Earth. Never venturing far from the solar glare, the planet sets shortly after the Sun or rises only a bit before sunrise, giving astrono-mers little opportunity to study it in telescopes. Its tiny size also inhibits seeing much detail. When it is closest to Earth, Mercury also passes between the Earth and the Sun, meaning that the hemisphere pointed toward Earth is not illuminated by the Sun.

259 **As recently as the 1960s, we had little idea what the surface of Mercury really looked like.** Hand-drawn maps of Mercury only showed faint shadings of light and dark with little agreement from one set of sketches to another. Photographs revealed

The U.S. spacecraft *Mariner 10* flew past Mercury and Venus in the early 1970s. (NASA/JPL)

little more. Astronomers couldn't even agree on the length of Mercury's day.

260 **In 1974 and 1975, our knowledge of Mercury suddenly jumped a million-fold.** On three separate occasions during these years, the *Mariner 10* spacecraft made a rendezvous with Mercury and transformed our view of the planet from a tiny shadowy blob into a world of incredible detail. Approaching to within 200 miles of the surface, *Mariner*'s camera eyes sent back hundreds of images of dramatic and varied terrain.

261 **Mercury is like the Moon but different.** At first glance, Mercury looks a lot like the Earth's Moon. It is covered with thousands upon thousands of craters. But there are some basic differences. Its cratered areas are not as densely packed as those on the Moon (possibly due to lava flooding from *volcanism* long ago). Mercury's craters are also flatter than those on the Moon, probably the result of Mercury's stronger gravitational field. Mercury does not have a great number of basaltic plains, or maria, as we see on our Moon. One notable exception is the Caloris Basin, the largest known *impact basin* on Mercury, formed when a large object from space collided with Mercury long ago. The basin is ringed by concentric chains of mountains, making it look like a giant bull's-eye about 800 miles across.

Photomosaics of the cratered terrain of Mercury taken by *Mariner 10*. (NASA/JPL)

262 **Mercury also has some *weird terrain.*** This is not a judgment call but what the region is actually called. It gets its name from the crisscrossed and jumbled pattern it displays. Not coincidentally, this region lies directly opposite the part of the planet where we find the giant Caloris Basin. In effect, when the object from space that formed the Caloris Basin struck Mercury, it sent seismic waves around and through the entire planet—waves that converged at the opposite point on the planet and created the weird terrain.

263 ***Scarps* are one type of feature found on Mercury but not on the Moon.** Mercury's scarps are cliffs that reach heights of almost 2 miles and run for up to 300 miles, cutting directly across craters. The largest known scarp is Discovery Rupes.

264 **The scarps on Mercury were probably formed in a different way than the cliffs on Earth.** On Earth, cliffs are frequently formed as the result of movements in our planet's crust that continue to this day. Mercury, however, doesn't have any such *plate tectonics.* Instead, its scarps are probably the result of activity that took place deep inside the planet. Scientists theorize that when Mercury was young, it had a molten core. In time, however, the core solidified and shrunk. The shrinkage resulted in some selected collapsing and buckling of the planet's crust far above and hence to the creation of Mercury's impressive scarps.

265 **Mercury has a weak magnetic field.** It is about 100 times weaker than the Earth's magnetic field but is strong enough to suggest that Mercury may have a large core of iron 1,100 miles across buried at its center. This makes Mercury, on average, much denser than our Moon.

266 **A day on Mercury is twice as long as its year.** Mercury only takes 88 Earth days to travel once around the Sun. And so, 88 Earth days is the length of a year on Mercury. But Mercury also slowly spins on its axis, rotating once every 59 Earth days *relative to the stars.* The combination of these two motions means that for any particular spot on Mercury, the elapsed time from one "high noon" to the next is 176 Earth days, so a day on Mercury is indeed twice as long as its year.

267 **Mercury has some pretty strange sunrises.** Mercury wobbles a bit on its axis as it orbits the Sun and also travels around the Sun in a rather elliptical orbit. The combination results in some pretty odd sunrises. In places on Mercury, the Sun comes up in the morning, then stops and goes back down again, only to finally rise a second time and then proceed slowly across the sky.

268 **Mercury may harbor clues to the origin of the solar system.** Recently, radar was bounced off of Mercury, revealing bright patches in some of the craters near Mercury's poles. Scientists speculate that the bright reflections may come from ice that was deposited in the craters when comets crashed there long ago. The ice would have remained protected in the perpetual shadows of these deep craters and hence not melted or evaporated away. Such material represents the primordial stuff out of which the solar system was created—elements and compounds that have remained virtually unchanged in the 4.6 billion years since the solar system's formation and which, therefore, can offer clues about conditions that prevailed at the time. (Recently, similar patches of ice were also discovered deep in a crater on the Moon).

269 **Mercury is a place where life could never have existed.** Mercury's small mass would have made it difficult for this tiny planet to have retained much of an atmosphere during its early years. Furthermore, its closeness to the Sun would have caused the intense heat and the strong solar wind to drive off any early atmosphere into space. Without air and water, Mercury, like our Moon, has always been a world devoid of life.

Venus

270 **Venus is our most common and beautiful "morning and evening star."** Because Venus travels around the Sun in an orbit that is interior to that of Earth, it appears to us to swing forever back and forth on either side of the Sun. Thus, Venus alternately appears in our southwestern sky for an hour or two after sunset and in our eastern sky for a few hours before dawn. Frequently during these times, with the exception of the Sun and the Moon, it can also be the brightest object in the sky, shining like a brilliant diamond in the purple glow of twilight.

271 **Venus comes closer to Earth than any other planet in the solar system.** Venus is the second planet out from the Sun and travels in a nearly circular orbit about 67 million miles from the solar surface. About every nineteen and a half months, Venus swings past the Earth and at such times is at the astronomically minuscule distance of only 26 million miles. Mars, on the other side of Earth, only approaches us to within 35 million miles, thus making Venus our closest planetary neighbor.

272 **For many years, Venus was called the Earth's "twin" or "sister" planet.** The diameter of Venus is only 408 miles smaller than the diameter of the Earth. This and the fact that the two planets lie on such close orbits led many to speculate that Venus may not be too unlike the Earth but with its own exotic touches. Early science fiction writers created visions that ran from a globe totally covered with water to a jungle planet inhabited by dinosaurs to a world boasting great civilizations. As scientific data on Venus gradually accumulated, however, our view of this neighboring world changed completely and we learned that size was just about the only thing the two planets have in common.

---

273 **Rather than a twin of Earth, Venus is "hell in space."** Early telescopic observations of Venus revealed only the most subtle of markings that seemed to continuously change. Soon it was realized that Venus was totally covered in silvery clouds that completely hid the surface from view. This veil helped fuel speculation about what kind of world might lie below. In time, however, scientific investigations revealed that instead of being abundant in life-giving oxygen, Venus's atmosphere was composed almost entirely of carbon dioxide. Realizing this was a powerful greenhouse gas, scientists immediately held out little hope that Venus might be very Earthlike.

274 **The clouds of Venus have subtle patterns.** The upper Venusian clouds have a subtle banded appearance more reminiscent of jet streams on Earth than the curling patterns associated with storms and fronts. These upper cloud decks are indeed powerful jets that speed around the planet at 250 miles per hour, or once every four days.

275 **Both Russian and American spacecraft have painted an ugly portrait of our sister planet.** When Russian and American spacecraft visited Venus in the 1970s and 1980s, they discovered a thick choking atmosphere 90 times denser than Earth's and

Bands of clouds are revealed in the upper atmosphere of Venus in this computer-enhanced image from *Mariner 10.* (NASA/JPL)

surface temperatures, both day and night, of around 850°F—increasing at times to nearly 900°F (even hotter than Mercury). Furthermore, Venus's clouds were found to be laced with droplets of sulfuric acid stronger than the acid in a car battery. Moved by its brilliant light, the ancient Romans named Venus after their mythological goddess of beauty and love. In reality, with its sulfurous fumes and ovenlike temperatures, Venus comes far closer to the biblical description of hell.

276 **Venus has acid mists.** Temperatures in Venus's clouds are so high that rain is impossible, but mists of powerful sulfuric acid do occur and are the chief component of the clouds (just as Earth's clouds are made of water vapor). Recent spacecraft observations suggest that the acid content of the atmosphere may have even been on the increase lately. Powerful volcanoes that appear to still be active and periodically vent huge quantities of sulfurous chemicals into the air are probably the source of the acid.

277 Venus is the *greenhouse effect* gone wild. The reason Venus is so hot is fairly easy to understand and serves to illustrate the classic greenhouse effect. Venus's clouds allow some light from the Sun to filter down to the surface. When this light falls on the surface of the planet, it is absorbed by the rocks and soil and reradiated back up into the atmosphere as heat. But the carbon dioxide gas in Venus's atmosphere absorbs this heat radiation. Some of it is then scattered out into space, but some is also scattered back down toward the ground. Thus, the atmosphere serves to trap some of the heat and this, in turn, has allowed surface temperatures to skyrocket. The high density of the atmosphere helps to transport the heat around the entire planet and creates little temperature variation from day to night or even from the equator to the poles.

278 Venus's clouds extend much higher above the planet than they do on Earth. Whereas most clouds on Earth lie within 7 or 8 miles of the ground, the cloud deck that blocks our view of our neighboring planet towers over 40 miles into the Venusian sky. Below may lie two or three additional cloud and haze decks, but below about 19 miles altitude, the sky becomes clear. Still, because of the sulfur in the air and the fact that little sunlight filters through, days on Venus are never bright and probably have a distinctly yellow cast.

279 Radar has allowed us to peek beneath Venus's obscuring veil. For centuries, the clouds of Venus prevented astronomers from getting even a glimpse of what the surface of the planet might be like. With the development of high-powered radar, however, all of that changed. Radar beamed from Earth and from spacecraft in orbit around Venus have allowed us to pierce the obscuring clouds. From the shape of the return echoes, detailed pictures of the entire surface of the planet have been created that reveal objects as small as a football stadium. In addition, a few Russian spacecraft have parachuted to the surface of Venus and survived the temperatures and pressures long enough to radio back pictures of rocks in the immediate vicinity of their landing points.

280 Venus has some fascinating surface features. In many places, the surface of Venus is covered by vast rolling plains. Elsewhere, however, there are several large "continents" of elevated terrain. One, Ishtar Terra (the Land of Ishtar), is about the size of the United States or Australia and the mountains of Maxwell Montes rise nearly 7 miles, higher than Mount Everest. Another highland region, Aphrodite Terra (the Land of Aphrodite), is about the size of Africa. Two smaller mountainous

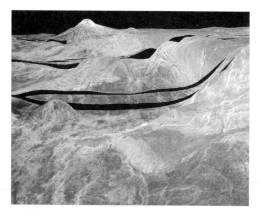

A computer made this image of a portion of the surface of Venus from radar echoes returned to the *Magellan* spacecraft in orbit around the planet. Temperatures here hover around 850°F day and night. (NASA/JPL)

sphere and hit the surface over an elongated area, rather like buckshot. The thick atmosphere appears to vaporize all small pieces of space debris before they reach the ground, resulting in no craters less than about 2 miles in diameter.

282 **Overall, there are far fewer craters on Venus than Mercury or the Moon, even though Venus is considerably larger.** While Venus has a very thick atmosphere, erosion due to weathering is not a major influence in slowly obliterating craters as it is on Earth. This is because winds at the surface of Venus are only about 4 miles per hour. Instead, Venus's craters are obliterated due to the periodic eruption of its volcanoes, which release lava that essentially repaves the surface of the planet.

regions are called Alpha Regio and Beta Regio. They appear to be sites of many volcanic peaks, some of which may periodically be active to this day.

281 **Venus also has its share of craters.** Some of Venus's craters are the result of impacts from objects in space, but most are volcanic in origin. The volcanoes are of a type known as *shield volcanoes,* which are due to molten lava pushing up, or *upwelling,* through a weak spot in a planet's crust. (On Earth, the Hawaiian Islands were created in this way.) Two large shield volcanoes on Venus, Sif Mons and Gula Mons, are each over 1 mile high. Impact craters include Cleopatra, which is near Maxwell Montes and measures over 60 miles across. Smaller craters are frequently somewhat elongated on Venus, possibly because the incoming debris from space broke up into pieces on its way through the thick atmo-

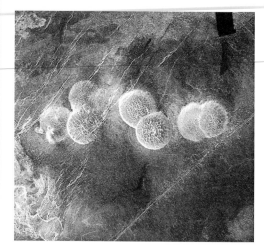

These unique features on Venus are known as "pancake volcanic domes." They were probably created when molten lava erupted onto the surface and was flattened by the enormous surface pressure or the lava withdrew back below ground. After flattening, the domes solidified. (NASA/JPL)

283 There are also unique "pancake volcanic domes" on Venus. These flattened domes, each about 15 miles across, were probably formed when molten lava oozed out of cracks in the Venusian crust and then withdrew, causing the dome to deflate like a fallen soufflé.

284 Venus evolved very differently than Earth. Venus appears to have always been too hot for liquid water to form on its surface. As a result, Venus never had oceans or seas. Earth's first atmosphere probably had lots of carbon dioxide in it the way Venus's atmosphere does today. (In both cases, the carbon dioxide was released in outgassing from numerous volcanic eruptions early in the planets' histories.) On Earth, much of this carbon dioxide was ultimately absorbed by the seas, thereby preventing a *runaway greenhouse effect.* On Venus, however, this could never happen and so the carbon dioxide was forever left in the atmosphere, where it created a runaway greenhouse effect that roasts the planet to this day. Radar studies also reveal nothing that looks like midocean ridges or other features on Venus that would be characteristic of large-scale plate tectonics.

285 In some ways, Venus resembles a young Earth. Many geologists believe that the surface of Venus in some ways resembles that of the young Earth at an age of perhaps 1 billion years. At that time in the Earth's past, volcanic activity had begun, but the Earth's crust was still relatively thin and the vertical motions of molten material (known as *convection*) below the crust, which, in turn, drive plate tectonics, had not yet begun. (Venus, of course, is the same age as the Earth and all the other planets in the solar system—about 4.6 billion years.)

286 Like Mercury, Venus has no moons. Both Mercury and Venus have no natural satellites of their own. For a time in the 1800s, a few astronomers reported the discovery of a satellite in orbit around Venus. They even named it Vulcan (to which Mr. Spock would undoubtedly say, "Fascinating!"). Before long, however, the astronomers found out that the satellite was an illusion caused by sporadic reflections of Venus's brilliant disk within the optics of the telescope. These reflections made it appear that a second object was located near Venus, an object that appeared to travel with the planet as it moved through space. Other telescopes with better optics failed to show the satellite and made scientists realize it was never really there at all. Fascinating indeed.

287 Once in a while Venus and Mercury pass directly between the Earth and the Sun, creating "minieclipses." Mercury passes between the Earth and the Sun about every four months, while Venus does the same thing about every nineteen and a half months. Astronomers call such events *inferior conjunctions,* as in "Venus is in inferior conjunction today," meaning that Venus is passing between the Earth and the Sun. But because the orbits of Mercury, Venus, and the Earth are not exactly in the same plane (that is, they're tilted a bit to each other), Mercury and Venus do not pass across the face of the Sun every time they are in inferior conjunction. Once in a while, however, such a *transit* takes place and, with the right eye protection (see the section on the do's and don'ts for safely viewing solar eclipses on page 83), you can enjoy these minieclipses. Mercury appears as a tiny round dot that slowly crosses the face of the Sun. Venus, because it is larger and closer, appears as a larger disk. On average, about 13 transits of Mercury occur each century. Transits of Venus take place in pairs separated by over a century. The last ones occurred in 1874 and 1882. The next transits of Venus will occur on June 8, 2004, and June 6, 2012.

Mars

288 **Many cultures named Mars after their god of war.** The Greeks called it Ares; to the Persians it was Nergal; to the Romans, Mars. The fourth planet from the Sun was frequently associated with a culture's god of war. The reason, it seems likely, was the planet's reddish hue, which apparently reminded many of blood. In reality, the color comes from a far less sanguine source—the presence of iron oxides in the Martian soil. Mars, simply put, is rather rusty.

289 **Unlike Venus, Mars has a thin atmosphere that is normally quite transpar-**

ent. When astronomers first looked at Mars in telescopes, they weren't nearly as frustrated as when they looked at Venus. The Martian air was typically quite clear and only occasional clouds would be seen, so the astronomers were able to see all the way down to the surface of the planet. What they saw were orange (not really red) deserts, interspersed with dark continent-sized markings and two gleaming white polar caps.

290 **Some of the first things we learned about Mars made it seem pretty similar to the Earth.** Being able to see its surface features, scientists were soon able to determine the length of a day on Mars: 24 hours, 37 minutes, a little more than half an hour longer than a day on Earth. Astronomers were also able to tell that, like Earth, Mars's axis was tilted to the plane of its orbit. Such a tilt produces seasons and the greater the tilt, the greater the severity of the seasons. Mars, it was discovered, has a 25.2° tilt—again, incredibly similar to the Earth's tilt of 23.5°. Thus, Mars, like Earth, experiences fall, winter, spring, and summer, although each season is about twice as long because it takes Mars about twice as long to orbit the Sun.

291 **Seasonal changes could be seen from month to month across Mars's disk.** Evidence for the changing Martian seasons was found in its polar caps and dark markings. As spring begins in each hemisphere, the appropriate polar cap gradually shrinks in size

Mars at Opposition · February 1995
Hubble Space Telescope · Wide Field Planetary Camera 2

PRC95-17 · ST ScI OPO · March 21, 1995 · P. James (U.Toledo), NASA

Valles Marineris Region
60° Longitude

Tharsis Region
160° Longitude

Syrtis Major Region
270° Longitude

These three images of Mars were taken by the Hubble Space Telescope in 1995 and together show the entire surface of the planet. Dark surface markings, lighter deserts, and a white polar cap are visible. (NASA/STScI)

as a "wave of darkening" spreads through the dark markings down toward the Martian equator. Some observers considered this to be evidence for the seasonal melting of the polar caps and the subsequent rejuvenation of Martian vegetation as the life-giving waters from the spring melt-off spread toward the equator. This assumption was wrong.

292 **The orbit of Mars is pretty egg-shaped.** The orbit of Mars is quite elliptical. While its mean distance from the Sun is about 141.7 million miles, the planet's actual distance from the Sun varies by about 26 million miles during the course of a Martian year (as compared to only about 3 million miles for the Earth).

293 **Eyeballing Mars from Earth—the good times and the bad.** The Earth orbits the Sun faster than Mars (because it is closer to the Sun) and swings by Mars every couple of years. When this happens, astronomers say that Mars is in *opposition* because it lies opposite the Sun in our sky. At such times, Mars is in the sky all night long and at its closest point to Earth for that particular orbit, making these the best times to observe the Red Planet. But because the orbit of Mars is so egg-shaped, all oppositions are not equal. The most favorable opposition occurs when the Earth happens to pass Mars at a time when our planet is at its *farthest* point from the Sun and Mars is at its *closest* point to the Sun. When this occurs, the distance between the two planets is a little less than 35 million miles. When the opposite situation holds, we say it is an *unfavorable opposition.* At such times, the two planets come no closer than about 64 million miles and Mars looks little more than half as large in a telescope.

294 **In 1877, exciting news came from Mars via Italy.** In that year, an Italian astronomer named Giovanni Schiaparelli reported seeing a network of fine dark lines crisscrossing the bright Martian deserts during a favorable opposition of Mars. Schiaparelli said some of the lines ran from one large dark area to another. Others came together at various "hubs" in the middle of a large stretch of desert. Schiaparelli called the features *canali* in his native Italian (meaning "channels") but refused to speculate on their nature. Before long some other astronomers were seeing the *canali* as well. One, an American named Percival Lowell, became so intrigued that he built the Lowell Observatory near Flagstaff, Arizona, and dedicated himself to studying Mars.

295 **In Percival Lowell's eyes, the *canali* of Mars quickly became "canals."** Taking into account the shrinking each spring of the Martian polar caps, the subsequent wave of darkening, and the pattern of the *canali* linking dark areas or converging at

dark "oases" in the deserts, Lowell quickly championed the notion that the *canali* were indeed broad strips of vegetation that bordered a planet-wide system of irrigation canals built by an intelligent race of Martians. Noting that he could see no reflections of sunlight anywhere on Mars, Lowell concluded that Mars was indeed largely a desert world and the Martian engineers had created the vast system of canals to thus channel the only precious water supply they had, namely the seasonally melting polar caps, to temperate zones of the planet and, therein, made the deserts bloom.

296 **It was intriguing to think that there might be other forms of life right next door on Mars.** The notion that there might be advanced sentient beings on a neighboring planet proved extremely intriguing to many and works of science fiction from the late nineteenth and early twentieth centuries played out scenarios of every possible type of Martian society. Madison Avenue was quick to jump on the celestial bandwagon. A turn-of-the-century newspaper ad proclaimed: "Mars is peopled and they want Kirk's Soap."

People tried to devise ways to send greetings to our Martian neighbors and let them know that Earth was inhabited, too. Ideas ranged from lighting huge signal fires in the Sahara Desert to giving thousands of people handheld mirrors that could be used to flash reflected sunlight up to Martians in a kind of Morse code. In 1892, the French Academy of Sciences even offered a prize of 100,000 francs to anyone who successfully contacted the inhabitants of another planet. The prize could not be claimed for contact with Mars, however, because this task was considered too easy to earn such a large prize.

---

297 **October 30, 1938: the night the Martians came to call.** In 1898, H. G. Wells published *The War of the Worlds,* in which the inhabitants of a dying desert world turned envious eyes on a greener Earth. On Halloween Eve in 1938, another Welles named Orson broadcast an adaptation of the novel on his radio program and panicked Americans by describing Martians landing in New Jersey packing poison gas. To many listeners that night, the thought of life on this nearby world seemed very credible.

---

298 **When it came to Martians, other people had their doubts.** Throughout the 1940s and 1950s, the debate over the possibility of life on Mars continued. While some reputable astronomers saw the canals, other equally qualified scientists with similar telescopes claimed they never saw a single one. Scientific analyses of Martian temperatures and the nature of the Martian atmosphere also began to spread doubts.

The Martian atmosphere was found to be painfully thin (the surface pressure is less than 1 percent that of Earth) and made up almost entirely of carbon dioxide. Indeed, the air was so thin that H_2O, if it existed on Mars, would go directly from a solid to a va-

por without ever becoming live dependent water. The tenuous nature of the Martian atmosphere also meant that the carbon dioxide present couldn't create any significant greenhouse effect to help compensate for Mars's greater distance from the Sun.

299 **The Martian weather report didn't hold out much hope for a picnic.** Measurements showed that while temperatures near the Martian equator in midsummer might top 70°F, the thin air could do little to retain the heat of the day. By sunset, the same spot would be down near freezing and computer models suggested that by the wee hours of the morning, readings could plummet to minus 100°F. During the long winter nights, polar regions might see temperatures hover near minus 150°F for months.

300 **In the 1960s and 1970s, we finally got to pay Mars a call.** In the mid-1960s, the *Mariner 4* spacecraft sped past Mars and snapped some pictures. Radioed back to Earth, they produced sadness and disappointment, for they showed a cratered Mars that looked more like the Moon than the Earth.

In the 1970s, other spacecraft went into orbit around the Red Planet and painted a far more complete picture. Ironically, the first of these spacecraft, *Mariner 9,* initially looked down on a planet totally devoid of detail. As fate would have it, a planet-wide dust storm was raging and obliterated everything in sight.

Gradually, the dust cleared and *Mariner 9*

(and later spacecraft) looked down on a world of fascinating geological wonders.

301 **The largest geological feature on Mars is the Tharsis Bulge.** Roughly the size of North America, this large plateau rises over 6 miles above the surrounding plain. The Tharsis Bulge is the youngest region on the planet (about 2 to 3 billion years old) because it is the site of a number of enormous volcanoes.

302 **There are some great national parks just waiting to be explored on Mars.** While Mars only has half the diameter and less than 30 percent of the area of Earth, the planet boasts a remarkable array of natural wonders. The first things to poke out of the settling dust storm for *Mariner 9* were the tops of several enormous extinct volcanoes. The largest, Olympus Mons (Mount Olympus), would cover the entire state of Col-

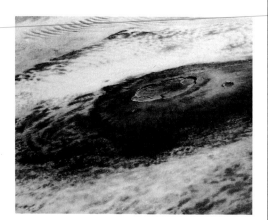

From high above, the *Viking* orbiter captures Olympus Mons wreathed in clouds. The largest extinct volcano in the solar system, it towers over 15 miles above the surrounding plain. (NASA/JPL)

orado and towers over 15 miles above the surrounding plains. Its *caldera,* or vent hole, alone is 55 miles across.

303 **Martian volcanoes are higher than those on Venus or Earth because the gravity is weaker on Mars.** The height of a volcano is typically limited by the strength of the gravity of the planet it happens to be on. This is because the volcano's height depends upon its ability to support its own weight. Earth and Venus are practically the same size and mass, so they have about the same gravity and the volcanoes of both planets reach the same approximate height. The gravity on Mars is only about 38 percent of Earth's gravity and its volcanoes are about two and a half times as high. The giant Martian volcanoes are known as *shield volcanoes.* They have gentle slopes and are formed from the rapid flow of lava. The volcanoes probably grew as large as they did because Mars seems to have no *plate tectonic movements.* On Earth, the Hawaiian Islands are examples of shield volcanoes, but a plate, moving over a hot spot, produced a series of volcanoes. On Mars, with no moving plates, lava was able to vent through the same place in the crust for millions upon millions of years, another factor that contributed to the gigantic size of the volcanoes.

304 **The "real" Grand Canyon is on Mars.** Mars has enormous canyons, including one system named the Valles Marineris (Mariner Valley). Up to 20,000

Close-up of a portion of the Valles Marineris, the "Grand Canyon of Mars." If brought to Earth, it would stretch from New York to California. (NASA/JPL)

feet deep and 120 miles from rim to rim, if brought to Earth, it would stretch from New York to California! Arizona's Grand Canyon is only as big as one of its minor

Wide-angle view of Valles Marineris. (NASA/JPL)

tributaries. The Valles Marineris, however, was not carved by water erosion like Arizona's Grand Canyon. Instead, it is the result of geological faulting—a collapse of the subsoil underneath.

305 **Some portions of Mars are heavily cratered.** About a quarter of Mars is covered by impact craters, especially in the highlands of the southern hemisphere. Indeed, this happened to be the region that was photographed by the first spacecraft to fly by Mars, so our first impression of Mars was of a far more Moonlike than Earthlike planet. The Martian craters, however, are far shallower than lunar craters and very few of the craters are smaller than about 3 miles in diameter—both signs of significant erosion by wind and weather.

Mars also has vast areas of cratered terrain, such as this area in the Meridiani Sinus region. These Martian craters are shallower than those we see on the Moon because they have been worn down by erosion. (NASA/JPL)

306 **Mars is the *lumpiest* planet in the solar system.** When it comes to overall differences in elevation, Mars is about as far from Kansas as you can get, Toto. From the depths of a great basin called Hellas to the tops of the giant extinct volcanoes that sit atop the Tharsis Bulge, the difference in elevation is a staggering 21 miles. On Earth, from the bottom of the Marianas Trench to the top of Mount Everest, the total difference in elevation is only a little over 12 miles.

307 **But what about the famous canals of Mars?** With the exception of the Valles Marineris, which does correspond to a "canal" on some of Lowell's maps, the spacecraft that visited Mars sent back no pictures of any canals. The canals, it turned out, had only been optical illusions, existing in the minds of those astronomers who *wanted* to see them. As Carl Sagan once observed, when it came to the Martian canals, it was a matter of "which side of the telescope the intelligence was on." The case of the Martian canals is a powerful one because it shows how even trained scientists can fool themselves into seeing something they want (or expect) to see.

308 **And what about Mars's "wave of darkening"?** The seasonal waves of darkening that spread down from the shrinking polar caps and were seen by virtually all astronomers also lent weight to the hope of finding life on Mars. Spacecraft and later Earth-based reconnaissance showed, however, that while the wave of darkening was real, it had nothing to do with vegetation returning to life in the Martian spring. Instead, scientists were seeing the result of seasonal hemispheric winds that would alternately cover dark surface rocks with lighter-colored dust and remove it.

309 **Mars is a very dry place.** Today, the Martian atmosphere is so thin that liquid water cannot exist. The polar caps are made of both water ice and frozen carbon dioxide, but in the Martian spring the water ice doesn't melt. Instead, it turns directly into a vapor (rather like mothballs do on Earth). Ice fogs occasionally form over the polar caps and canyons, but no Martian rains currently fall. In short, Mars is drier than the Sahara Desert.

310 **In 1976, scientists nevertheless went in search of the Martians.** In spite of the harsh data, some scientists still held out the hope that some life might exist on Mars. Life on Earth, after all, had proven to be very tenacious, often surviving in Antarctic ice or baths of boiling acid. Perhaps primitive forms of life might somehow survive in the Martian soil. In the summer of 1976, two American spacecraft, *Viking 1* and *Viking 2,* touched down on the rusty plains of Mars. (*Viking 1* landed on July 20, 1976, seven years to the day after the first astronauts landed on the Moon.) Each came equipped with a robot arm that could scoop up Mar-

Rocks and dunes spread out to the horizon on the Chryse Planitia (the Plains of Gold) in this image taken by the *Viking 1* lander. In reality, the plains aren't gold but rusty red from iron oxides in the Martian soil. (NASA/JPL)

tian soil and deposit it inside the spacecraft, where a miniature laboratory performed several experiments to test for life. No positive results were found.

311 **Could there be life elsewhere on Mars?** While *Viking 1* and *Viking 2* were made to touch down at places on Mars that scientists felt would prove the most favorable for life, the *Viking* landers were not mobile and thus could only examine their immediate surroundings. Thus, it can be argued that they hardly were able to do a very thorough job of searching for life on the planet as a whole. Furthermore, they also only performed experiments designed to search for life that would closely resemble the chemistry of Earth life. Nevertheless, the *Viking* probes failed to find any organic compounds on Mars that scientists believe are necessary to life of any kind. So, overall, the negative results are generally accepted as significant, at least for surface and near surface life on Mars today.

312 **There is evidence, however, that Mars was a wetter and warmer place in the ancient past.** While extinct today, Mars's huge volcanoes would have created a substantial carbon dioxide atmosphere through enormous amounts of outgassing earlier in the planet's history. A thicker carbon dioxide atmosphere, in turn, would have created higher global temperatures and made the presence of liquid surface water possible. Indeed, spacecraft images of the Martian plains reveal features called *runoff* and *outflow channels* that look very much like they were carved by water erosion. In some cases, the size of the channels suggests that the water flow may well have equaled that of major rivers on Earth.

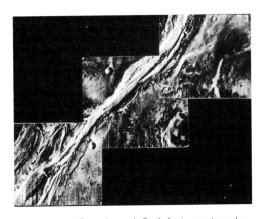

A Martian outflow channel. Such features strongly suggest that significant quantities of water flowed on Mars at least once in the past. (NASA/JPL)

313 **Where has all the water gone on Mars?** While the permanent portion of Mars's northern polar cap appears to be made mostly of water ice, the amount isn't nearly sufficient to account for all the water that seems to have flowed on Mars in the past. Most scientists think that much of the remaining water may be locked in the soil over large expanses of the planet in the form of *permafrost.*

314 **The areas surrounding some Martian craters lend credence to the argument that large amounts of permafrost may lie just beneath the surface.** The material surrounding Moon craters like Copernicus looks like a dry mixture of rocks, soil, and dust, but the regions surrounding some Martian craters give the appearance of liquid that has flowed out from around the crater. This is precisely what would be expected if subterranean ice had been liquefied by the heat of the impact.

315 **Like Earth, Mars may go through periodic *ice ages.*** The same mechanism that has led to ice ages on Earth, including gradual changes in the shape of our planet's orbit, is believed to have created periodic global freezes and thaws on Mars as well. If so, the current frigid conditions on Mars may simply be due to the fact that the planet is presently in the grips of an ice age. Someday, should Mars's orbit again significantly change, the climate could warm and liquid water may once again flow across the rusty plains of Mars.

316 **What about the "face on Mars"?** Among the thousands of images taken from orbit by the two *Viking* spacecraft is one that shows a rock that looks something like a human face staring up into space. Unfortunately, much has been made of this coincidence in nonscientific circles. In short, just as you can see specific shapes in clouds if you look at enough clouds from enough different angles, so you can with rocks. Given the number of rocks on Mars and the different angles of the Sun at different times of the day (producing different shadows), it would be odd if people didn't see a rock that looked like something familiar from some angle at some time of day. A rock formation known as the "Old Man of the Mountains" in New Hampshire presents a rather lifelike profile of a bearded man when viewed from a certain direction. Between Santa Fe, and Taos, New Mexico, lies Camel Rock, which from the

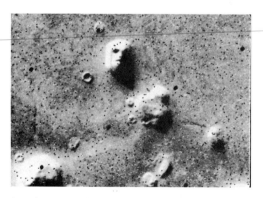

A rock outcropping on Mars, when viewed from a particular angle with the Sun at a certain point in the sky, is seen by some to resemble a face. The feature, however, is a natural result of erosion and is no more the work of intelligent beings than the "Old Man of the Mountains" is in New Hampshire. (NASA/JPL)

right direction does indeed resemble a camel. Many other examples exist around our world, none of which were made by intelligent beings. They were carved by nature and refined by the human imagination. The same is true for the "face on Mars."

317 **The years from 1997 to beyond 2000 will see a significant new phase in Martian exploration.** Two U.S. spacecraft arrived at the Red Planet in 1997, the first in 20 years. The Mars *Global Surveyor* went into orbit around Mars in September 1997 and is sending back images of the planet that resolve surface details down to less than five feet across. It is also surveying the overall mineralogy of Mars from orbit and analyzing where mascons may be located under the Martian surface in the same way that scientists did studies of the Moon with lunar-orbiting spacecraft.

318 **Another spacecraft touched down on Mars and released a rover.** A separate spacecraft, *Pathfinder,* landed on the surface of Mars on July 4, 1997, and released a tiny robot called *Sojourner* that's about the size of a laser printer. Controlled by scientists back on Earth, *Sojourner* had a pair of cameras for stereo imaging and was able to make little excursions to photograph and study the soil and rocks and send the data back to Earth. *Pathfinder* also took daily weather measurements from its landing site. Daily pictures from the Mars *Global Surveyor* and *Sojourner* and daily weather reports from *Pathfinder* were made available in homes and classrooms via the Internet. In the coming years, other U.S. and possible Russian missions to Mars are planned, with a possible manned expedition being loosely discussed for the year 2018.

319 **Some rocks on Earth may be from Mars.** No spacecraft from Earth has ever returned from Mars with rock samples like the Apollo astronauts brought back from the Moon. Yet scientists believe that a dozen or so rocks in their possession may possibly be from Mars. Scientists learned from analyses done by the *Viking* landers 20 years ago that Martian rocks have different percentages of certain elements than similar Earth rocks. Gases trapped in one of the rocks held by the scientists match gas ratios in the Martian atmosphere. How did the rocks get here? They may have been literally blasted off Mars when a chunk of material from space crashed into the planet long ago. (Such bombardments were common in the solar system's ancient past and such impacts can still occur today.) Flying through space, they ultimately arrived at the Earth, where they entered our atmosphere and finally crashed on our planet. A long shot to say the least but a very lucky opportunity for science. We'll have more on rocks from space that fall from the sky in the next chapter (see page 144).

320 **If these rocks are from Mars, they may give us some insight into past conditions there.** Some of the rocks show evidence of interaction with water and most have substances like salts and clays that offer further evidence of weathering. In all, they help confirm deductions drawn from the fact that Mars has huge volcanoes and flow channels, namely that Mars had a thicker atmosphere and running water at some time in the past. In 1996, a team of scientists reported that certain microscopic structures within some of these rocks may be due to the presence of very primitive life on Mars long ago. Further tests are being independently conducted. If proved to be true, it will be an *extremely* important scientific discovery.

---

321 **Mars is the Red Planet because it's "rusty."** The *Viking* landers did extensive studies of the Martian surface soil and found that it was high in iron oxides, or rust. Hence Mars's reddish coloration. The subsoil rocks, however, are likely not to be red in color because they would not have been exposed to the atmosphere. Overall, Mars probably doesn't have a higher percentage of iron than the Earth and, like on Earth, most of Mars's iron probably sank to the center of the planet when it was still forming and in a molten state.

---

322 **Martian winds are normally light, but once in a while things change quickly.** Surface winds on Mars normally don't get much above a gentle breeze, but when summer occurs in the southern hemisphere, extra heating from the Sun can cause an abrupt change in the weather. Winds suddenly and dramatically increase, lifting the fine desert dust into the air. Within days, a raging dust storm is in progress and spreading rapidly. Within weeks, winds of over 150 miles per hour have created billowing clouds of dust that tower 20 miles into the sky and have spread to encompass the entire planet! It was precisely such a storm that was at full fury when *Mariner 9* arrived on Mars in 1971. However, while such Martian winds can be very high, they wouldn't knock you over because the air is so thin.

323 **Sometimes the daytime sky on Mars isn't the familiar blue we find on Earth.** When the first color pictures came back from the surface of Mars via the *Viking* landers, scientists were somewhat surprised at the color of the sky. Knowing the air was very thin, they had long expected the daytime sky to be a very dark blue like we would see from a high-flying jet aircraft on Earth. Instead, the Martian sky was something between a yellowish brown and pink. Fine particles of the "rusty" Martian desert dust— suspended in the air by Martian winds—had colored the sky. When dust storms are not near, however, the Martian sky can appear dark blue.

324 **Recent studies suggest Mars may have lost much of its early atmosphere because of the Sun.** Evidence strongly suggests that Mars once had a far denser atmosphere than it does today. It had been long assumed that in time the atmosphere had escaped into space because the weak Martian gravity could not hold it down. Recently, however, some scientists have suggested that the Sun may have also played a significant role. Unprotected by a magnetic field, the solar wind can directly impinge on the Martian atmosphere and, in effect, may have helped "chip it away" over the eons.

325 **The Hubble Space Telescope recently found significant changes in the Martian climate.** With its new corrected optics, the Hubble Space Telescope was able to give us the best views of Mars since the *Viking* spacecraft surveyed the Red Planet back in the late 1970s. The new images allowed scientists to have their first detailed look at the general Martian weather picture in nearly 20 years and they found significant changes during that time. In particular, the Mars of the 1990s is a cloudier, colder, and drier Mars than just a few decades ago.

326 **Mars has two tiny companions.** In its journey around the Sun, Mars is accompanied by two tiny moons, Deimos and Phobos. The moons take their names from two characters in Greek mythology whose names translate as Terror and Panic—fitting companions for the god of war. Looking like cratered potatoes, Deimos and Phobos are irregular chunks of rock no more than 10 and 18 miles across, respectively. The two satellites were discovered in 1877 by Asaph Hall. The largest crater on

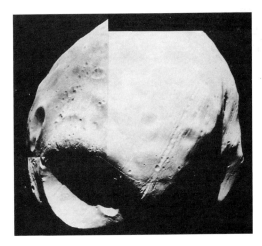

A close-up of Phobos, the larger of Mars's two tiny moons. In the foreground is the crater Stickney. (NASA/JPL)

Phobos is named Stickney, in honor of Angelina Stickney, Hall's wife, who encouraged him to make the search.

327 **From the surface of Mars, its two moons appear to travel in opposite directions.** The smaller of the Martian moons travels in an orbit about 12,000 miles above the surface of Mars, while Phobos is much closer at only about 3,500 miles above the surface. The closer a satellite is to a planet, the faster it must travel to maintain its orbit and Phobos travels around Mars so fast that it completes an orbit in less time than it takes Mars to rotate once on its axis. The net result is that from the surface of Mars, Deimos rises in the east and sets in the west, while Phobos rises in the west and sets in the east.

CHAPTER 6

THE OUTER SOLAR SYSTEM: REALM OF THE GIANTS

Jupiter

328 **Jupiter is a big planet.** Jupiter is the largest planet in the solar system. It's so big that if it were a goldfish bowl, it would take over 1,200 marbles the size of the Earth to fill it up. Put another way, Jupiter is so large that if you ground up all the other planets, they would fit inside Jupiter with space left over.

Jupiter: the largest planet in the solar system. Just below the center of the disk can be seen the Great Red Spot. An image of the Earth is included for size comparison. (NASA/JPL)

329 **Jupiter and the other giant planets (Saturn, Uranus, and Neptune) are very different from Mercury, Venus, Earth, and Mars.** With the exception of tiny Pluto, the solar system consists of four relatively small planets—Mercury, Venus, Earth, and Mars—followed by four giant ones. The little planets are all rocky worlds with atmospheres (if they exist) that extend only a few dozen miles from their surfaces. By contrast, the giant planets that lie beyond have far greater masses and huge atmospheres that extend downward from the cloud tops for thousands and thousands of miles.

330 **It's always a cloudy day on Jupiter.** When we gaze at Jupiter through a telescope, all we see are the tops of enormous systems of clouds. The clouds come in a variety of shadings and, because of

Jupiter's rapid spin, are frequently stretched out into great parallel bands that completely encircle the planet, giving it a striped face.

331 Chemically speaking, Jupiter is a lot more like the Sun than the Earth. While Jupiter has an iron core as does the Earth, about 85 percent of the giant planet is made up of hydrogen with most of the other 15 percent consisting of helium. Less than 1 percent is made up of all the other elements. Jupiter's strong gravitational field allowed it to retain all the elements in the original cloud out of which the solar system formed. Earth, with its much weaker gravitational field, lost most of its light elements (hydrogen and helium) into space long ago.

332 Jupiter's clouds are a riot of color. Unlike Earth's clouds, which are all simply white, Jupiter's clouds also contain brilliant splashes of color: shades of yellow, orange, red, and brown. The colors are believed to be due, in part, to the presence of complex molecules, including ethane and phosphine. The colored clouds lie between brighter white clouds. These bright bands, called *zones,* are places where gas is rising in the Jovian atmosphere. The alternate dark bands are referred to as *belts* and are places where gas is descending to lower depths. The belts in general are a bit lower in the atmosphere than the zones, so they are also slightly warmer. Jet streams cut back and forth across Jupiter's face between the zones and the belts, sometimes racing past each other at hundreds of miles per hour.

333 Days go by faster on Jupiter than on any other planet in the solar system. Although Jupiter is the largest planet, its size doesn't keep it from spinning the fastest of all. A day on Jupiter is less than ten Earth hours long. But just how long a day lasts depends on where you are. Like the Sun, Jupiter is not solid, so different latitudes on the planet rotate at different speeds. Near the equator, a Jovian day lasts about 9 hours and 50 minutes, but a Jovian day near the poles is about 9 hours and 56 minutes long. So on Jupiter, unlike Earth, a point that's due south of you today won't be due south of you tomorrow. If you want to go on vacation to a different latitude on Jupiter, just wait until it's passing by at your longitude. You'll save on airfare.

334 The top of Jupiter's atmosphere is a seething caldron of storms. Within Jupiter's multicolored clouds, telescopes and spacecraft have revealed enormous swirling systems of clouds that pack incomprehensible amounts of energy. As jet streams rush past each other, they create ribbons of turbulent gas tens of thousands of miles long and propel hurricanes the size of Earth's continents. In addition to the striped belts and zones, Jupiter also has *spots.* The largest, known as the Great Red Spot (GRS), is an oval-shaped maelstrom of clouds more than twice the size of Earth that has been raging unabated on Jupiter for over 350 years.

A *Voyager* image showing a close-up of the region around the Great Red Spot. The GRS is over twice the size of Earth and has been seen for over 350 years. Smaller white storms nearby are about 45 years old. Above and to the left of the Great Red Spot, clouds take on the shape of a series of waves crashing on a beach. (NASA/JPL)

335 **Jupiter has some of the wildest weather in the solar system.** The temperatures of Jupiter's upper clouds are cold—about −240°F. Here it likely snows ammonia and ice crystals in blizzards larger than our entire planet. As we descend, temperatures rise steadily and snowstorms give way to torrents of water and ammonia rain driven by winds of over 300 miles an hour. Enormous thunderstorms may well up within Jovian clouds that tower tens of thousands of feet high and discharge superbolts of lightning that would vaporize entire cities on Earth.

336 **The weather on Earth and Jupiter is driven by very different things.** On Earth, the main energy source that drives the planet's weather systems is the Sun. The Sun heats the Earth's land and oceans, causing evaporation and currents of warm air to rise and spin through the atmosphere. Jupiter is much farther from the Sun, however, so the Sun can deliver far less heat energy to the giant planet. Yet Jupiter has weather systems whose size and energy dwarf anything found on Earth. In Jupiter's case, the main source of heat energy that drives its weather doesn't come from above but rather from below. While temperatures at Jupiter's cloud tops are around −240°F, temperatures deep in Jupiter's core measure in the tens of thousands of degrees F. This tremendous heat welling up from below and the planet's rapid rate of spin combine to produce Jupiter's Herculean storms.

337 **Jupiter has no solid surface.** Unlike the Earth and its closest neighboring planets, Jupiter does not have a solid surface. If you were to descend through the Jovian clouds for several thousand miles, you would eventually reach a place where heavy rains merge with a vast ocean. This ocean is like none found on Earth, however, for instead of being composed of water, Jupiter's ocean is made up of hydrogen that is compressed under pressures millions of times greater than we find at the surface of the Earth into a liquid that conducts electricity. Scientists refer to this strange "soup" as *liquid metallic hydrogen*. Below this oddest of oceans lies a molten core of iron and silicates as massive as 20 Earths compressed under 100 million Earth atmospheres of pressure.

338 **Because it is still collapsing, Jupiter has infernolike central temperatures.** All of the planets in the solar system collapsed out of a giant cloud of gas and dust about 4.6 billion years ago. While some of Jupiter's internal heat is due to energy released by radioactive decay, there must be another source of heat at work to produce such high temperatures. Scientists suggest that the extra heat may be due to the fact that Jupiter's interior is probably still collapsing.

339 **It doesn't take much collapse to generate a great deal of heat.** A "settling down" of only a few inches per year in Jupiter's massive core would generate enough additional energy to create Jupiter's incredible central temperatures. As a result, Jupiter actually radiates two and a half times as much heat as it receives from the Sun.

340 **Will Jupiter someday ignite as a star or did it somehow just miss becoming a star?** In both cases, the answer is a simple and very definite no. The more mass an object has as it forms, the higher its central temperature will rise. In order for Jupiter to have become a star, it would have to have formed with almost 100 times more mass than it has. An object needs a central temperature of over 1 million°F to qualify as a star and generate its own energy by converting hydrogen into helium through nuclear reactions. Jupiter's mass isn't going to increase and neither will its central temperature, which is in the range of tens of thousands of degrees and thus far less than the 1 million°F readings you need to ignite thermonuclear fusion. Thus, Jupiter failed to become a star by a wide margin and will never be able to become one.

341 **Jupiter has a thin dusty ring.** Scientists didn't really expect the giant planet to have a ring, but the *Voyager 1* spacecraft was programmed to look for one nonetheless. Not to disappoint, *Voyager* found a small faint ring around Jupiter's equator about 35,000 miles above the cloud tops. The ring looks much brighter when seen from the nightside of Jupiter (looking back toward the Sun) than from the dayside (as seen from Earth). (Just like a dusty windshield appears much more brightly lit if you're in the car driving into the Sun than

The *Voyager* spacecraft. *Voyager 1* flew past Jupiter and Saturn. *Voyager 2* made flybys of Jupiter, Saturn, Uranus, and Neptune between 1979 and 1986. On the right arm of the spacecraft are the two cameras. The images and other data were transmitted to Earth by radio signals sent from *Voyager's* large antenna. On the central part of the spacecraft can be seen a cover over a record that contains sounds and pictures from Earth. (NASA/JPL)

Broken into almost two dozen pieces by a close encounter with Jupiter in 1992, the fragments of Comet Shoemaker-Levy 9 (SL-9) flew in formation like a squadron of planes, each with its own tail. To keep track, scientists assigned each a letter designation. In July 1994, these fragments returned to slam into Jupiter at over 130,000 miles per hour. (NASA/STScI)

it does from outside the car.) This, in turn, indicates the particles that make up Jupiter's ring are small, like dust. A thin disk of even finer dust seems to extend from the rings down to the cloud tops.

342 **Jupiter has a large, powerful, and deadly magnetic field.** Jupiter's molten iron core causes it to behave as though it has a gigantic bar magnet buried inside. The result is that Jupiter is surrounded by a powerful magnetic field that is 20,000 times stronger than Earth's. Like Earth's magnetic field, Jupiter's field is shaped like a giant doughnut that encircles the planet. Jupiter's field, however, is 1 million times the volume of Earth's field and actually extends beyond the orbit of Saturn. Jupiter's magnetic field also interacts with the solar wind and traps trillions upon trillions of charged particles in its own enormous version of Earth's Van Allen radiation belts. These charged solar particles create auroras on Jupiter similar to those found on Earth. Several of Jupiter's satellites orbit within these zones of intense radiation, which are so powerful that they would prove instantly fatal to a human being.

For this reason, unmanned spacecraft, instead of humans, will probably always have to explore these worlds.

343 **In July 1994, over 20 mountain-sized pieces of a comet named Shoemaker-Levy 9 crashed into Jupiter at 130,000 miles per hour.** In the summer of 1994, telescopes around the world, the Hubble Space Telescope in Earth orbit, and the *Galileo* spacecraft en route to Jupiter, all zeroed in to

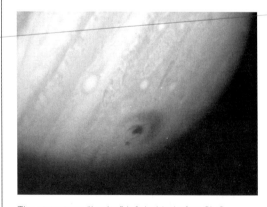

The enormous "bruise" left behind after SL-9 fragment G exploded in Jupiter's upper cloud deck. The area of destruction, including the impact zone and the crescent-shaped shock wave, is larger than the Earth. (NASA/STScI)

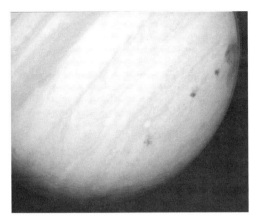

A series of SL-9 impacts leave a succession of scars on Jupiter's battered face. The discolorations lasted for more than a year. (NASA/STScI)

cloud decks, releasing the energy of millions of tons of TNT. The explosions created shock waves the size of Earth and dredged up huge dark clouds of material from the lower clouds that persisted for over a year.

344 **In late 1995, a new chapter in the exploration of Jupiter began.** On December 7, 1995, after a six-year trip, the *Galileo* spacecraft reached Jupiter. But unlike *Pioneer* and *Voyager* spacecraft that just flew past the planet back in the 1970s, *Galileo* was designed to go into orbit and send back at least two years' worth of data. First on the agenda was the release of a 746-pound probe that became the first man-

watch the unprecedented event. The nearly two dozen impacts occurred over a five-day period and detonated in Jupiter's upper

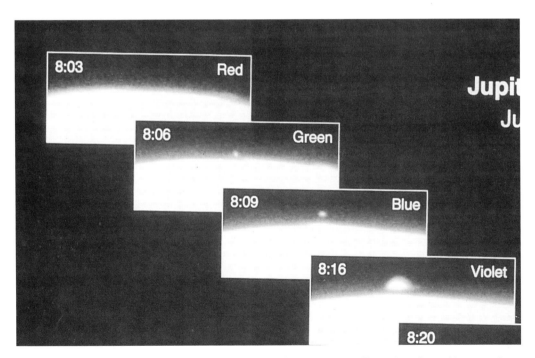

A series of images from the Hubble Space Telescope taken through various filters show the rapid progression of events as a mushroom cloud from SL-9's W fragment impact rises over Jupiter's limb. (NASA/STScI)

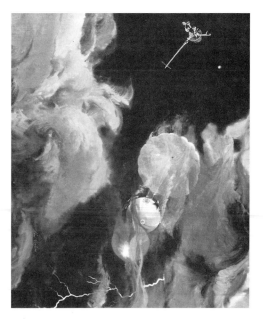

An artist's conception of the *Galileo* probe descending into the upper clouds of Jupiter as the main spacecraft orbits above. In reality, *Galileo* found much less lightning in Jupiter's upper atmosphere than expected. (NASA/JPL)

for its electronics. The information received by anxious scientists back on Earth painted a surprisingly different picture of Jupiter than they had anticipated. Before the probe's entry, scientists had clocked winds in Jupiter's upper atmosphere at over 200 miles per hour, but the probe recorded gusts of over 335 miles per hour.

346 **The probe also found Jupiter to be "drier" and less chemically diverse.** The probe found much less water vapor entering its sensors than scientists expected. Absent entirely at these levels was a rich variety of complex organic molecules, such as ethane and phosphine, that scientists had assumed gave Jupiter some of its richer coloration. The discovery of such molecules and water would have been very exciting, for these are the same ingredients that are believed to have been present when life first developed on Earth. Such a discovery would not have said anything about life on Jupiter but would have shown that the chemistry that may have led to life on Earth was not unique to our planet and indeed might be widespread throughout the universe. The probe also detected much less lightning in the atmosphere than anticipated and less helium at these levels, which could ultimately give us insight into exactly how Jupiter and the other planets formed. In all of this, however, we must remember that the probe only entered and analyzed the Jovian atmosphere at one point during one hour of one particular day, and thus hardly

made object to ever enter the clouds of Jupiter. The main spacecraft then fired retro-rockets and established a long looping orbit that alternately carried it in for repeated close looks at Jupiter and out for encounters with three of the planet's largest satellites: Ganymede, Callisto, and Europa. *Galileo*'s images showed far more detail than those sent back from *Voyager*.

345 **The *Galileo* probe found a windier Jupiter than scientists expected.** The probe, which descended into Jupiter's upper cloud deck, had no cameras onboard but sent back data for almost an hour before high temperatures and pressures were too much paints a complete picture or even perhaps a typical one.

347 **Jupiter's four largest satellites were discovered by Galileo back in 1610.** When Galileo first looked at Jupiter through a telescope, he discovered four little starlike objects arranged in a straight line on either side of the planet. Watching them from night to night, he saw that they changed position but always followed Jupiter. He correctly surmised that they were satellites of Jupiter—the first satellites to be found in orbit around another planet. Knowing where his bread was buttered, Galileo initially named the moons in honor of his patrons, the wealthy Medici family. Since Galileo's time, however, these four worlds have been referred to as the *Galilean satellites*.

348 **The Galilean satellites of Jupiter are actually bright enough to be seen with the naked eye.** These moons are so large that they are actually brighter than the faintest stars that can be seen with the naked eye. They remained undiscovered until the invention of the telescope, however, because the glare from Jupiter blinds us to their fainter light in the same way that it would be impossible to see a firefly crawling on the rim of a searchlight shining directly into our eyes.

349 **The Galilean satellites take their names from Greek mythology.** In Greek mythology, Zeus (the king of the gods) was the equivalent of Jupiter (the father of the sky). And so, in time, Jupiter's four largest satellites, always hovering about it, were appropriately named after several of Zeus' many mistresses and consorts: Callisto, Ganymede, Europa, and Io.

350 **Jupiter has a large and colorful entourage.** In the years since Galileo first looked to the sky, astronomers have discovered twelve additional satellites orbiting Jupiter. Indeed, there are so many that the planet and its moons look a bit like a miniature solar system. Twelve moons travel around Jupiter in one direction, while the outermost four (probably captured asteroids) orbit in the opposite direction. The moons vary in size from little more than oversized rocks to worlds larger than Mercury and Pluto. Several of the moons have been explored at close range by spacecraft. The amazing images sent back by the spacecraft show that these satellites vary as much as the planets do.

351 **Callisto is a big "dirty ball of ice."** The outermost of the four Galilean satellites is Callisto. About 3,000 miles in diameter, it is considerably larger than the Earth's Moon. But whereas the Earth's only satellite is made of rock, Callisto is mostly frozen water with a smattering of metals and silicates. The surface of Callisto is peppered with thousands of bright ice craters that were caused by the impact of smaller objects from space. Callisto's largest and

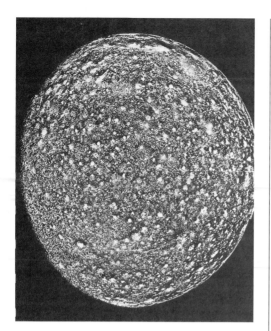

The surface of Jupiter's satellite Callisto is peppered with thousands of impact craters. (NASA/JPL)

caused either by expansion of the satellite's interior as it solidified or by tidal forces from giant Jupiter. In 1995, astronomers discovered a very tenuous ozone atmosphere surrounding this moon.

353 **Europa has the topography of a billiard ball.** Traveling inward from Ganymede, we encounter Europa. Almost 2,000 miles across, Europa is orange in color and appears to be covered with a crust of ice that makes this world the smoothest in the solar system. The elevation differences across its entire surface are less than 600 feet.

most impressive feature is a giant bull's-eye-shaped formation called Valhalla. About 1,800 miles across, it consists of ten concentric rings of shattered ice and is undoubtedly the result of a huge boulder from space striking Callisto long ago.

352 **Ganymede is the largest moon in the solar system.** With a diameter of about 3,280 miles, Ganymede is not only the largest satellite in the solar system, it is larger than both Mercury and Pluto. Like Callisto, it is made of water ice that encases a rocky core. Ganymede's surface also has craters, but the landscape is striped here and there with strange-looking grooved terrain. These features, known as *wrinkle ridges,* are probably the result of stresses in Ganymede's crust

354 **When scientists finally found the Martian canals, they were on Europa.** Europa's surface is crisscrossed in many places with lines that run in different direc-

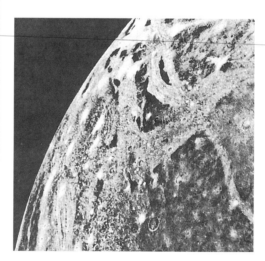

From a distance of 152,000 miles, Ganymede, the largest satellite in the solar system, shows broad wrinkle ridges, signs of internal stress. (NASA/JPL)

Jupiter's Europa, smooth as a billiard ball, is covered with features that look like cracks. The surface is likely a thin layer of ice that encases a worldwide ocean. (NASA/JPL)

tions for hundreds of miles. When scientists saw similar features on Earth in sea ice that had partially melted and refrozen several times, it led some to speculate that Europa's icy surface crust may be quite thin and encase a worldwide ocean. Temperatures above the freezing point would be maintained in this alien ocean by heat delivered through *tidal friction* from Jupiter.

355 **Europa is also a colorful place with exciting possibilities.** Europa's orange coloration has also led scientists to speculate that this moon might contain the same kinds of complex organic molecules that were the precursors to the development of life on

Earth. Someday, a spacecraft may chip off a piece of Europa's ice and return it to Earth. A future spacecraft/submarine may send back some fascinating data and pictures from this very intriguing moon as it explores the exotic depths of this alien ocean.

356 **Io looks like a pizza.** The most bizarre and exotic satellite in all the solar system has to be Io. As the *Voyager* spacecraft approached this moon, scientists looked down on a landscape like none they had seen before. Not a single crater could be found. Instead, Io's blotchy surface was a cosmic palette of color—reds, yellows, oranges, black, and white—that made it look not unlike a pepperoni pizza with anchovies.

357 **Io's strange appearance is caused by an extraordinary phenomenon.** Why does Io look so strange? In a *Voyager* image of Io's dark side that was purposely overexposed (to show the stars and make sure *Voy-*

Io, which looks like a pizza, is the most volcanically active world in the solar system. (NASA/JPL)

A giant volcano erupts along Io's limb, blasting sulfurous material high into the sky. (NASA/JPL)

ager was headed in the correct direction), scientists found the answer. There, rising above the satellite's limb, was a plume from an active volcano that was blasting material nearly 200 miles into the Ionian sky. In short order, over half a dozen other active volcanoes were found, all spewing enormous sprays of molten sulfur. Each volcano was the size of

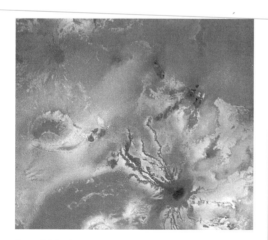

From high above, we peer down the throat of an Ionian volcano. The caldera appears as a dark oval. Outflowing material creates a radial pattern in all directions. (NASA/JPL)

France. Io has no craters because it is constantly resurfacing itself with matter from within. Io is the most volcanically active world in the solar system—a world that is literally turning itself inside out.

358 **Io's volcanism is due to Jupiter.** Io's proximity to massive Jupiter is the reason for Io's incredible volcanoes. Just as the Moon exerts tidal forces on Earth, so Jupiter does on Io. But Jupiter is so massive that its tidal effects on Io are extreme. Like a tennis ball constantly flexed in a mighty grip, Io's surface is forced to bulge up and down by as much as 300 feet on every orbit. As a result of the constant friction, Io is little more than a giant drop of molten sulfur encased in a thin surface crust. Breaking through holes and cracks in the crust, the material inside literally explodes into the near vacuum of space, creating giant volcanic plumes. The sulfur, in different chemical states and at different temperatures, accounts for the variety of colors on Io's surface.

359 **Io continues to be volcanically active.** While the *Voyager* spacecraft discovered the volcanoes of Io back in the late 1970s, the Hubble Space Telescope has allowed us to continue to keep an eye on this remarkable world. In July 1995, a new yellow-and-white splotch appeared on Io. Nearly 200 miles across, it is almost certainly the outflow from another giant volcano.

360 **Io has a large iron core.** In May 1996, the *Galileo* spacecraft discovered variations in Jupiter's magnetic field that were attributable to Io having an iron core about 1,000 miles across—nearly half the diameter of the satellite itself. Io thus becomes the first satellite known to have a magnetic field of its own.

361 **Io is a bit of a litterbug.** As Io travels around Jupiter spewing its sulfurous fumes, it "pollutes" its orbit by leaving a cloud of sulfur atoms in its wake. Over time, this has formed a complete doughnut-shaped cloud around Jupiter that traces Io's entire orbit.

362 **Jupiter's tidal forces create a significant pattern of differences among its largest satellites.** The densities of the four largest satellites of Jupiter increase as we get closer to Jupiter and the temperatures of their interiors also rise. Callisto and Ganymede are little more than "dirty balls of ice," while Europa probably is an ocean world under a thin crust of ice. And Io is almost completely made of molten sulfur and iron with no craters and virtually no water or ice at all. Similarly, the surface of Callisto is probably quite ancient and shows no evidence of internal activity. By contrast, Ganymede's wrinkle ridges demonstrate that some geological activity has gone on there. Europa's crisscrossed ice pattern may indicate repeated meltings and refreezings, perhaps even to the present, and its thin or slushy ice crust has obliterated most signs of cratering in the past. Finally Io, through its erupting volcanoes, is having its surface continuously "repaved." Indeed, Io has the "youngest face" in the solar system.

363 **Jupiter itself is the reason for the primary differences between its major satellites.** Simply put, the closer a satellite is to Jupiter, the stronger its tidal forces and the greater its tidal heating. Thus, the far-out worlds (such as Callisto and Ganymede) are frozen solid and have been for a long time, while worlds closer in are warmer. This, in turn, accounts for whether a moon is solid or liquid or molten and determines whether the original water the satellite may have had remains frozen in place or has evaporated into space long ago, leaving heavier substances behind to give that moon a higher density.

S aturn

364 **Saturn is the lord of the rings.** Twenty years ago, astronomers wondered why, of all the giant planets, only Saturn had rings. Today, we know that all four giants (Jupiter, Saturn, Uranus, and Neptune) have rings of some sort, but none can compare to the magnificent rings of Saturn. Shining brilliantly, Saturn's rings span over 200,000 miles (nearly the distance from the Earth to the Moon) and can be seen in even a small amateur telescope.

365 **Saturn's rings were first seen by Galileo, who couldn't figure out what he was looking at.** When Galileo first looked at Saturn through a telescope, he noticed that something about this planet was unusual. He reported that "the sixth planet is three" (that is, a *triple* planet) and described how Saturn (an old man in Roman mythology) apparently needed a servant on either side to help him move about the heavens. Galileo knew that what he saw in his telescope wasn't round like Jupiter, but to him it looked as though the planet had a smaller planet on either side. He also described the objects at other times as looking like handles on a teacup or like "ears."

other astronomer, Christiaan Huygens, realized that what Galileo had seen was actually an unattached ring that completely encircled the planet. The fact that Galileo failed to discern a ring may have been due to the inferior optical quality of his early telescope, but it may also have been due to the fact that no one had ever seen a ring around a planet before, so Galileo's brain was not prepared to explain to him what his eyes were seeing.

366 **Christiaan Huygens solved the mystery of the sixth planet.** Years later an-

367 **When viewed in a telescope from Earth, Saturn is a three-ring circus.**

Beautiful Saturn with three of its satellites: Tethys, Dione, and Rhea. (NASA/JPL)

Seen from Earth, Saturn appears to have three rings. They are simply called the A-ring (outer), the B-ring (middle), and the C-ring (inner). The B-ring, the broadest, is separated from the A-ring by a gap named the *Cassini Division* (for the astronomer who discovered it). It's wide enough to drop our Moon through and can be seen in medium-sized amateur telescopes. The C-ring is also referred to as the *Crepe ring* because of its gauzelike or semitransparent appearance.

368 **When it comes to the rings of Saturn, there's a lot more than meets the eye.** As the *Voyager* spacecraft approached Saturn in 1980 and 1981, finer structure in the rings began to be revealed. What appeared to be only three rings from Earth turned out to be first hundreds, then thousands of "ringlets." From close up, Saturn's rings began to look like a phonograph record. In addition, the *Voyager* spacecraft discovered rings never seen from Earth, including one that mysteriously braids and unbraids like a girl's hair.

369 **While they may look like a racetrack or CD, Saturn's rings are not solid.** It has long been known that the rings and ringlets are not solid disks but instead are made up of millions of pieces of "dirty ice." They range in size from particles the size of grains of sand to icebergs the size of a small house. Each object in this blizzard behaves like a tiny moon in its own separate orbit. Like the planets orbiting the Sun, the closer a ring particle or boulder is to Saturn, the faster it travels. Some race around at speeds

of up to 50,000 miles per hour. To our eyes, all these minisatellites blur, like the blades of a fan, into the beautiful adornment we call the rings.

370 **Saturn's B-ring has *spokes*.** Flying high over Saturn's broad middle ring, *Voyager 1* discovered what appeared to be dark spokelike streaks. Scientists believe the spokes may be electrostatically charged dust suspended just above the ring that is trapped in Saturn's magnetic field and forced to orbit the planet as it spins.

371 **Gaps between the rings are the result of tugs-of-war between some of Saturn's satellites.** In addition to the 3,000-mile-wide gap known as the Cassini Division, other gaps are also visible within the rings. In time, astronomers realized that these separations were not only semipermanent features

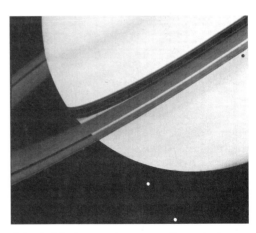

Saturn's thin rings can be seen casting a shadow on the cloud tops far below. The clouds can be seen shining through a major gap in the rings known as the Cassini Division while Tethys and Dione orbit nearby. (NASA/JPL)

From close up and high above, the rings are seen to subdivide into thousands of ringlets. Dark spokes are also visible in the broad central B-ring. (NASA/JPL)

of the ring system, but were the direct result of gravitational tuggings between several of Saturn's moons. The forces between certain moons act as invisible agents to clear ring material out of certain zones at particular distances from Saturn and create the gaps that are seen.

372 Other Saturnian satellites act like sheepdogs to keep rings from disappearing. *Voyager 1* discovered a very thin ring just beyond Saturn's A-ring that scientists were surprised could even exist. According to theory, the pieces of rock and ice that make it up should have dissipated and scattered into space long ago. Looking closer, however, they found two tiny satellites, one on either side of the ring. The two satellites, appropriately called *shepherd satellites,* act like sheepdogs and use their

gravitational pull to actually herd particles of ring material that begin to wander off back into the ring.

373 Once in a while, Saturn's rings disappear. A couple of years after Galileo first saw Saturn's mystifying shape, he was further mystified to discover that the planet's "servants" or "appendages" had completely *disappeared,* leaving only a single round planet in his telescope. Today, we realize that the rings seem to disappear every 15 to 17 years when they are positioned edge-on as seen from Earth. The celestial magic trick is made possible because, although the rings measure over 200,000 miles across, they are less than 100 feet thick! Trying to see them turned edge-on from the distance of Earth is like trying to see a phonograph record edge-on at a distance of 20 miles! The most recent time the rings were edge-on as seen from Earth was in 1995.

374 The rings lie within a region of strong tidal forces. Every object has an imaginary surface around it called the *Roche limit.* Within this surface, tidal forces created by the object are greater than the force of gravity between other objects placed there. Thus, within the Roche limit of Saturn, any material that has collected cannot come together to form a satellite but instead must remain individual pieces. In some cases, objects venturing closer than the Roche limit to an object can even be torn apart by these tidal forces, thus creating rings.

375 Saturn is made of such lightweight stuff that if you could find a bathtub big enough to hold it, Saturn would float. Like Jupiter, Saturn is a giant cloud-covered planet with an atmosphere that descends for thousands of miles. And, like Jupiter, Saturn is made up mostly of the two lightest elements in nature: hydrogen and helium. Saturn also contains a smattering of heavier and more complex chemicals, but its overall density is actually less than that of water. This means that, given a large enough bathtub, Saturn would literally bob around like a marshmallow in a cup of hot cocoa.

376 While Jupiter is a riot of color, Saturn looks more like a big lump of pale butterscotch pudding. From its Great Red Spot to its belts of orange and brown, Jupiter's atmosphere displays dramatic swirls and splashes of color. Saturn, by comparison, is far more subdued. Its pastel yellowish brown cloud bands are interspersed with white. The reason seems to be a combination of two factors. First, there is a high haze layer in Saturn's atmosphere that makes it seem as though we are looking at the planet through frosted glass. Second, Saturn has a more thorough mixing of weather systems, which makes large single-colored cloud features quite rare.

377 *Voyager* found strange hurricanes on Saturn. *Voyager*'s cameras spied pinwheel-shaped swirls of clouds, where it rains ammonia, that are likely hurricanes the size

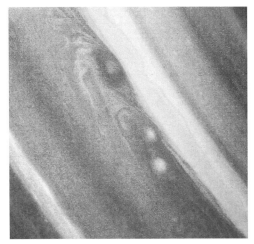

Storms on Saturn. (NASA/JPL)

of Asia. Such meteorological phenomena are nothing, however, compared to a mind-boggling event that may occur like clockwork on Saturn every 30 years.

378 Saturn may have the biggest blizzards in the solar system. Each time Saturn is closest to the Sun, the additional warming it receives triggers a huge upwelling of clouds from deep within the planet. As copious amounts of ammonia vapor rocket upward into the Saturnian stratosphere, they are turned into trillions of snowflakes. Caught by jet streams that blow at over 1,000 miles per hour, the storm quickly explodes into a giant ammonia blizzard that encompasses an area several times the size of Earth. Seen as an enormous white cloud that covers millions of square miles, the blizzard rages for weeks before slowly subsiding.

379 Like Jupiter's Callisto and Ganymede, many of Saturn's moons are made of ice. Like the rings themselves, many of Saturn's satellites are made of ice. At nearly 1 billion miles from the Sun, however, the ice is so cold it behaves quite differently than ice does on Earth. Losing virtually all of its elasticity, the ice of the moons of Saturn can be as strong as steel yet shatter like glass.

380 **The moon called Tethys contains a massive canyon.** On Tethys, there is a canyon called Ithaca Chasma that stretches two-thirds of the way around this world. Ithaca Chasma may be a crack that opened up when Tethys's interior cooled, froze, and expanded (not unlike what happens to your car's cooling system when you put too little antifreeze in your radiator).

381 **Mimas looks like Darth Vader's Death Star.** Tiny Mimas is a little sphere of ice barely 300 miles across, yet it sports an impact crater named Herschel that

Saturn's moon Mimas is only about 300 miles in diameter but sports a crater, Herschel, that is over 65 miles wide. Its appearance reminds some astronomers of the Death Star. (NASA/JPL)

is over 65 miles across and has a mountain peak 3 miles high in its center. Not only does this colossal feature make Mimas look like the Death Star in the film *Star Wars,* it shows just how big a crater can be on a moon this size. Had the object that plowed into Mimas been any larger, it would have shattered the moon into pieces.

382 **Mimas also illustrates just how small a world can be and still be round.** All of the moons in the solar system that are larger than Mimas are more or less spherical like the planets. But satellites smaller than Mimas are typically irregular in shape. Mimas-sized moons and larger objects had enough mass when they formed to have become molten for a time. When something is molten, the force of gravity naturally molds it into a sphere. Smaller satellites (as well as asteroids and comets) never went through a molten phase and so are irregular in shape.

383 **We have found Camelot and it's in orbit around Saturn.** Many of the surface features on Mimas have been whimsically named by astronomers after characters in the Arthurian legend, including Guinevere, Lancelot, Merlin, and, of course, Arthur himself.

384 **Enceladus is the giant's ice cream scoop.** Enceladus is another ice moon of Saturn with its share of craters. But it is also marked by a long and wide swath that looks as though a giant tongue came along

Saturn's icy moon Enceladus is a little over 300 miles across. Just below center a broad area of the satellite has been wiped clean—the likely result of liquid water gushing in torrents across Enceladus's surface at some time in the past and obliterating everything in its path. (NASA/JPL)

and licked it clean, obliterating all detail. The border of this area is even marked by craters that are partly intact and partly wiped away. Oversized extraterrestrials with a liking for cosmic-sized snow cones not being a likely hypothesis, scientists think this is where the moon's ice melted at least once in the past and gushed in torrents across the landscape. The heating source? Tidal tuggings from giant Saturn.

385 **Iapetus is a two-faced world.** Iapetus is one of the strangest worlds in the solar system. About 900 miles across, it has one hemisphere coated with ice as bright as newly fallen snow, while portions of the opposite hemisphere are darker than asphalt. Astronomers speculate that the dark material may be some sort of rich organic material (not unlike tar) that has somehow welled up from deep inside the moon because of tidal heating between Iapetus and Saturn.

386 **On Hyperion, no two days are alike.** Hyperion is about 160 miles across and looks like a cross between a hamburger and a hockey puck. Its far-from-spherical shape is probably due to a collision in the ancient past that knocked off one or more large chunks and blasted what was left into an egg-shaped orbit. Its odd shape and strange orbit give the moon such a chaotic rate of spin that the length of a day on Hyperion varies from one day to the next. If the long days always occurred on Saturdays and Sundays, it wouldn't be such a bad place to live.

387 **Titan is a world shrouded in natural gas.** Titan, Saturn's largest satellite, is the second-biggest satellite in the solar system (after Jupiter's Ganymede). With a diameter of 3,200 miles, Titan is larger than the planets Mercury and Pluto. It is not only a planet-sized world, it also has another feature typically characteristic of a planet: an atmosphere. In fact, Titan's atmosphere is two and a half times as dense as that of Earth. When the *Voyager* spacecraft flew past Titan, scientists hoped to see its potentially exotic surface, but instead all they saw was a globe enshrouded in a featureless orange haze. Rich in methane (commonly known as natural gas), Titan's atmosphere is acted on by the Sun's light to produce a natural hydrocarbon smog. Some scientists have suggested that over the years mists of organic compounds may have filtered down through the atmosphere and accumulated on Titan's surface in orange-colored drifts of sludge. Others imagine cloud decks where ethane rain falls into seas or lakes of liquid methane.

388 **In 1994, scientists got to peer through Titan's obscuring clouds.** Recently, astronomers have used the Hubble Space Telescope to cut through Titan's obscuring smog. At particular wavelengths in the infrared, the atmosphere becomes somewhat transparent, allowing views of the surface to be glimpsed. Thus far, light and dark features have been mapped, but from nearly 1 billion miles away, even the Hubble Space Telescope cannot resolve details well enough to allow us to figure out exactly what we are seeing.

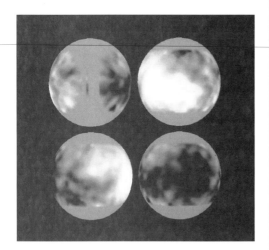

Astronomers used special filters on the Hubble Space Telescope to penetrate the thick obscuring atmosphere of Titan. These light and dark regions exist on the satellite's surface, but their exact nature is still unknown. (NASA/STScI)

389 **The next wave of Saturn exploration will occur shortly after the turn of the twenty-first century.** Early in the twenty-first century, a spacecraft named *Cassini* is scheduled to travel to Saturn. The main craft will go into orbit around the ringed giant. A probe called *Huygens* will dislodge from the mother ship and descend via parachute into Titan's exotic atmosphere. While no camera will be onboard *Huygens* (thanks to budget cuts), other instruments will tell us more about the weather and chemistry of this unusual moon.

390 **In the distant future, Titan could become an interesting place to live.** Nearly 1 billion miles from the Sun, it's not surprising that Titan's temperatures aren't exactly balmy. Observations and computer models suggest that surface readings probably hover near −250°F. Four or five billion years from now, however, things may be quite different. As our Sun grows older, it will someday expand to become a red giant star, engulfing and incinerating Mercury and Venus and boiling away the oceans of Earth. But the very changes that will make life on Earth impossible may cause Titan to "bloom." The makeup of Titan is scientifically fascinating, for Titan is a veritable organic chemistry lab, rich in many of the molecules that scientists believe were present on Earth when life first began here. Should Titan's temperature rise sufficiently, some interesting evolution (first chemical and later biological) may not be out of the question. Our distant ancestors may just find a new home in orbit around Saturn when our old one is no longer fit for life.

---

391 **Astronomers have known for years that Saturn has the largest number of moons in the solar system, but the family may be even larger than suspected.** Astronomers recently took advantage of a special event to go looking for more satellites around Saturn. For 15 to 17 years at a time, the Earth is positioned in such a way that we are able to see either the top side or the under side of Saturn's rings. Then, for the next decade and a half, we see the opposite side of the rings. But in between, for periods of a few weeks, the rings are seen virtually edge-on. At such times, the rings virtually disappear because they are so thin. With the rings' visibility greatly diminished, so is the glare of sunlight reflecting off of them. This, in turn, allows astronomers to search for tiny satellites that may have gone undetected. During the summer of 1995, astronomers took advantage of a *ring plane crossing* and used the newly sharpened eyes of the Hubble Space Telescope to ferret out what may prove to be even more family members. Future observations will be needed to tell for sure.

*U*ranus

392 **Uranus was the first planet to be discovered with a telescope.** In 1781, astronomer William Herschel (who wasn't a half-bad musician, by the way) spotted Uranus in his telescope. Unlike a star, Uranus had a small disk and Herschel at first thought his discovery might be a comet. Carefully noting its change in position over time, however, he was able to plot Uranus's orbit and found that it didn't follow the long looping path of a comet but the nearly circular path of a planet—in this case, a planet beyond Saturn.

393 **Under the right circumstances, you can see Uranus with the naked eye.** At maximum brightness, Uranus is actually bright enough to be glimpsed with the naked eye on a clear, dark, moonless night. Undoubtedly, over the centuries lots of folks did, but they failed to take note of the planet's slow telltale motion amid the stars from night to night and so didn't realize it was a planet.

---

394 **If Herschel had gotten his way, the seventh planet wouldn't have a name that makes junior high school students snicker no matter how you pronounce it.** Herschel, the first human being to officially discover a planet, felt he ought to have the right to name it. If Herschel had gotten his way, the planets (in order from the Sun) would have been: Mercury, Venus, Earth, Mars, Jupiter, Saturn, and . . . George. Although born in Germany, Herschel was living in England at the time of his discovery and, being a loyal subject of the crown, thought it would be a wise idea to name his discovery *Georgium Sidus* (Georgian Star) after the current monarch, George III. Had Herschel been successful in his efforts, the very man who just a few years before had lost the American colonies would have gained an entire new planet (though admittedly one that would have been difficult to tax). Other astronomers (especially those in France) somehow had objections to the English king getting his own planet and so, ultimately, the new world was named Uranus, the father of Saturn in Roman mythology.

---

395 Uranus is about as bland as a planet can get. In even the largest telescopes on Earth, Uranus looks like nothing more than a small bluish dot. It's not that the planet is really small (it's actually about four times the diameter of Earth), it's just that at an average distance of almost 1.8 billion miles, almost anything would look small. As *Voyager 2* approached Uranus in 1986, scientists hoped to begin to make out details. But while Uranus grew larger in *Voyager 2*'s cameras, it remained a virtually featureless disk.

396 *Voyager 2* gave us our first glimpses of detail on Uranus. In its closest approach, *Voyager 2* finally was able to spy the tops of what looked like giant cloud complexes similar in structure to giant thunderstorms on Earth (but probably lacking the thunder and lightning). Each was approximately the size of the United States.

In recent years, the Hubble Space Telescope has revealed similar cloud features on Uranus from time to time, but for the most part, Uranus remains a visual understatement. The presence of a high haze layer above Saturn's clouds gives Saturn a more muted appearance than Jupiter. In the case of Uranus, however, the situation is even more extreme. Indeed, in addition to haze, the Sun's ultraviolet rays have created a thick acetylene smog high in the Uranian clouds through which very few details can typically be seen.

397 Uranus rolls around the Sun on its ear. While the Earth's axis of rotation is tilted 23.5°, Uranus's axis is tipped over an incredible 97.9°. This means the axis of the planet lies almost in the plane of its orbit. As a result, Uranus travels sideways as it orbits the Sun. Many astronomers surmise that Uranus's odd orientation is the result of a collision with a massive object early in the history of the solar system—a collision that knocked Uranus on its ear.

398 Uranus has strange days, strange nights, and strange seasons. The fact that Uranus's axis is tilted 97.9° results in first one of its poles and then the other pointing toward the Sun for 42 Earth years at a time. In other words, from a point near either of Uranus's poles, day and night last 42 years each. Similarly, summer and winter last equally long periods. By a quirk of atmospheric circulation, however, winter is slightly warmer than summer, although temperatures at the cloud

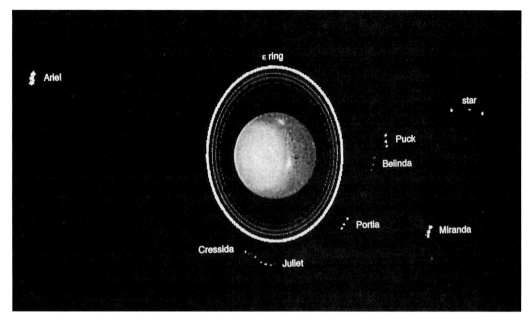

A composite of exposures by the Hubble Space Telescope show Uranus, its rings, and several of its moons. Multiple exposures to record the faint rings have resulted in multiple images of the satellites. Two rare cloud features on Uranus are 1,800 and 2,700 miles across. (NASA/STScI)

tops in both seasons rarely warm to more than −300°F.

399 **Uranus gets its blue color from methane gas in its atmosphere.** Like Jupiter and Saturn, Uranus is another giant cloudy world consisting mostly of hydrogen and helium. Its atmosphere also has trace amounts of methane gas, however. The methane absorbs the reds, oranges, and yellows in the Sun's light and scatters back the blue into our eyes, thus making the planet appear that color.

400 **In 1977, rings were discovered around Uranus without anyone actually seeing them.** In 1977, astronomers were observing Uranus, waiting for it to pass in front of a star.

From the way the star's light dimmed as it went behind the planet, the scientists would be able to deduce things about the structure of Uranus's upper atmosphere. Before the planet ever got in line with the star, however, the star's light blinked off and on several times. The astronomers correctly deduced that it had done so because a system of thin dark rings surrounding the planet had eclipsed the star's light. The process, as expected, repeated after the star passed behind the planet. From the number of blinks and their duration, astronomers estimated that Uranus was surrounded by nine very thin rings.

401 **In 1986, we actually got to see the rings of Uranus for the first time.** *Voyager 2* got close enough to Uranus to map its rings in great detail. Indeed, in a purposely planned reenactment of the type of event that had led to their discovery in the first place, scientists carefully positioned *Voyager* to watch another star pass behind the rings as it swept past Uranus. *Voyager*'s cameras confirmed the spiderweb-like thinness of the rings and added a particularly faint tenth ring to the nine previously discovered from Earth. *Voyager* also provided a good explanation why the rings had never been actually seen on Earth. Not only were they thin, they were also as black as charcoal. One possible explanation holds that the ring particles may be coated with methane ice, which darkens when exposed to sunlight. As with the rings of Jupiter and Saturn, the Uranian rings gird the planet's equator, so they travel around the Sun on their side, as does the planet itself.

402 **Uranus has a magnetic field that really knows how to swing.** With the

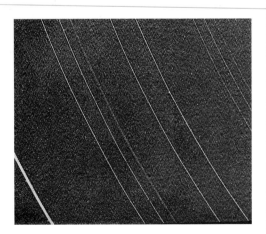

Voyager 2 provided our first direct views of the thin dark rings of Uranus. They are likely made of methane-coated pieces of ice darkened by the action of sunlight. (NASA/JPL)

Earth, Jupiter, and Saturn, the axis of each planet's magnetic field lies pretty close to the planet's axis of rotation. But this doesn't have to be the case, as Uranus demonstrates. When it comes to Uranus, the planet's magnetic axis is tipped almost 59° to its axis of rotation, so as Uranus spins around, we are repeatedly looking down one magnetic pole and then the other.

403 **When it comes to Uranus, just how fast it spins depends on what you're talking about.** Like Jupiter and Saturn, different latitudes on Uranus spin around at different rates, so just how long a day is depends on where you are. But Uranus is so featureless that astronomers frequently talk about its rate of spin by tracking the spin rate of its magnetic field instead. This is actually a way of saying how fast the core of the planet rotates, as the magnetic field is tied to the molten iron heart of the planet. While we cannot see into the core, instruments onboard *Voyager 2* called *magnetometers* were able to sense Uranus's magnetic field (trailing across it like a cat's whiskers trailing across your face) and map the magnetic field as well as its rate of spin.

404 **One spacecraft found twice as many moons around Uranus in a couple of days as astronomers on Earth had found since the planet's discovery.** Uranus's five largest satellites were discovered from Earth between 1787 and 1948. *Voyager 2*, in its flyby of the planet in 1986, spied ten

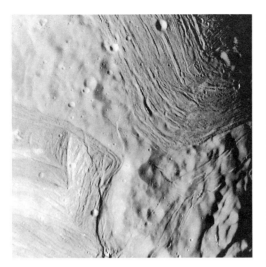

A close-up of some of the truly chaotic terrain found on Uranus's "patchwork quilt" moon, Miranda. Scientists believe that Miranda was broken apart by a powerful collision and then reassembled by gravity in a haphazard way. (NASA/JPL)

more ranging in size from about 70 to less than 20 miles across.

---

405 **Miranda is a "patchwork quilt" of geological wonders.** Of all the moons in the solar system, Uranus's Miranda is the greatest geological treasure trove. While Miranda is only about 300 miles in diameter, within the confines of this tiny world lie cliffs, canyons, and terrain that can only be described as a chaotic jumbled mess. Scientists believe that earlier in its life, Miranda was struck at least once by an object massive enough to break it into pieces. Pulled back by the force of their mutual gravity, the pieces reassembled, but not in their original arrangement, thus creating a moon that resembles a giant three-dimensional jigsaw puzzle where the pieces don't quite fit.

406 **The ice cliffs of Miranda are not only a great geological wonder, they'll also make a great theme park ride someday.** Few geological features rival the vertical ice cliffs of Miranda, the tallest of which, Verona Rupes, towers 9 miles high and represents the greatest sheer drop in the solar system. If you stood at the top and stepped off into space, you would find the gravity of this tiny world so slight that it would take you nearly half an hour to reach the canyon floor below.

A close-up of the 9-mile-high ice cliffs of Miranda—one of the true wonders of the solar system. (NASA/JPL)

$\mathcal{N}$eptune

407 **Neptune was discovered because Uranus was off-course.** Within a century of Uranus's discovery, the planet had traveled far enough around the Sun for astronomers to realize that it was not keeping to the orbit they had expected. A young English astronomer named John Couch Adams, who had recently graduated from Cambridge University, calculated that the planet's deviant behavior could be explained if another planet beyond Uranus was tugging on the planet with its gravitational pull. Adams even calculated where this unknown planet might be found, but no one in England bothered to look for it. A year later, a French astronomer named Urbain Leverrier independently made the same calculations and reached the same conclusion, but he couldn't get any of the French observatories to look for the planet, either. Frustrated, Leverrier took his calculations to German astronomer Johann Gottfried Galle at the Berlin Observatory in the fall of 1846. Galle pointed his telescope skyward and found the new planet the very first night.

408 **Neptune is a cobalt blue wonder.** Aptly named for the god of the sea, Neptune is colored in rich shades of blue. Like Uranus, its atmosphere contains methane gas, which gives this world a deep cobalt blue color by absorbing the red light in the Sun's spectrum and directing the rest back at

us. Unlike Uranus, however, Neptune's atmosphere is not shrouded in a thick haze but at times can display an incredible array of dynamic features.

409 **In 1989, *Voyager 2* flew past Neptune and photographed a huge dark blue "eye" staring right back.** Dubbed the Great Dark Spot, it was the size of the Pacific Ocean and is believed to have been a giant storm system that made a deep well in Neptune's upper cloud deck through which we were able to peer down to darker blue clouds far below. Time-lapse movies of the Great Dark Spot made from individual *Voyager* images showed its fluid nature. Over time, it actually changed shape, getting slightly longer and shorter, rounder and skinnier, and, in general, looking not unlike a "giant slug."

410 **The Wizard's Eye and the Scooter were two of the more interesting features of Neptune.** In addition to the Great Dark Spot, *Voyager 2* also discovered a smaller dark spot that scientists cleverly dubbed the Small Dark Spot. Further inspired by white clouds that perpetually gathered over the center of this dark blue well in the Neptune atmosphere, scientists nicknamed this feature the Wizard's Eye. The white clouds, made of methane ice crystals, were similar to the cirrus clouds (or so-called mare's tails) in the upper atmosphere of Earth. Another brilliant white cloud was nicknamed the Scooter because it raced around Neptune at a speed faster than any other cloud feature.

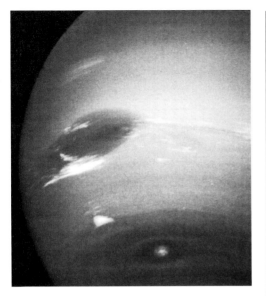

Neptune as it appeared to *Voyager 2* in 1989. Visible are the Great Dark Spot, the Wizard's Eye, and the Scooter. High methane ice crystal clouds show up as brilliant white just under the Great Dark Spot and in the center of the Wizard's Eye. (NASA/JPL)

411 **Neptune has blizzards of frozen natural gas.** *Voyager* also took pictures of long streaks of white that cast shadows on the blue cloud tops far below. They turned out to be enormous clouds made of methane ice crystals (frozen natural gas) that would stretch from New York to Paris.

412 **The fastest winds in the solar system are found on Neptune.** While jet streams on Jupiter have been clocked at over 300 miles per hour and some on Saturn exceed 1,000 miles per hour, Neptune's jets can rip along at an astounding 1,400 miles per hour—the fastest winds in the solar system. If Earth had jet streams that strong, storms would be propelled clear across the United States in barely two hours and commercial jets could fly coast to coast in even less time.

413 **The Hubble Space Telescope recently recorded remarkable changes in Neptune's weather systems.** In 1994, the refurbished Hubble Space Telescope provided the first close-up views of Neptune since *Voyager 2* flew past in 1989. The images revealed dramatic changes, including the complete disappearance of the Great Dark Spot in Neptune's northern hemisphere and the emergence of a comparably colossal dark storm in its southern hemisphere! Gone too were both the Wizard's Eye and the Scooter. More recent images taken with the Hubble Space Telescope suggested that Neptune can somehow undergo such tremendous changes in as little time as a few weeks.

414 **Even more recent images reveal a quieter, gentler Neptune.** HST images from 1995 and 1996 reveal the disappearance of all of the dark spots, great or otherwise, from Neptune's atmosphere with only high bright clouds remaining.

415 **Could there be a link between Neptune's weather and the sunspot cycle?** When *Voyager* reached Neptune in 1989 and found a very active atmosphere with several dark storms, the Sun was near the maximum in its sunspot cycle. By the mid-1990s, when Neptune's atmosphere was lacking in such storms, the sunspot cycle was at minimum. Could there be a link? More HST observations in the coming years will be needed to tell.

416 **Galileo may have actually observed Neptune without knowing it was a planet.** In Galileo's notebook from December 1612, there is a sketch of Jupiter and its four largest satellites. Also included is another object that Galileo thought was a star. Had he observed it more carefully, he would have seen it slowly move from night to night and probably would have realized it was a planet. Had this happened, the planet that is usually the eighth from the Sun would have been discovered before the seventh.

417 **Triton is an amazing moon at the edge of the solar system.** Given that Neptune and its satellites orbit nearly 3 billion miles from the stimulating warmth of the Sun, scientists hardly expected to find worlds alive with activity. Neptune, however, did not disappoint and neither did its largest satellite, Triton. Geologist Larry Soderblum of the Jet Propulsion Laboratory began the press conference following the Triton encounter by *Voyager 2* with the words: "What a way to end the solar system!" This pink and gray world nearly 1,700 miles in diameter has a varied landscape. In places, there are frozen lakes up to 300 miles across. Elsewhere, valleys

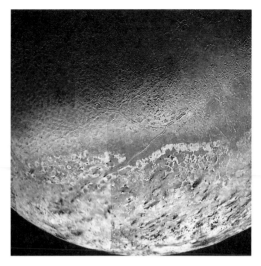

Neptune's largest satellite, Triton, as seen by *Voyager 2*. Along the top of the image is the mysterious topography dubbed "cantaloupe terrain" and crisscrossing grabens that reveal geological activity. Below these, dark parallel streaks reveal the presence of giant geysers. (NASA/JPL)

of ice crisscross through cratered terrain and an entire hemisphere was seen to be covered with a unique pattern that looks like the skin of a cantaloupe. Its origin continues to baffle scientists.

418 **Triton also has an "oil field" out of control.** In Triton's opposite hemisphere, there is a vast array of dark streaks that are dozens of miles long. The strange features puzzled astronomers until movement was seen within some of the streaks. *Voyager 2* had uncovered a large number of active geysers, each blasting Sun-darkened nitrogen spray 8 miles or more into the sky. The spray, carried by winds, then paints the landscape in long black streaks.

419 **Triton, the coldest spot in the solar system, is also one of the brightest.** With a temperature of around −400°F, Triton's surface is the coldest place found so far in the solar system. This icy surface also reflects more than 90 percent of the weak sunlight that falls on it, making Triton one of the brightest spots in the solar system.

420 **Scientists had to retrain the *Voyager 2* spacecraft to get the images they wanted from Neptune and Triton.** When *Voyager 2* was originally launched from Earth in 1977, its primary targets were Jupiter and Saturn. Scientists knew they could take advantage of a rare planetary configuration to send *Voyager* on to Uranus and Neptune, but

The surface of a large frozen lake on Triton. Similar in appearance to the lunar maria, it marks a place where a basin was flooded with a mixture of water, ammonia, and methane, which then froze in the frigid temperatures. (NASA/JPL)

they didn't really expect the spacecraft, built in the early 1970s with late-1960s technology, to still be working by the time it got to these outer planets. However, by 1986, as it approached Neptune, *Voyager* was still alive and well. There was a problem to be solved, however, before *Voyager* could send back any useful pictures from Neptune: Light levels at Neptune are only .0025 what they are on Earth because of the great distance of the Sun. To get good pictures, long time exposures had to be taken and cameras turned to track the planet as the spacecraft sped past.

421 **Fancy shooting was responsible for getting *Voyager 2* to Neptune.** When *Voyager 2* left Earth, it barely had enough fuel for its extended mission. But scientists took advantage of a rare configuration of the giant planets that wouldn't be repeated for over 170 years and utilized a technique known as *gravitational assists.* Gravitational assists work by bringing a spacecraft past a planet at just the right distance and at just the right angle and letting that planet's gravity redirect and boost the spacecraft on to the next planet.

Voyager was given just the right trajectory as it rounded Jupiter to allow the giant planet's gravity to redirect it so that when *Voyager* crossed the orbit of Saturn, Saturn was there to meet it. Saturn, in turn, performed the same trick to propel *Voyager* on to an encounter—five years and several billion miles later—with Uranus. Uranus did the same, getting *Voyager* to Neptune. All of this was accomplished with only short firings of a small rocket to make slight corrections to the flight path along the way. In some cases, the computer guidance programs were so on target that no engine corrections were needed at all. By this ingenious maneuvering, astronomers got to see four planets (and lots of satellites) instead of two. The skill of the aiming necessary to perform the feat is equivalent to making a four-ball, four-pocket shot on a billiard table 22 miles long.

$\mathcal{P}$luto

422 **Pluto was discovered the same way Neptune was.** In the mid-1840s, Uranus's odd wanderings led to the discovery of Neptune. By early in the twentieth century, however, astronomers determined that Neptune was also off its expected course, so it was suggested that yet another planet might lie beyond. After a long and systematic

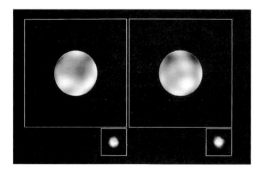

Computer-processed images from the Hubble Space Telescope reveal "features" on the surface of Pluto. (NASA/STScI)

search, tiny Pluto was discovered in 1930 by an American astronomer named Clyde Tombaugh at the Lowell Observatory in Flagstaff, Arizona. At first, Tombaugh simply referred to the object as Planet X. Many formal names were suggested by people around the world, but the astronomical community settled on Pluto. Pluto was actually the choice of a little girl in England, not because it's the name of Mickey Mouse's dog, but because Pluto was the god of the shadowy underworld in Greek mythology and thus seemed like the perfect name for such a dark and distant place.

423 **Pluto is the tiniest planet of them all.** With a diameter of only a little over 1,400 miles (less than the distance from New York to Denver), Pluto is the smallest planet in the solar system. In fact, Pluto is smaller than seven satellites or moons: the Moon, Io, Europa, Ganymede, Callisto, Titan, and Triton.

424 **The Hubble Space Telescope has given us our only glimpses of detail on Pluto to date.** Pluto is so tiny and far away that it looks like little more than a speck in even the largest telescopes on Earth. Perched above our planet's turbulent atmosphere, the Hubble Space Telescope has recently been able to make out some detail on the planet's surface. At this point, these large surface "features" can only be described as patches of light and dark. Their actual geological na-

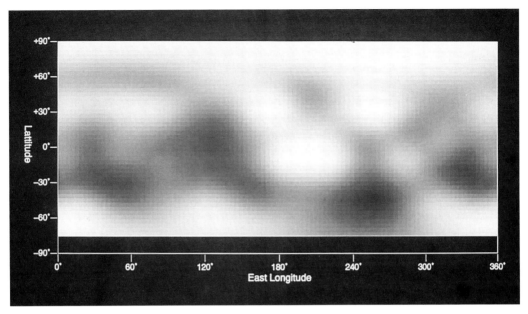

A map of light and dark areas on Pluto. Determining the exact nature of these "features" will likely have to wait for a spacecraft to visit Pluto sometime in the twenty-first century. (NASA/STScI)

ture will have to wait for a spacecraft to visit Pluto someday.

425 **While spacecraft have yet to visit Pluto, scientists think that Triton may have given them a sneak preview.** As statistics on Triton began to come in from *Voyager* in 1989, scientists couldn't help but notice some basic similarities between Triton and Pluto. Triton is about 1,700 miles in diameter, while Pluto is a little over 1,400 miles across. The densities of Triton and Pluto are similar (about twice that of water). Furthermore, both worlds average about the same distance from the Sun over significant portions of their orbits. All of this has led some to suggest that if we send a spacecraft to Pluto someday, we might find a world not too different from Triton. Ironically, until *Voy-*

ager's Triton encounter, scientists had more or less written off Pluto as a pretty dull place. But now that Triton has proven to have such a fascinating and varied landscape with both active geology and weather, Pluto is seen by many as a potentially more interesting place. Several plans to send a small spacecraft to Pluto are on the drawing boards. By the second decade of the twenty-first century, we may get our first close-up glimpse of this distant world.

426 **Naming the planets in order from the Sun sometimes requires you to change your tune.** If you asked most folks who went to school before the 1970s to name the planets in order from the Sun, they would probably answer: "Mercury, Venus, Earth, Mars, Jupiter, Saturn, Uranus, Neptune, and

Pluto." From time to time, however, this answer is wrong. This happens to be one of those times. The orbit of Pluto is so elliptical that it periodically brings the tiny planet closer to the Sun than Neptune. Since 1977, Pluto has been closer to the Sun than Neptune and will continue to be so until 1999, when Neptune will again take over the dubious distinction of being second to last. So in 1999, the little mnemonic "My Very Educated Mother Just Sent Us Nine Pizzas" will again help you to remember the names of the planets in their correct order from the Sun. (It should be noted, however, that the orbits of Neptune and Pluto do *not* intersect. Thus, a modern collision between these worlds is impossible.)

427 **Pluto is neither as big nor as bright as once thought.** Astronomy textbooks published before the late 1970s typically list Mercury as the smallest planet in the solar system. But this was based on an incorrect assumption about Pluto. Astronomers had assumed that Pluto had no satellites and so attributed all of the brightness they saw coming from this speck in the sky to Pluto alone. In 1978, however, a satellite was discovered orbiting Pluto. Scientists realized that some of the brightness (and size) that had been attributed to Pluto now had to be attributed to its moon. Pluto had been named for the god of the underworld, so its satellite was appropriately called Charon for the pilot of the boat that transported departed souls across the River Styx.

428 **Pluto and Charon look more like a miniature double planet than a planet and its satellite.** With diameters of about 1,430 and 745 miles, Pluto and Charon are closer in size than any other planet and its satellite in our solar system. Furthermore, they are less than 12,000 miles apart, making them look like a little barbell with an invisible connecting shaft tumbling end over end through space.

429 **Is there a tenth planet beyond Pluto?** Soon after the discovery of Pluto, some scientists argued that its small mass was insufficient to account for all of the off-course wanderings that had been observed in the orbital motions of Uranus and Neptune. They suggested a tenth planet must lie beyond Pluto. Since the ninth planet now had a name, this would-be tenth world was dubbed the new Planet X and the search was on. For decades, various astronomers in various parts of the world combed the skies for the elusive object without success. Finally, in 1992, they learned why. A scientist at the Jet Propulsion Laboratory in California reexamined the old Uranus and Neptune data and found that it had not been analyzed correctly. Fixing the mistake, he realized that the presence of Uranus, Neptune, and Pluto accounted for all the wanderings of the outer planets. No one had found the tenth planet for a very good reason—it had never existed in the first place.

430 **Nevertheless, some scientists believe that there may be many small icy objects beyond Pluto.** Such objects may be at least tens if not hundreds of miles in diameter and may resemble worlds like Pluto and Charon. Whether you would call these objects *minor planets* (a name now reserved for the asteroids), "icy midgets," or something else may be more a matter of semantics than science. If they exist, Pluto may someday be seen as simply the largest and closest of this different breed of solar system nomad. In 1996, one such object dubbed 1996TL66 was found. It had a diameter of about 300 miles. The discovery of many more will likely follow.

CHAPTER 7

ASTEROIDS, COMETS, METEOROIDS, AND SPACE DUST

$\mathcal{A}$steroids

431 **There is a gap in the procession of planets in our solar system.** In 1772, a professor of mathematics named Johann Titius noticed a curious mathematical progression in the distances of the planets from the Sun. Specifically, he started with the number 0, added 3, and then kept doubling the number. In this way, he got the numbers 0, 3, 6, 12, 24, 48, 96, 192, and so on. Titius added 4 to each of these numbers and then divided each of those numbers by 10. The result was the series of numbers 0.4, 0.7, 1.0, 1.6, 2.8, 5.2, 10.0, and 19.6. Curiously, if you divide the distance of each of the planets out to Uranus by the distance of the Earth from the Sun, you get the series 0.4, 0.7, 1.0, 1.5, 5.2, 9.5, and 19.2. In short, the actual distances of the planets Mercury through Uranus from the Sun are very close to the numbers in his series. There's probably no science going on here, just numerical coincidence. However, Titius noticed that his series included the number 2.8, but there was no planet to be found at 2.8 times the distance of the Earth from the Sun. According to Titius's "law," there should have been a planet between Mars and Jupiter. Instead, there seemed to be a gap in the solar system.

432 **On the first night of the nineteenth century, the planetary gap was filled.** On the night of January 1, 1801, quite by accident, an Italian astronomer named Giuseppe Piazzi ran across a faint little object in his telescope. From night to night, it slowly moved amid the stars and soon its orbit could be calculated: 2.8 times the distance of the Earth from the Sun. Piazzi named it Ceres after the Roman goddess of the harvest and the ancient deity of his native Sicily, where the discovery was made.

433 **Ceres filled Titius's gap but not in a very substantial way.** Ceres was at the "right distance" to be the missing planet between Mars and Jupiter, but it was clear from the start that Ceres wasn't much of a planet. Not even bright enough to be seen by the naked eye, Ceres turned out to be less than 600 miles across—about a quarter of the diameter of the Earth's Moon. Piazzi referred to the newly discovered object as an *asteroid,* meaning "starlike," because it was too small to be seen as anything more than a point of light in his telescope. Today, astronomers still use this term but also refer to such objects as *minor planets.*

434 **Today, we know that Ceres has lots of company.** Within a few years of the discovery of Ceres, other asteroids were discovered and were given names like Pallas, Vesta, and Juno. Each was smaller than Ceres—on average less than 300 miles across. Today, over 20,000 asteroids are known and dozens more are discovered each year. The actual total, if you include stuff down to the size of refrigerators, probably exceeds 100,000, most of which (but not all) swarm in a broad swath of space between Mars and Jupiter. The asteroids do not all occupy a single orbit.

435 **There are a few large asteroids and lots of little ones.** Only fifteen asteroids are more than 150 miles across. Most are less than a few miles in diameter. Many are no larger than a football stadium. The reason is that, over the billions of years since our solar system formed, there have been lots of collisions between these minor bodies, resulting in fragmentation.

436 **Many asteroids have a number and a name.** Asteroids carry a number that indicates the order in which they were discovered. The larger asteroids also have names. Thus, the first four asteroids to be discovered are officially known as 1 Ceres, 2 Pallas, 3 Juno, and 4 Vesta, though frequently the numbers are dropped. (These asteroids are also the largest to be discovered.) With so many asteroids out there, exotic names from mythology were soon exhausted. Today, there are asteroids that are officially called Chicago, California, and Bertha.

437 **The asteroids are not the pieces of a planet that blew up.** While this was a popular notion at one time, neither any evidence nor any plausible mechanism has ever been found to, in any way, substantiate such an idea. Indeed, if you were to glue all the known asteroids together, you would wind up with an object with less than a tenth of the mass of the Earth's Moon. In reality, there was simply never enough mass present between Mars and Jupiter to have a planet form there in the first place. The asteroids are the fragments of a planet that never was.

438 **Most of the asteroids are not round.** From observations of the satellites of the planets, we see that all such bodies more than about 300 miles in diameter are quite spherical but that the smaller satellites are invariably far from spherical. There are two factors at work. When an object of sufficient mass first forms, the heat generated by the pieces of the object falling together under the force of gravity (aided by the heat of radioactive decay) is sufficient to make the object molten for a time, thus it forms a sphere. An object of less than the critical amount of mass (and hence size) never becomes molten, so it never is molded by gravity into a sphere. Collisions are the second factor. While large objects frequently get cratered in collisions, small objects get broken into smaller irregular pieces.

439 **Asteroids are pieces of material left over from the formation of the solar system.** The solar system formed out of a giant cloud of gas and dust about 4.6 billion years ago. Gradually, grains of dust began to stick together and, in time, these tiny bits of matter stuck to each other as a swarm of debris circled the new Sun. Some of these objects continued to grow larger and more massive, gathering, or *accreting,* more matter to themselves through the force of their growing gravitational pull. Eventually, the largest became the planets and the satellites of the planets, but smaller pieces happened to avoid accretion or were prevented due to gravitational tuggings from giant Jupiter. These, in turn, underwent numerous collisions with each other, which fragmented them more and more. Among these objects are what we today call the *asteroids.*

440 **On its way to Jupiter, the *Galileo* spacecraft gave us our first close up looks at two—make that three—asteroids.** In October 1991, the *Galileo* spacecraft flew

The asteroid Gaspra as imaged by the *Galileo* spacecraft on its way to Jupiter. It measures about 12.5 × 7.5 × 7 miles in size. (NASA/JPL)

past an asteroid named 951 Gaspra. As anticipated, it proved to be an irregularly shaped object (looking a bit like a large cratered potato) about 12 miles long. In August 1993, *Galileo* had a second planned asteroid encounter, this time with one named 243 Ida. To everyone's surprise, although Ida was only a little over 1 mile across, it had a tiny object in orbit around it. While some were initially

The *Galileo* spacecraft also flew past the asteroid Ida and discovered a tiny satellite, which has been named Dactyl. (NASA/JPL)

tempted to name this baby satellite Ho (as in "Ida-Ho"), astronomers ultimately settled on the name Dactyl.

Galileo's brief encounters with the asteroids showed some significant differences between these objects. While both Gaspra and Ida are irregularly shaped and have cratered surfaces, scientists were able to determine that the surface of Gaspra is significantly older than that of Ida. The coloration of the two asteroids is also somewhat different, indicating differences in chemical composition. This, in turn, reflects measurements made from Earth of various asteroids that indicate a rather wide range in surface brightness and coloration. In short, asteroids may run the gamut from objects made of stony materials to others that contain significant quantities of iron, as we see in the meteorites that fall to Earth. In addition, asteroids that orbit closer to the Sun may show the influence of the Sun's heat in their formation, while more distant asteroids may not.

441 **Another asteroid was imaged by radar from Earth.** The object in question is known as 4179 Toutatis. From the images that were returned, this tiny asteroid appears to be two objects that are barely stuck together. One is about 3 miles across and the other is about 1.5 miles across and they are connected at a narrow "neck." In short, 4179 Toutatis looks like a peanut.

442 **Some asteroids share an orbit with Jupiter.** Not all asteroids orbit the Sun between Mars and Jupiter. Two groups of asteroids known as Trojan asteroids actually share Jupiter's orbit. One group lies 60° ahead of the giant planet, while the other group trails Jupiter by 60°. These points are not arbitrary but are known as *Lagrangian points* after the French mathematician who first worked out the theory. The Lagrangian points are places of increased gravitational stability between Jupiter and the Sun.

443 **Other groups of asteroids come much closer to home.** Three such "families" exist: the Amors, whose orbits can carry them between Earth and Mars; the Atens, whose orbits can carry them between Earth and Venus; and the Apollos, whose orbits *cross* that of Earth. The latter are clearly of more than academic interest, for while none of the orbits of any of the Apollos currently actually intersect the orbit of the Earth, these orbits are constantly changing because of the gravitational pull of the various planets. Ultimately, any of the more than 1,000 known Amor-Aten-Apollo (or AAA) Objects could prove a threat to Earth. Indeed, the cratering that our planet has experienced in the past has been due in part to such objects striking the Earth.

444 **Some scientists are advocating surveillance of all such objects.** In recent years, some concerned scientists have suggested to Congress that funding be provided for a detailed and systematic search for all such potentially threatening *near-Earth ob-*

jects. Currently, three observatories keep vigils, but someday a computer-automated network of telescopes could effectively scan the heavens nightly, tracking all known objects and discovering new ones—whose orbits could then be calculated. While the actual chances of a sizable object striking the Earth are slim over the near term, such surveillance is now technologically possible. Furthermore, given enough warning, we could also hope to deflect, fragment, or destroy the object with an intercepting spacecraft (that would latch on to it and push it away like a tug) or by bombardment with nuclear weapons. It is an interesting irony to consider that the same weapons that are capable of destroying all life on this planet might someday be used to save it.

445 How often do Apollo asteroids hit the Earth? On average, estimates indicate, about three Apollo asteroids strike the Earth during the course of every million-year interval.

446 There have been a couple of reasonably close calls (in astronomical terms) in recent years. In 1968, an Apollo asteroid named Icarus came within about 3,750,000 miles of Earth; in 1989, another, code-named 1989FC, sailed within 500,000 miles; and in 1991, a third asteroid called 1991BA missed us by about 106,000 miles—less than half the distance to the Moon.

447 How big are the near-Earth asteroids? Most Apollo asteroids are less than 1 mile across, but one has been found that is about 6 miles across.

448 How big would an asteroid have to be to create significant destruction on the Earth? The Barringer Meteor Crater near Winslow, Arizona, is about 4,200 feet across and was created by an object about 160 feet across. An asteroid fragment only 500 feet in diameter could explode upon impact with enough energy to destroy a major metropolitan area. Chunks of Comet Shoemaker-Levy 9 that probably measured a few miles across created shock waves on Jupiter in 1994 that were as large as the Earth. If an asteroid of that size ever strikes our planet, it will probably destroy all life.

449 Jupiter is actually the culprit in all this. The immense gravitational pull of Jupiter easily disrupts the paths of asteroids venturing near. Indeed, since the formation of the solar system, Jupiter has been altering the orbits of asteroids (and comets) and, for the most part, redirecting them into the inner solar system. This activity helped create a great *period of bombardment* in the early history of the solar system that produced many of the craters we see today on the Moon, Mercury, and Mars. Other redirected asteroids account for the existence of the current crop of Amor-Atens-Apollo Objects that still plague us.

Comets

450 **Comets are big frozen clumps of mud.** The *heart,* or nucleus, of a comet is a dark irregular chunk of ice, rock, and dust that looks a bit like a blackened potato and can range in size from hundreds of feet to several miles across. The ice is typically a combination of frozen water, frozen ammonia, and solid carbon dioxide.

451 **If you discover a comet, you get your name in the sky.** A curious tradition has grown up in astronomy whereby if you discover a comet, the scientific community names it after you. This "find a comet and get your name in lights" policy is responsible for many an amateur astronomer purposely scanning the skies for fuzzy patches of light that aren't supposed to be there. Some have been so successful that they have several comets to their credit. Sometimes a small group of people will find a comet together, as was the case with Shoemaker-Levy 9, the comet that plowed into Jupiter in 1994. It was named after its codiscoverers, Caroline and Eugene Shoemaker and David Levy, who stumbled upon this newsmaker completely by accident. The 9 signifies that this was their ninth joint comet discovery.

452 **There once was a man who went looking for comets but became fa-** **mous for something else instead.** Charles Messier was a comet hunter in the late 1700s. As Messier scanned the skies from night to night in search of comets, he would occasionally come upon objects that looked somewhat fuzzy like comets but did not move slowly among the stars. For a serious comet hunter like Messier, observing these objects was a waste of time. In frustration—and to assist others who were also hoping to become famous by discovering a comet or two—Messier compiled a list of 103 of these objects so they could be easily avoided. Messier did go on to discover several comets in his lifetime, but ironically, he is far better known today for this list of 103 things to avoid. Indeed, his list consists of nothing less than some of the most astronomically important and beautiful objects in all the heavens, from giant clouds of gas and dust where new stars are born to the remnants of exploded stars to entire galaxies of hundreds of billions of stars. Today, each is still cataloged with the letter M (for Messier), followed by its numerical entry in his list. We shall have occasion to encounter and talk about many of them later in this book.

453 **The most famous comet of all is Halley's Comet, but Halley didn't discover it.** Halley's Comet returns to Earth's skies every 75–76 years, having visited in 1910 and again in 1985–86. In 1705, English astronomer Edmund Halley was the first to apply a new form of mathematics developed by Sir Isaac Newton to plot the orbits of comets by tracking their changing positions in the sky. In so doing, Halley discovered that a bright comet that had been seen in 1682 had a long elliptical orbit that carried it in close to the Sun and then out beyond the orbit of Neptune. He also discovered that the orbit of this comet of 1682 was virtually identical to those of other bright comets seen in 1607 and 1531 and suggested that all three sightings were the same comet. Extrapolating ahead the appropriate number of years, Halley predicted that the comet in question would again appear in 1758. He didn't live to see it, but on Christmas night, 1758, the comet returned right on schedule. In his honor, it was named Halley's Comet. Until Halley's time, astronomers thought that all comets were seen only once. Now it was realized that at least some were "repeaters." Armed with this revelation, astronomers have been able to trace recorded observations of Halley's Comet back to before the time of Christ.

454 **All comets are divided into three parts.** The *heart* of a comet is its nucleus, the frozen mud clump itself. When close enough to the Sun, the nucleus is surrounded by a nearly spherical cloud of gas and dust known as the *coma.* Streaming out from these is the most recognized part of the comet, the comet's *tail.*

455 **Most comets spend most of their lives in the low-rent realms of the solar system.** Most comets travel in huge orbits well beyond those of Neptune and Pluto. It has been suggested that some may inhabit a disk-shaped zone known as the Kuiper Belt about eight times farther from the Sun than Pluto, while trillions more may orbit in a vast spherical region called the Oort Cloud that may stretch from well beyond the Kuiper Belt halfway to the nearest star.

456 **If comets typically dwell so far from the Sun, what causes some to visit the innermost reaches of the solar system and be seen by us here on Earth?** Nearby stars and massive planets like Jupiter are primarily responsible. Their gravitational pull can be sufficient to periodically nudge a comet nucleus in just the right way to send it on a long slow journey toward the inner solar system. As it approaches the orbits of the inner planets, the comet speeds up, ultimately swinging around the Sun. Then, depending on how much its orbit has been changed, the comet either heads back to the depths of interstellar space, never to be seen again, or follows a new smaller orbit that results in its periodic return to the inner solar system to become a repeat comet for us on Earth.

457 **Some comets come really close to the Sun.** Occasionally, a comet will plunge right into the Sun and destroy itself in a fiery burst of light.

458 **Comets may have small hearts, but they have big tails.** Comet nuclei are typically less than 15 miles across—mere specks on an astronomical scale. But from a tiny heart, an impressive comet can grow. As the nucleus tumbles in toward the inner solar system, radiation from the Sun begins to warm its ices and dust, which are released as powerful geysers into the vacuum of space. This dusty fog becomes the comet's coma and can grow to more than 100,000 miles across. The solar wind and radiant energy from the Sun then interact with the liberated gas and dust in the coma, pushing it out far into space to form the comet's long and beautiful tail. The tail of a major comet can stretch 100 million miles.

459 **Some comets brighten quite steadily as they approach the Sun, while others are less predictable.** The action of the Sun grows a comet's coma and tail, so many comets become increasingly more spectacular as they approach our solar system's central star. But the mixture of ices and dust in the nucleus of most comets is not very homogeneous. The result can be sudden outbursts of dust or gas that, in turn, can lead to sudden increases in brightness or equally dramatic fizzles. A great example of a dud was Comet Kohoutek, which appeared in the winter of 1973–74 and greatly underperformed as it neared the Sun. With this in mind, astronomers frequently are cautious in predicting whether or not an incoming comet might become the "Comet of the Century."

460 **Why was Halley's Comet so disappointing to many last time around?** While Halley's Comet provided opportunities for great scientific success during its last appearance, many people had heard how spectacular the comet was in 1910 and were disappointed at its seemingly lackluster performance this time around. Had the stories about the comet in 1910 been exaggerations? Well, perhaps a little. But there were several other important factors. First, the comet came much closer to the Earth in 1910, so it looked larger. Second, while people living in the northern hemisphere had the best view of Halley in 1910, the southern hemisphere was better positioned in 1985–86. This kept the comet quite low in the sky for most of Europe and North America, where there was more obscuring water vapor. Next, the comet was a bit smaller in the 1980s than it was back in 1910. (Since a comet's coma and tail come from its nucleus, the comet gradually wears away over time.) The biggest reason for the difference in appearance between Halley's first and second visit in the twentieth century, however, wasn't the comet but rather the Earth itself. There were many more city lights by the 1980s—city lights that shine up into the sky and obliterate all but the brightest stars. From the South Seas, the Australian outback, or the plains of Africa, Halley's Comet was still an impressive sight this last time around, but in or near places like New York or London, it was simply invisible in skies awash with urban glow.

461 **The tail of a comet always knows where to point.** Because it is pushed into space by the pressure of sunlight and the solar wind, a comet's tail always points away from the Sun. As a comet approaches the

Halley's Comet during its most recent swing past Earth on March 16, 1986. The comet's two tails can clearly be seen—the dust tail appearing smooth, while the gas (ion) tail takes on the appearance of a cascading waterfall. (NOAO)

ASTEROIDS, COMETS,
METEOROIDS, AND SPACE DUST

Sun, its tail streams out behind the nucleus. But as a comet withdraws from the Sun back into the outer solar system, its tail leads the way. If this seems strange, just picture a woman with long straight hair walking with the wind and then against the wind.

462 **Some comets grow two tails.** Comets are typically made of ices and dust. Sometimes the liberated gas forms one tail and the dust creates a separate tail. The dust tail tends to be very smooth in appearance, while the gas tail can exhibit a more "textured" look that is not unlike a cascading waterfall.

463 **Some comets have more than one head.** Comet nuclei are made of pretty crumbly stuff. Indeed, if a comet ventures too close to a massive object like Jupiter or the Sun, tidal forces can actually break the nucleus into two or more pieces. In 1992, as Comet Shoemaker-Levy 9 flew close to Jupiter, it broke into over 20 pieces that subsequently flew in formation like a squadron of planes before slamming into the giant planet in 1994.

464 **Comet Shoemaker-Levy 9 and Jupiter made awesome celestial fireworks together.** Over a one-week period in July 1994, more than 20 mountain-sized chunks of Comet SL-9 collided with Jupiter at over 130,000 miles per hour. The result was a series of spectacular explosions in Jupiter's upper atmosphere. Each released the energy of millions of tons of TNT and left black Earth-

sized "scars" that lasted for over a year. (See pages 114 and 115.)

465 **In the spring of 1996, people were treated to one of the brightest comets in years.** Comet Hyakutake was discovered by Japanese amateur astronomer Yuji Hyakutake in late January, 1996. By mid-March, it had been seen by millions of people around the world. Its bright coma was even visible in urban areas and under dark country skies, its tail was seen by some observers to stretch halfway across the sky! Making the show easy to catch for most people was the fact that the comet became visible during the evening hours and was high in the sky.

466 **Professional astronomers also paid attention to Hyakutake.** Scientists used telescopes and spacecraft to confirm the presence of frozen water, ammonia, and methane, as well as other compounds, including ethane, carbon monoxide, and methyl alcohol, in the comet. These substances have been found in other comets as well. For the first time, however, X rays were discovered coming from a comet. In this case, they were being emitted from a region in the comet's inner coma on the sunward side. Astronomers think the X rays may be due to particles in the coma getting tangled in the Sun's magnetic field and being accelerated until they give off the high-energy radiation. Comet Hyakutake was certainly the "Great Comet of 1996," and one of the great ones of the century. In late winter and early spring 1997, we were treated to an even more spectacular comet, Hale-Bopp.

---

467 **Throughout time, many comets have collided with the Earth.** Few worlds with solid surfaces in our solar system are without craters, the vast majority of which are due to collisions with meteoroids and comets. The Earth is no exception, for while erosion and continental drift have erased many of our planet's scars, impact craters can still be found from the Arizona desert to the Australian outback. In 1908, a huge explosion over sparsely populated Siberia was probably due to a small comet or meteoroid vaporizing in the Earth's upper atmosphere. While such events were far more common in the solar system's ancient past, Comet Shoemaker-Levy 9 proved that a rare but real threat exists today. Indeed, the media attention to SL-9 prompted some astronomers to push for the creation of a network of telescopes that would systematically scan the heavens and warn us about any objects of dangerous size that might be heading toward Earth.

---

468 **In 1986, an armada of five spacecraft flew by Halley's Comet—the only such encounter in history.** While one Japanese spacecraft, *Suisei,* made a wide sweep of Halley's Comet at a distance of 93,800 miles and studied the chemistry of the cloud of material engulfing the nucleus, a second Japanese spacecraft, *Sakigake,* studied the solar wind around the comet. At the same time, two Russian craft, *Vega 1* and *Vega 2,* closed in on Halley, coming within 5,000 miles of the nucleus. *Vega 1* and *Vega 2* produced detailed images of the comet's inner structure and relayed their data to the European Space Agency to help target its spacecraft, *Giotto,* which made a run for the nucleus itself. Encountering fierce dust storms, *Giotto* came within 400 miles of Halley's heart and revealed an irregularly shaped velvety black object, 5 by 9 miles in size, spouting giant geysers of material into space.

469 **Although comets usually travel at several miles per second, they appear to hang almost motionless in the sky.** Unlike meteors or *shooting stars,* comets don't streak rapidly across the sky. This doesn't mean, however, that comets don't travel fast. Within the inner solar system, comets can tear along at more than 150,000 miles per hour. Their relatively great distances from Earth, however, make comets only seem to move a little from night to night against the background of stars. So, as you look at a comet, it appears to hang quite motionless in the sky and can usually be seen for several nights (or even weeks) in a row.

470 **In the course of history, comets have had more than their share of glory and infamy and deserved neither.** Throughout time, no other class of celestial phenomenon has inspired such awe, wonder, superstition, and fear as comets. They have been credited with signaling the birth of great leaders, including Julius Caesar. They have been said to foretell victory or defeat in battle (depending, of course, on which side you were on). And they have been blamed for every conceivable form of human misery, including pestilence, famine, and the plague. A prayer uttered by Christian soldiers during the Crusades went: "From the sword, the Turk, and the comet, dear Lord, deliver us." And if such notions seem naive by twentieth-century standards, consider the fact that in 1910, when an astronomer innocently announced that Halley's Comet contained traces of substances that were poisonous, people scrambled to buy gas masks and "comet pills" sold by hucksters who were more than willing to relieve them of their money.

471 **Comets are like Rosetta stones from which we can read the early history of our solar system.** While the Earth and the other planets have evolved tremendously over the 4.6-billion-year history of our solar system, comets have remained virtually untouched by time. The reason is that most comets have spent the vast majority of their existence in the frigid outer reaches of our solar system where temperatures hover near absolute zero and no chemical changes can take place. Thus, comets provide snapshots frozen in time of the chemistry and physical conditions that were present at the time the Sun and planets were being born.

472 **Comets may be responsible for both the birth of life on Earth and its periodic demise.** Comets appear to be rich in water and organic molecules, both of which were important for the development of life on Earth. Some have even suggested that the countless comets that bombarded the early Earth seeded our planet with some of the water and organic molecules that played a crucial role in the ultimate development of life on Earth. Ironically, however, the collision of large comets and meteoroids with the Earth in subsequent eras may well have been responsible for many of the massive extinctions our planet has experienced, including the "death of the dinosaurs."

473 **Comets are cosmic litterbugs.** Because comets are made of rather crumbly stuff and are essentially sand blasted by the Sun as they fly through space, it's not surprising that they leave bits and pieces of themselves behind. Indeed, the orbits of comets are literally strewn with this cosmic debris.

$\mathcal{M}$eteoroids, Meteors, and Meteorites

474 **Space debris that burns up in Earth's atmosphere turns into celestial fireworks.** As the Earth orbits the Sun, it plows into debris left behind by many old comets. Microscopic particles of comet dust filter down through the atmosphere, but pieces the size of grains of sand and larger burn up due to friction. Meteors or *shooting stars* are the result.

475 **It doesn't take much space debris to make an impressive shooting star.** As pieces of cometary debris are swept up by the Earth, they streak down into our atmosphere at speeds of up to 45 miles per second. The heat generated as the debris incinerates is so great that a piece the size of a grain of sand can create a shooting star bright enough to be seen from the ground on a clear night. Larger pieces produce extremely bright shooting stars, which are also called *fireballs* or *bolides*. Some of these can outshine the brighter planets or even rival the brilliance of the Moon.

476 **Bright shooting stars might seem to be traveling at very low altitude, but they usually aren't.** The human eye can frequently be fooled and interpret something that is bright as also being very close. In reality, most of the space debris that burns up as shooting stars does so while it is still 30 to 40 miles above the surface of the Earth.

477 **Meteors can play some funny tricks.** Some meteors can be seen to change color. The different colors are due to the burning of different chemical compounds within the piece of debris, as well as the changing temperature of the object as it travels through our atmosphere. Because this material can be quite crumbly in some objects, some meteors break into pieces as they are buffeted by the air at very high speeds. Some

meteors even leave smoke and vapor trails behind that last for several seconds before dispersing.

478 **Many shooting stars come in *showers*.** While some shooting stars occur as the result of random pieces of space debris entering the Earth's atmosphere, most occur in showers. The Earth's orbit actually intersects the debris paths of many comets during the course of the year. As the Earth passes through one cluttered part of its orbit after another, a series of *meteor showers* take place. Because the Earth is at the same point in its orbit at the same time each year, the dates of the showers remain the same.

479 **The meteors from a meteor shower all appear to radiate from a point in the sky, just like the spokes of a wheel radiate out from the hub.** The point in question is appropriately called the *radiant point* of the shower and is simply the point in space toward which the Earth is heading on that particular night as it orbits around the Sun. The effect is just like what you observe if you drive through a snowstorm at night. No matter which way you drive, the snowflakes appear to radiate from a point in front of you that is simply the point toward which you are driving at that moment.

480 **A meteor shower's radiant point moves around the sky during the night.** As the Earth rotates on its axis, the stars and constellations move around the sky during the course of the night. Thus, a meteor shower's radiant point moves around as well. Typically, the radiant point rises in the eastern part of the sky early in the evening and gets progressively higher in the sky as the night draws on. This, in turn, means that more meteors are seen late at night (after midnight) than earlier.

481 One of the best meteor showers of the year typically occurs in August. Each August, the Earth passes through debris left behind by Comet Swift-Tuttle. The result is the annual Perseid meteor shower, which peaks on the night of August 11 (going into August 12) when the Earth passes through the densest part of the comet's debris swarm. The shower is called the Perseid meteor shower, or simply the Perseids, because its shooting stars all appear to radiate from a point in the sky within the constellation Perseus. Of course, the meteors don't really come from Perseus. This is simply the direction in space toward which the Earth is heading in mid-August.

482 Another good meteor shower can be seen each December. The Geminid meteor shower, or the Geminids, peaks each year on the night of December 11 (going into December 12). As you have probably guessed, the radiant point for this shower is the constellation Gemini. Temperature contrasts between August and December, however, usually make the Perseids a bit more comfortable to watch.

483 How many shooting stars can you see at night? On a typical night when no meteor shower is taking place, a person scanning the whole sky might expect to see two or three shooting stars in the course of an hour. On average, during the Perseid or Geminid meteor showers, this number can climb to thirty to fifty per hour.

484 The number of meteors you see depends on many factors. As with stars, there are lots of faint shooting stars and few really bright ones. So just how many you see on any particular night depends on how clear and dark your sky is. Even a covering of high thin clouds can greatly reduce the number of stars and shooting stars you see. The darkness of the sky also makes a big difference. City lights shine up into the sky, scatter off water vapor and pollutants, and, in general, make the sky much brighter than it is from rural locations. Even the light from a natural object, namely the Moon, can have a big effect. (Just compare the number of stars you can see on a night when there's no Moon in the sky with the number you see on nights around the time of full Moon.) So, if you want to increase your chances of seeing shooting stars, pick a clear dry night when the Moon is not a factor and get as far away from city lights as you can.

485 Most meteor showers have good years and bad years. Even showers like the Perseids and Geminids have good and bad years. This is partly due to the phase of the Moon and partly due to the meteor swarm itself. When a full Moon occurs around August 11–12, for example, it is not going to be a good year for the Perseids because the Moon will be up all night long and "wash out" all but the very brightest meteors. But even when moonlight is not a factor, a particular shower can offer many meteors or just a few. This is because the debris is typically not distributed evenly along the comet's orbit. In years when the Earth travels through a particularly dense swarm of debris, many shooting stars are seen. Little debris in other years, of course, produces fewer meteors.

486 **On rare occasions, meteor showers can be spectacular.** In most years, a November meteor shower known as the

The Leonid meteor shower over Kitt Peak National Observatory in Arizona on the night of November 18, 1966. For a short time, tens of thousands of meteors per hour were seen. (NOAO)

Leonids produces a few dozen meteors per hour at most. In 1966, however, observers in portions of the western United States saw over *100,000 shooting stars per hour!* In 1833, the Leonids reportedly produced a similarly dazzling display over portions of Europe. The reason for such outbursts lies in the fact that the debris can be clumped very close together over small portions of the comet's orbit. However, because the size of the space that contains these very high concentrations is usually small, such *superpeaks* frequently last only a few hours and, therefore, are only seen by people that happen to be in the right place on Earth (the right couple of time zones) at the right time. These outbursts are also not easy to predict. In 1993, some predicted that the Perseids would produce a superpeak over Europe, but it failed to materialize.

487 **No special equipment is needed to observe a meteor shower.** You don't need a telescope or binoculars to watch a meteor shower. In fact, since the meteors radiate outward across a large part of the sky, telescopes and binoculars greatly cut down on the number of shooting stars that you see because they severely limit how much of the sky you see. The best way to watch a shower is to just find an open spot and lie down on a blanket or reclining lawn chair so that you can see as much of the sky as possible and not get a stiff neck.

488 **What's the difference between *meteors, meteoroids,* and *meteorites?*** Meteors are synonymous with shooting stars, that is, streaks of light that flash through the sky. A small piece of space debris, while it is still in space, is a meteoroid. Upon entering the Earth's atmosphere, meteoroids incinerate and create streaks of light called meteors. If a meteoroid is too large and does not completely burn up, the piece that crashes into the Earth is called a meteorite.

489 **Meteorites generally fall into three categories.** Most meteorites are generally classified as either *stones* or *irons*. Stony meteorites are made mostly of silicates. Irons, as the name suggests, have a high iron content (usually about 85 percent to 95 percent) with the rest being mostly nickel. A third category, the *stony-irons,* which are very rare, share the characteristics of both.

490 **Stony meteorites are more common, but we find more irons that have landed on Earth.** Most stony meteorites are found when people actually see the object fall to Earth. Such instances are extremely rare, but over 90 percent of meteorites recovered this way turn out to be stony meteorites. There are two problems, however, with discovering stony meteorites that have already fallen to Earth. First, they look a lot like common rocks from Earth, so they don't stand out in the crowd. Second, they are subject to the forces of erosion, so they break down in relatively short periods of time. Iron meteorites, although much rarer, don't erode

quickly and because of their shiny black and metallic appearance are more easily identified as "different."

491 **One of the best places to go meteorite hunting is Antarctica.** Meteorites stand out well against the snow, making Antarctica a choice spot to look for them. Furthermore, over the centuries, glaciers have even helped to "herd" meteorites by gradually carrying them to the edge of the continent for easy pickings by informed observers.

492 **Our Earth bears the scars of many meteorite impacts.** From the Arizona desert to the woodlands of eastern Canada to the Australian outback, craters created by the fall of meteorites can be found. The Barringer Meteor Crater near Winslow, Arizona, measures 4,200 feet wide and is over 600 feet deep. In its vicinity, over 30 tons of iron meteorite fragments have been found. The New Quebec Crater in Canada was discovered from aerial photographs in 1950 and is about twice as large.

493 **What and where are the largest meteorites ever found?** The largest meteorite ever found on Earth is known as the Hoba West meteorite. It was found near Grootfontin in South Africa. It has a volume of 9 cubic yards and an estimated weight of more than 50 tons. The largest meteorite in a museum is the Aneghito meteorite, which is on display at the American Museum of Natural History in

New York. It was discovered by Admiral Robert E. Peary in Greenland in 1897 and weighs 34 tons. The largest meteorite to be recovered in one piece in the United States is the Willamette meteorite, which is on display at New York's Hayden Planetarium. It was discovered in Oregon's Willamette River Valley (from whence it got its name) but may have actually fallen in western Canada before it was dragged by advancing glaciers during the last Ice Age into what would become the United States.

494 **How many meteorites have been found worldwide?** To date, over 3,000 meteorites have been cataloged and placed in the collections of various museums around the world.

495 **What should you do if you find a rock that you have good reason to believe is a meteorite?** Experts in *meteoritics* may be found at some universities and in some natural history museums. These experts would be able to examine your treasure and tell you if it is a meteorite (or a meteor-wrong). If the object really did come from outer space, they may even be interested in buying it from you.

496 **There is only one documented case of a person being struck by a meteorite.** In 1954, Mrs. E. Hewlett Hodges of Sylacauga, Alabama, was struck a glancing blow by a meteorite that crashed through the roof of her house and left a nasty bruise. There are several instances of parked cars being struck by falling meteorites but no known reports of moving cars hitting parked meteorites.

497 **A few meteorites may have actually come from Mars.** Based on their unique chemical composition, a small number of meteorites that have landed on Earth are believed to have come from Mars. Scientists theorize that the impact of a large asteroid-sized object with Mars at some point in the past may have had sufficient energy to catapult rock fragments all the way from one planet to the other. If this is true, we have Mars rocks on Earth, even though no spacecraft has gone to Mars to retrieve any. In 1996, teams of scientists reported finding microscopic structures in at least one of these meteorites which some believe are indicative of the one time presence of very primitive life. If proven, the discovery would be of great significance.

498 **Other objects called *tektites* may also have an exotic origin.** Tektites are glassy objects that tend to be roughly spherical or teardrop in shape. They have been found in various parts of the world. Some scientists believe that they are made of silicates from the Earth's crust that were fused in the heat of impacts from material originally blasted off the surface of the Moon.

Space Dust

499 **_Micrometeorites,_ the runts of the litter.** Many very tiny meteorites are encountered by the Earth all the time. Known as micrometeorites, they are typically about 0.00004 of an inch across and are so lightweight they simply float down through our atmosphere. It is estimated that the Earth accumulates 50 to 100 tons of this material every day!

500 **The _zodiacal light_ is the glow of interplanetary dust.** Micrometeorites make up what is known as _interplanetary dust._ Each particle travels around the Sun in its own orbit and, like the solar system itself, the distribution of dust is shaped like a thin disk with the Sun at the center. By knowing just where and when to look, we can view this dust, or more correctly, sunlight scattering off it (like we do when we see "sunbeams" in a darkened room). Known as the zodiacal light, it appears as a faint glow arcing up into the southwestern sky an hour or two after sunset and the southeastern sky an hour or two before dawn. (In the predawn sky, it is also known as the _false dawn.)_ To see the zodiacal light, however, you need to be under very clear and very dark skies, which are getting harder and harder to find.

501 **What is a star?** Stars are gaseous objects that, for most of their lives, generate their own energy by means of internal nuclear reactions.

502 **Just how many stars you can see at night depends very much on where and when you look.** How many stars can you see with the naked eye at night? The answer depends on the darkness and clarity of the sky and just where you happen to be stargazing from. From downtown New York or Tokyo, you would be lucky to make out more than a dozen of the brightest stars. But from a rural location on a clear moonless night, a person with good vision can make out more than 3,000 stars with the naked eye.

503 **Since neolithic times, people have tried to create a sense of order in the heavens.** At first glance, the night sky can seem like a pretty big mess with a jumble of stars scattered helter-skelter across the heavens. A first reaction might well be to wonder how any astronomer can possibly find his or her way around and make sense of it all. At second glace, however, the situation isn't quite as hopeless. The eyes and mind begin to work together almost unconsciously to trace patterns among the stars—in effect, playing a simple mental game of "connect the dots." As it turns out, people have probably been doing this since before the dawn of recorded history. Today, we refer to these patterns or grouping of stars as *constellations* (from the Latin *cum* and *stella,* meaning "together" and "stars").

504 **To many early peoples, the patterns of the stars had very practical significance.** Early peoples were quick to recognize that the stars moved around the sky from hour to hour and that the same star patterns could be found in the sky at the same time of year, year after year. Thus, the sky became both the first clock and the first calendar and told those who knew how to recognize the star patterns when to plant, when to harvest, and when to hunt migratory or seasonal animals. For early peoples, knowledge of the sky could literally mean the difference between life and death.

505 **In some cultures, having knowledge of the sky meant having power.** Much of the economy of ancient Egypt was dependent upon the annual flooding of the Nile and the subsequent refertilization of the land. Every year the Egyptian priests were able to come to Pharaoh and accurately predict when the all-important flooding would occur. Indeed, they seemed to possess sacred foreknowledge imparted to them by the gods themselves. In fact, the priests were simply careful sky watchers who had noticed that every year the Nile would flood shortly after the day when the bright star Sirius rose just before dawn. So every year they carefully watched the predawn sky and, by so doing, maintained the illusion of having magical power over the very lifeblood of the kingdom and hence over Pharaoh himself.

506 **Modern astronomers divide the sky into 88 constellations.** The entire sky is now typically divided into 88 regions called constellations. Most people picture the constellations as stick figures (rather like "connect the dots" pictures) with the stars as dots. To astronomers, however, the constellations are more like state boundaries. Thus, just as the contiguous United States is divided into 48 states and any town can be unambiguously described as being in one state or another, so the entire sky is divided into 88 constellations and any particular star can be described as being in a particular constellation. There is no science to the constellations. They're all just arbitrary artificial boundaries in the sky. But if an astronomer is talking about studying a particular object that happens to be located in a particular constellation, other astronomers can use the designation to get a general idea of what part of the sky the astronomer is talking about.

507 **Some things that are commonly thought of as constellations really aren't.** Ask people to name a constellation and the Big Dipper might be the first thing that comes to mind. It would surprise most folks to know, however, that in the complete list of all 88 constellations, the Big Dipper does not appear. Neither does the Little Dipper. Each, however, is *part* of a constellation. The Big Dipper is part of Ursa Major, the Great Bear, while the Little Dipper is part of Ursa Minor, the Little Bear. Neither Dipper, however, is a constellation in itself. Instead, these subgroups of stars that form recognizable patterns in the sky are known as *asterisms*. In Chinese tradition, the sky isn't divided into 88 constellations but instead into over 300 asterisms. Some are small groupings of stars, while others are pairs or even individual bright stars.

508 **The *zodiac* is probably the most famous group of constellations.** In Western tradition, the zodiac is a group of 12 constellations that form a band that completely encircles the sky. The zodiac consists of Pisces, Aries, Taurus, Gemini, Cancer,

Leo, Virgo, Libra, Scorpius, Sagittarius, Capricornus, and Aquarius. The word *zodiac* comes from the Greek and means "band of animals." Although most of the zodiacal constellations are indeed animals, Gemini (the Twins), Virgo (the Virgin), Aquarius (the Water Carrier), and Libra (the Scales) are not and Sagittarius (the Archer) is typically pictured as half-man and half-beast.

509 **The zodiac has been significant to both astronomers and astrologers.** The zodiac became famous simply because this is the region of the sky through which the Sun, Moon, and all the visible planets appear to move. For astrologers, this region was significant because they saw these celestial objects as favoring these particular constellations with their presence. For astronomers, it afforded insight into the shape and physics of the solar system. The fact that the zodiac occupies a relatively narrow band in the heavens says a lot about the shape of the solar system, for it means that the solar system must be relatively *flat*. Put another way, it means that the orbits of all the planets and our Moon lie in approximately the same plane. Indeed, with the exception of Pluto, the orbits of all the planets are within a few degrees of each other. In other words, the solar system is shaped like a pancake with the Sun at the center. This is a natural consequence of a law of nature known as the Conservation of Momentum. Other solar systems that we have observed in the process of formation around other stars also seem to have a disk-like shape.

510 **Some constellations are ancient, while others are relatively modern.** Some constellations, like Leo (the Lion), date back to the time of the ancient Pharaohs of Egypt. Other constellations, primarily in the southern hemisphere, received Western names around the year 1600 from two Dutch travelers, Pieter Keyser and Frederik de Houtman, who created a series of fanciful animals in new parts of the sky they saw on their global journeys—parts of the sky that were not visible from Europe. Over a century later, Nicolas de Lacaille decided to honor some of the developing tools of the scientific and industrial revolutions by grouping some additional southern stars into such constellations as the Furnace, the Air Pump, and the Microscope. Of course, the native peoples of the southern hemisphere had long ago traced their own patterns among the stars and created their own constellations.

511 **Among the peoples of the Andes, some of the constellations are dark instead of light.** At certain times of the year in the Andes Mountains of South America, the bright band of the Milky Way can be seen to arc high overhead. It is mottled in places, crossed with dark pockets and ribbons where obscuring dust hides the light of more distant stars. Curiously, natives of the Andes not only created constellations by connecting imaginary lines between the stars but also fashioned a series of animals out of the shapes of these dark regions. Hence, in places we find such "dark" constellations as the Fox and the Hen. One constellation, known as the Llama, actually consists of a dark region and two bright stars, Alpha and Beta Centauri, which mark the Llama's eyes.

512 **Sometimes the same pattern of stars is seen as different objects by different groups of people.** What is seen as the Big Dipper by many non-Native North Americans is called the Plough by people in England and is fashioned into a wagon pulled by three horses by the Chinese. To some Native American tribes, the bowl of the Dipper was envisioned to be a bear, while the stars of the handle were hunters in pursuit. (People with good vision are able to see that the middle star in the handle of the Dipper is actually two stars that are very close together in the sky.) And so, to many Native Americans, the fainter of these stars is a cooking pot being carried by one of the hunters. To native peoples of Polynesia, the curved sweep of stars that Westerners fashion into Scorpius (the Scorpion) is instead seen as a giant fish hook thrown skyward in joy by the god Maui after he used it to pull up great clumps of soil from the ocean depths to form the many beautiful islands of the Pacific.

513 **The North Star isn't the brightest star in the sky.** Ask people to name the brightest star in the sky and many will answer, "The North Star." This is a common misconception, however. In a list of the brightest stars, Polaris, the North Star, barely makes the Top 50. The North Star's fame doesn't stem from its extraordinary bright-

ness but from its unique position in the sky. At this point in history, Polaris is the star that is closest to the point in the sky astronomers call the *North Celestial Pole.* The North Celestial Pole lies directly above the North Pole on Earth, so all the stars in the sky appear to turn slowly about the North Celestial Pole as the Earth turns on its axis. As a result, the stars appear to turn around a point that is very close to Polaris. Thus, the North Star remains almost stationary in the sky and its location always points north.

514 **Just as the Sun appears to move across the sky during the daytime, so the stars appear to move across the sky at night.** Every day the Sun appears to move across the sky from east to west. This motion, of course, is only *apparent,* for it is really due to the spinning of the Earth on its axis. At night, of course, the stars also appear to travel across the sky for the same reason. It's like being on a giant merry-go-round and convincing yourself that you are not turning but that the world is turning around you. Indeed, the illusion is so good many ancient peoples were convinced that the Earth was stationary and everything in the universe traveled around it.

515 **The apparent motion of the stars at night may at first seem a bit more complex than the motion of the Sun across the daytime sky.** Go out on a clear night and find a star near the eastern horizon. If you go out an hour or so later, the same star will have climbed a bit higher in the sky (just as the Sun climbs higher in the eastern sky in the morning). During that hour, stars in the western sky will have moved lower on their way to setting. And stars in the southern sky will have traveled across the sky from left to right. All of this motion of individual stars, however, is simply the *apparent* motion of the sky due to the Earth's rotation. Because the Earth is really spinning on an axis that passes through a point in the sky near Polaris, the North Star, all the stars in the sky appear to be traveling in circles around Polaris.

516 **You can capture the hourly motion of the stars by using a simple 35-mm camera.** All you need are a simple 35-mm camera and a tripod to see the apparent motion of the stars in the sky. You can point your camera anywhere in the sky, but you can get one of the neatest effects by pointing it at the North Star. Use reasonably sensitive film (with an ASA or ISO number of 200). Pick a clear moonless night without a lot of wind. Clamp the camera down tightly, open the diaphragm wide, set the lens at infinity, set the shutter speed for "time," and take a one- to two-hour time exposure. When the film is developed, you will see that each star has left a curved trail (cleverly called a *star trail*) on the film. Each is part of a circle swept out by the star as the Earth turned on its axis. If you aim your camera south or toward the eastern or western horizon, you can capture the apparent motions of the

stars in these parts of the sky as well. If you use color film, you may pick up the different colors of the brighter stars.

517 **Polaris hasn't always been the North Star, nor will it always be.** Polaris is our planet's current North Star because of its current proximity to the North Celestial Pole. Thus, Polaris is the star toward which the Earth's axis currently points. Over time, however, the Earth slowly wobbles on its axis (like a top wobbles as it spins). This motion, called *precession,* is very slow: One complete wobble takes about 26,000 years. As a result of this motion, Earth's axis traces out a large circle in the sky over the course of the same 26,000 years. At the moment, the Earth's axis points close to Polaris, but in the past and in the future, other stars that lie on or near the *precession circle* in the sky take turns being our North Star. When some of the ancient Pharaohs ruled Egypt, a star we call Thuban in the constellation Draco, the Dragon, was the North Star. Around A.D. 14,000, a bright blue-white star called Vega, now high overhead in our summer sky, will be the North Star. And, in about A.D. 26,000, Polaris will once again serve in this capacity.

518 **If Polaris is currently the North Star, what's currently the *South* Star?** Just as the Earth's axis extends from the center of the Earth out through the North Pole and up into the sky to a point near Polaris, so too our planet's axis extends out through the Earth's South Pole and on into the sky. But if you were to journey to a land "down under," you

Circumpolar star trails. (©Anglo-Australian Observatory)

STARRY, STARRY NIGHTS:
BEYOND THE SOLAR SYSTEM

wouldn't find a South Star because there simply isn't a bright or semibright star at or near that appropriate point in the sky at this time. Folks in places like Australia and South America have some pretty stars to look at at night, but, alas, no South Star.

519 **The North Star and your fist are all you need to tell your latitude—unless you are south of the equator.** Because the Earth's axis points almost directly at the North Star, it turns out that the height of the North Star above the horizon is just about equal to the latitude from which you observe it. Thus, from New York, for example, which lies at about 41°N, Polaris remains almost fixed in the sky at a point about 41° above the northern horizon. A fist held at arm's length is about 10° across, so for a New Yorker, Polaris may be found about "four fists" up in the northern sky. By comparison, from Miami, where the latitude is about 26°N, Polaris lies only 26°, or about "two and a half fists," above the horizon. From the North Pole, at latitude 90°N, Polaris lies directly overhead at the top of the sky; from the equator, Polaris lies on the northern horizon; and, south of the equator, Polaris remains forever out of sight below the horizon.

520 **People at the same latitude see the same stars.** Because the height of the North Star is determined by the latitude from which you observe it, people at the same latitude anywhere in the world see the same stars at the same local time on the same night.

Thus, people in New York, Madrid, Ankara, and Beijing (who all live at about the same latitude) see the same stars on the same nights of the year, even though they live many miles apart.

521 **People who live at very different latitudes see many different stars at night.** Santa Claus at the North Pole and scientists at the South Pole see entirely different hemispheres of stars at night. Folks in between typically only view a portion of the night sky seen by folks at other latitudes. If people from a place like New York, for example, vacation in Rio or Australia, they will not see the Big and Little Dippers or other stars near Polaris that they are used to seeing at home because these constellations will be below the northern horizon. Instead, however, new stars and star patterns never visible from home (such as the Southern Cross) will populate the southern sky. In addition, some familiar constellations typically found in the southern sky when they are in New York will instead be found over in the north *upside down.*

522 **When it comes to looking out on the universe, there are definitely some "cheap seats."** Not all stargazing sites on Earth are created equal. From the North Pole, for example, the North Star always lies fixed at the top of the sky. Since all the other stars circle around the North Star, each star stays at its particular height above the horizon all the time. In short, from the North Pole, no stars rise and no stars set. This is also

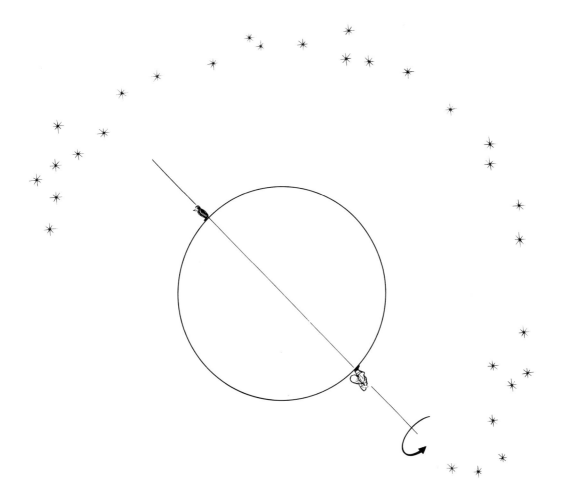

People living at different latitudes see the stars in different positions. Santa Claus and penguins see entirely opposite halves of the universe.

true from the South Pole, but the stars we would see from there would be an entirely different group of stars because all the stars seen from the North Pole would be below the horizon. Hence, from the Earth's poles, we look out on opposite hemispheres of the universe around us, so an observer only sees half the stars that surround the Earth from either location. These are the "cheap seats."

523 **Where are the "best seats" for viewing stars?** The "best seats" for viewing the stars are on the equator. That's because the North Star lies on the northern horizon and all the stars visible from both hemispheres slowly parade into view during the course of the year. From the equator, every naked-eye star within sight of Earth can be seen at one time of the year or another and no stars remain permanently below the horizon.

---

524 **The Earth's motion around the Sun also affects which stars we see at night.** Imagine placing a lamp in the middle of a room to represent the Sun and yourself walking in a circle around the lamp to represent the Earth traveling around the Sun during the course of the year. At any point in time, half of your body would be exposed to the light while the other half would be in shadow. In this way, you would be simulating the fact that half of the Earth is always experiencing daylight while the other half has night. If there were stars painted on the walls of the room and the lamp was very bright, you would see only half the stars at one time because the glare of the lamp would prevent you from seeing the stars in the same direction as its glow. In the same way, the nighttime side of the Earth looks out on different stars during the winter than it does during the summer and different stars in the fall than it does in the spring. As a result, we see different constellations at different times of the year and different constellations are typically associated with the different seasons.

525 **Some constellations are always in the sky.** Because all the stars in the sky appear to circle around the North Star, some stars remain in the sky all year around. From New York City, for example, at a latitude of 41°N, the North Star remains quite fixed about 41° above the northern horizon. Thus, all those stars that lie *within* 41° of the North Star never reach the horizon as they circle around the North Celestial Pole. They remain up in the sky all year long and are known as *circumpolar* stars (from the Latin, meaning "around the pole"). From Miami, at a latitude of 26°N, only stars within 26° of Polaris are circumpolar. Going to extremes, from the North Pole *all* the stars are circumpolar, while from the equator *none* of the stars always remain above the horizon. Instead, all stars rise and all stars set.

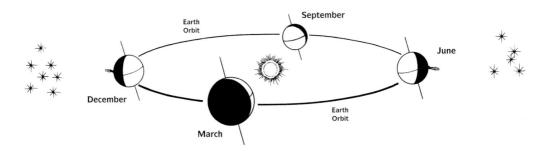

The Earth's passage around the Sun also means that at different times of the year, the nighttime side of the Earth looks out on stars in different directions. Thus, many stars and constellations are traditionally associated with different seasons.

526 **Astronomers name the stars in different ways.** Many of the brightest stars in the sky actually have individual proper names. Many of these come down to us from a period about a thousand years ago when Arab astronomers made significant contributions to astronomy and produced the best star charts of the day. Some of the Arabic names come down to us intact, while others have become corrupted over the centuries through Western usage. In either event, it has resulted in many of the bright stars having strange and even exotic-sounding names. For example, the star Deneb (from the Arabic, meaning "tail" or "tail feathers") marks that part of the anatomy of Cygnus, the Swan. A star in the constellation Pisces (the Fish), is called Alrischa (Arabic for "knot") and appropriately marks the knot in the rope that ties the two fish together in the autumn sky. Other star names just sound good coming off the tongue. From left to right, the three stars in the belt of Orion, the Hunter, go by the wonderful names of Alnitak, Alnilan, and Mintaka.

527 **Other star-naming schemes are less exotic.** Another naming scheme is typically used by professional and amateur astronomers alike. The rule is actually very simple and even logical. The brightest star in any particular constellation is typically called by the first letter of the Greek alphabet (*alpha*) followed by the Latin possessive form of the name of the constellation itself. Thus, for example, the brightest star in Taurus, namely Aldebaran, is also known as Alpha Tauri (quite literally meaning "first of Taurus"). The brightest star in nearby Orion goes by the funny-sounding name Betelgeuse but is also known as Alpha Orionis. The second-brightest star in a particular constellation bears a name that consists of the second letter of the Greek alphabet (*beta*) followed again by the Latin possessive of the constellation. Hence the brilliant blue-white star Rigel in Orion's left foot is also known as Beta Orionis ("second of Orion").

528 **Unfortunately, there are only 24 letters in the Greek alphabet and there are a lot more than 24 stars in each constellation.** Sometimes when two or more stars lie rather close together in the sky, they are each given the same Greek letter designation, but a little superscript is added. Thus, the double-double in Lyra is known as epsilon 1 (ϵ^1) and epsilon 2 (ϵ^2) Lyrae. Variable stars, as noted elsewhere, are usually given conventional alphabet designations, such as RR Lyrae. But such schemes weren't enough as telescopes got bigger and bigger and allowed us to see more and more stars. In the process, star catalogs were created that simply give each star a number. A star might be called HD 213468, meaning that it is star number 213468 in the *Henry Draper Catalog* compiled at Harvard University early in this century. Or a star might be designated as SAO 347981 in its listing in the star catalog of the Smithsonian Astrophysical Observatory. Indeed, because there are so many of them, most stars really do just have numbers and the same star will have different numbers in different catalogs. It may be quite impersonal, but it gets the staggering job done.

529 **There aren't really any stars named Jenny Smith or Uncle Albert.** In recent years, a few companies and even institutions have made a significant amount of money "naming stars after people." In many cases, for a set fee, you get a nice-looking certificate and a picture or map of some part of the sky showing "your star." It may be a nice gesture and a "unique gift," but do astronomers then actually refer to those particular stars by those names in their studies or the scientific papers that they write? The answer is no and it never will be anything but no. A piece of paper with your name "attached" to a particular star may reside somewhere in a Swiss bank vault, but it will never reside in any catalog or publication used for research in any observatory or university. If you want to get your name in the sky, go out and discover a comet. Lots of amateur and professional astronomers have and, if you discover one, the professional astronomical community really will officially name it after you. But when it comes to stars, you can't just buy your way in.

530 **The sky looks two-dimensional, but it's really four-dimensional.** When we look up at the night sky, we can get the very strong impression that the stars are dots attached to or projected on a dome spread over our heads. Indeed, many ancient cultures imagined something like this to be the case and modern planetariums create a very cred-

ible reproduction of the sky in this way. In reality, of course, the stars are not all the same distance from Earth but spread out in three-dimensional space. In addition, because light takes a finite time to travel from each of the stars to us on Earth, gazing out into the sky also means gazing through the fourth dimension, namely time. More on this in a minute, but first let's think about the fact that . . .

531 **On Earth, we sometimes see things in the present but hear them as they were in the past.** If you sit out in the bleachers at a baseball game and observe the game carefully, you will notice that you *hear* the batter hit the ball *after* you *see* him hit the ball. Similarly, you typically see a bolt or flash of lightning before you hear the thunder. The reason in both cases is that sound travels considerably slower than light. At room temperature, sound travels at about 1,100 feet per second, whereas light, in a vacuum, travels at 186,000 miles per second (and almost as fast through air). This means that you hear distant things the way they were a short time in the past, but you see things virtually as they happen—at least on Earth.

532 **When we look into space, we see things the way they were in the past.** Objects in space are so much farther away from us than objects we see on Earth that light can take a significant time to get to our eyes. So we actually see different objects in space the way they were at various points in time in the past. For example, it takes light approximately one and a half seconds to travel the nearly quarter of a million miles from the Moon to the Earth. And so, when we look up and see the Moon in the sky, we aren't seeing it as it *is* but as it *was* about one and a half seconds ago. At a distance of 93 million miles, the Sun looks to us the way it was about 8 minutes and 20 seconds ago. (If the Sun somehow magically disappeared, we wouldn't know about it on Earth for another 8 minutes and 20 seconds.) The various planets appear to us as they were minutes or even hours ago and we see the stars as they were years in the past and galaxies the way they were millions or even billions of years in the past. Thus, when we look out into space, we are also actually looking back in time!

CHAPTER 9

STAR SECRETS

533 Figuring out the distance to the nearer stars is as straightforward as blinking your eyes. How far away are the stars? It's a simple question but one that seems impossible to answer. After all, you can't just run a tape measure out to a star and then read the answer. To understand how the feat is done, hold your index finger out in front of your face at a distance of about a foot and look at it alternately through your left eye and then your right eye, blinking back and forth. As you do this, you will notice your finger appearing to jump back and forth relative to background objects, that is, objects that are significantly farther away. The apparent jumping of your finger is an effect known as *parallax* and is simply due to the fact that your two eyes each look at your finger from slightly different directions. The amount of shifting or parallax depends on the distance of your finger from your face. Move your finger closer to your face and the apparent shift increases; move it farther away and the parallax decreases.

534 Astronomers extend the basic concept of parallax in measuring the distances to the stars. In this case, using a telescope, a picture of a nearby star is taken against the background of much more distant stars. Six months later, when the Earth is on the opposite side of the Sun, a second photograph is taken. Comparing the two images, the nearby star can be seen to jump back and forth relative to the background stars. Again, the nearer the target star, the greater the amount of jump, or parallax. Since the distance from one side of the

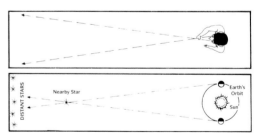

Looking at your finger while blinking each eye in succession causes your finger to "jump" back and forth relative to more distant objects. Astronomers apply the same concept in determining the distances to nearby stars.

Earth's orbit to the other is a lot farther than the distance between your eyes, parallaxes of many stars can be measured in this way. Nevertheless, the stars are so far away that all parallaxes are tiny.

535 **The parallax of even the nearest star is very, very small.** Alpha Centauri, the nearest star system to Earth, has a parallax of only 0.76 seconds of arc or about 0.004 percent of the diameter of the full Moon.

536 **Astronomers usually measure the distances to the stars in *light-years*.** You could measure the distance from New York to Paris in inches, but using a larger unit of measure, the mile, is far more practical. Similarly, the distances to the stars are so enormous that miles are impractical. Instead, astronomers typically use a unit of measure called the light-year as the yardstick of the universe. A light-year is the distance that light travels in one year or a little under 6 trillion miles. The distances to many of the stars we see with the naked eye are measured in tens to hundreds of light-years.

537 ***Parsecs* are also used to measure distances in space.** In addition to light-years, distances in space are also sometimes given in units called parsecs. One parsec is equal to 3.26 light-years or about 20 trillion miles.

538 **The Alpha Centauri star system contains the closest stars to Earth beyond the Sun.** What looks like a single star to the eye is really three stars that all orbit around each other in the southern constellation Centaurus. Two of the stars are similar to our Sun, while the third is a tiny red star. These stars are the closest to the Sun and lie at a distance of 4.3 light-years or about 25 trillion miles. (Which of the three stars is the closest to us at this time? The little red one known as Alpha Centauri C or Proxima Centauri, whose name means "closest of Centauri.")

539 **Some other bright stars are close, while others are quite far away.** The brightest star in the sky, Sirius, in the constellation Canis Major, is one of the nearer stars to Earth at a distance of 8.8 light-years. Nearby Canis Major is the constellation Orion, the Hunter, which contains the bright stars Betelgeuse and Rigel. In contrast to Sirius, Betelgeuse lies at a distance of about 590 light-years, while Rigel is even farther from Earth at nearly 900 light-years. Thus, the light that enters your eyes from Betelgeuse tonight left the star before Columbus discovered America and Rigel shines with light that left it while Europe was just emerging from the Dark Ages. In the summer sky, we find a pattern of stars known as the Summer Triangle, which consists of the bright stars Vega, Deneb, and Altair. Vega lies 27 light-years away, while Altair is 16 light-years distant. Deneb is an astonishing 1,600 light-years from Earth, shining with light that left it before the fall of the Roman Empire.

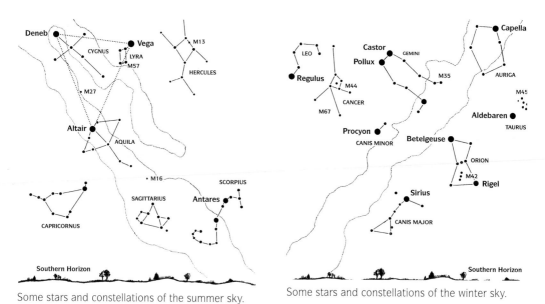

Deneb

Vega

CYGNUS

LYRA

M57

M13

HERCULES

M27

Altair

AQUILA

M16

CAPRICORNUS

SAGITTARIUS

Antares

SCORPIUS

Southern Horizon

Some stars and constellations of the summer sky.

LEO

Regulus

Castor

Pollux

GEMINI

M35

Capella

AURIGA

M44

CANCER

M67

Procyon

CANIS MINOR

Betelgeuse

ORION

M42

Rigel

M45

Aldebaren

TAURUS

Sirius

CANIS MAJOR

Southern Horizon

Some stars and constellations of the winter sky.

PERSEUS

CASSIOPEIA

h & chi Persei

ANDROMEDA

TRIANGULUM

M31

CYGNUS

PEGASUS

PISCES

EQUULEUS

Some stars and constellations of the fall sky.

540 **Astronomers use a machine called a** *measuring engine* **to measure stellar parallaxes.** The instrument consists of a large slab of granite with a hole in the center. The image of a portion of the sky is taken on a plate of glass, which is placed over the hole. Light is shined through the plate and an image is created on a large screen. The star or stars whose parallax is to be measured are then positioned in a crosshair and their position(s) are recorded. Over time many plates of the same region are measured and compared with the appropriate calculations done by computer. Measuring engines are kept in climate-controlled spaces because expansion and contraction of the plates and the granite slab can greatly affect the measurements.

The distances to over 6,000 stars have been measured in this way. Most of these stars lie within a few hundred light-years of Earth because, beyond that, the stars simply have parallaxes that are too small to measure. For stars beyond this limit (which, astronomically speaking, is still very close to Earth), other ingenious techniques must be used.

541 **In referring to the brightness of the stars, astronomers use the term** *magnitude.* The ancient Greeks Hipparchus and Ptolemy grouped the stars they saw into general categories based on their apparent brightnesses. The brightest stars were called stars of the *first* magnitude, while the faintest stars that could be seen were termed stars of the *fifth* magnitude with the other stars falling in between. Later, when sensitive instruments were developed that could accurately measure differences in brightness, it was discovered that the stars of one particular magnitude were about 2.5 times brighter than those of the stars of the next higher magnitude. It was also realized that not all the stars that the ancient Greeks had placed in the first magnitude category were really of equal brightness, so the brightest had to be redesignated as being of *zero* magnitude or even having magnitudes with negative numbers.

542 **When you think about a star's brightness in terms of its magnitude, you have to think backward.** Just remember, if one star is *brighter* than another, its magnitude is represented by a *lower* number, not a higher number. In addition to getting your magnitudes straight, this may also give you some insight into how astronomers think.

543 **Today, astronomers extend the magnitude scale a lot farther in both directions than did the ancient Greeks.** When the brightness of Sirius, the brightest star in the sky, was accurately measured, it was seen to be significantly brighter than other stars of the first magnitude with which it had been grouped by the ancient Greeks. Indeed, Sirius was sufficiently brighter to warrant a magnitude designation of –1.4. Some of the planets, at times, can shine even brighter than Sirius and so have even lower magnitudes.

Venus, for example, can occasionally shine at magnitude −4.4, making it appear three magnitudes (in other words, about 2.5 × 2.5 × 2.5 or more than 15 times) brighter than Sirius. In comparison, the full Moon has a magnitude of −12.7 and the Sun shines at magnitude −26.7. With the invention of the telescope, stars far fainter than can be seen with the naked eye were discovered and so star catalogs began to fill with objects fainter than fifth magnitude. Indeed, today's giant telescopes have recorded stars and galaxies as faint as magnitude +30.

544 **There are *apparent* magnitudes and there are *absolute* magnitudes.** Obviously, if we say the Sun is millions upon millions of times brighter than some newly discovered stars and galaxies, we are only referring to how bright these various objects *appear* to us and *not* how bright they really are. With this in mind, astronomers refer to the *apparent* brightness of an object by its *apparent* magnitude but also speak in terms of an object's actual, or intrinsic, brightness by referring to that object's *absolute* magnitude. Somewhat arbitrarily, the absolute magnitude of an object is the magnitude it would have if it was placed at a distance of 10 parsecs (32.5 light-years) from Earth. If the Sun was 32.5 light-years away, it would have a magnitude of +4.84, which means that it would barely be visible to the naked eye. By comparison, Rigel, the lower right-hand star in Orion, has an absolute magnitude of −8.1. At a distance of about 1,400 light-years, it's one of the brightest stars in our sky, but at 32.5 light-years, it would be visible in broad daylight.

545 **In general, the closer a particular star is to us, the brighter it looks.** Just as streetlights appear to get brighter as we approach them, so it is with the stars. How much brighter? You might think that if there were two identical stars but one was twice the distance from us as the other, it would look half as bright. In reality, the more distant star would only look a *quarter* as bright as its twin. If one star was three times the distance of a twin, it would appear only *one-ninth* as bright. Scientists refer to such a relationship as an *inverse square* law because as the one variable goes *up* (in this case, the distance of the star), the other variable goes *down, not* in proportion to the distance but rather in proportion to the distance *squared.* Twice as far, one-quarter as bright. Three times as far, one-ninth as bright. Four times as far, one-sixteenth as bright, and so on.

546 **In general, the hotter a star, the brighter it looks.** The hotter the star, the more energy it radiates per square inch. Thus, the hotter the star, all other things being equal, the brighter the star really is.

547 **All other things being equal, the bigger a star is, the brighter it will appear.** Stars shine because they radiate energy from their surfaces out into space. If two stars have identical colors (see page 184), they also have identical temperatures, so they radiate the same amount of energy from every square inch of their surfaces out into space each second. But if one of the stars is larger than the other, it also has *more* surface area, that is, *more* square inches of surface through which to radiate its energy. And so, if two stars of the same temperature but different sizes are the same distance from us, the *larger* one will look *brighter.*

548 **Not all of the brighter stars in the sky are among the closest stars.** Sirius, the brightest star in our sky, may be found in the southern sky in winter. Located at a distance of about 8.8 light-years in the constellation Canis Major, the Great Dog, it is one of the closer stars to Earth. Sirius is a bit larger than the Sun but appears as bright as it does to us because of both its high temperature and its closeness. A star of identical brightness to our Sun, however, if placed at the same distance as Sirius, would appear only as bright as an average star in our sky. Nearby Sirius, in the constellation Orion, the Hunter, lie two stars almost as bright as Sirius known as Betelgeuse and Rigel. They are far more distant than Sirius but compete with it in apparent brightness because they are both giant stars and thus have a lot more surface area through which to radiate their energy into space.

549 **The apparent brightness of a star is also affected by how "clean" the space is between us and the star.** Most people think of outer space as a pretty empty place. Sure, there are planets and moons and stars out there, but space is empty in between all these things, right? Well, not really. There is matter between the stars—gas and other stuff that astronomers refer to as *interstellar dust.* This interstellar dust isn't like the stuff you find on the dresser or under the bed if you haven't cleaned in a while but, in truth, astronomers aren't exactly sure just what it's made of. But whatever it is, it lies in clumps and patches throughout the plane of our galaxy (and many other galaxies, for that matter) and acts something like fog on a gloomy night. We all know that headlights of approaching cars don't look as bright on a foggy night as they do on a clear one because the fog scatters some of the light out of the beam on its way to our eyes. In a similar way, interstellar dust dims starlight passing through it.

550 **Interstellar dust also affects the apparent colors of stars.** In addition to dimming starlight, this "space dust" also *reddens* the light of distant stars because it scatters more of the blue light in the star's radiation out of our line of sight than the red light. So the more interstellar dust between us and a particular star, the *fainter* and *redder* the star will appear.

551 **Stars come in a wide variety of natural colors.** Our Sun is yellow, but not all stars are. There are also red stars and orange stars, white stars and blue stars. There are stars the color of a robin's breast and others as violet as amethyst.

---

552 **The color of a star tells us its temperature.** Imagine taking an iron poker and sticking it in a very hot furnace. If you pulled it out after a few minutes, the tip might have become hot enough to be glowing a dull red. A thermometer would register its temperature at about 5,000°F. Stick the poker back in the furnace and remove it a few minutes later and you would find the tip to be glowing a bright yellow with a temperature of about 11,000°F. Several more minutes of heating would produce a poker that was white-hot, its temperature topping 20,000°F. With a very hot furnace and further heating, the poker might be made to shine brilliantly blue-white and, like a welder's arc, register a temperature of 30,000°F or more. In the same way, the surface temperature of a star is revealed by its color. The coolest stars are red, while the hottest are blue. Yellow stars like the Sun have temperatures that are in between. Match the color of a star to that of a hot poker and you will learn its temperature even though you could never journey to the star with a thermometer in hand.

553 **The colors of some stars are quite obvious.** At first glance, all the stars might appear to be white or, in other words, have no color at all. But if you look carefully at some of the brighter stars, you may be able to tell that they really are different colors.

Low in the summer sky, for example, lies Antares, the brightest star in the constellation Scorpius (the Scorpion). Translated from the Greek, the name means "rival of Mars" and is derived from the star's reddish color. High overhead at the same time of year shines brilliant blue-white Vega. In the winter sky, we find the constellation Orion, the Hunter. Betelgeuse, which marks the hunter's right shoulder, is noticeably red, while Rigel in his left foot is decidedly blue.

554 **Binoculars and telescopes can help you see the colors of stars.** The eyes see colors in proportion to how much light enters them. This is why colors are typically vibrant on a bright sunny day, more muted on a cloudy day, and almost nonexistent under the darkness of night. We can frequently discern different colors among the brightest stars in the sky with the naked eye, but the fainter stars all look uniformly white. Telescopes and binoculars focus all the light that comes down their tubes into the eye and so concentrates it, making the colors of all stars more obvious and even quite vibrant. The second-brightest star in the constellation Cygnus, the Swan, for example, looks quite colorless to the naked eye. In only a modest-sized telescope, it is resolved into two stars: one a brilliant gold, the other a deep bluish-purple.

555 **A device known as a *spectroscope* is an incredibly powerful tool that can tell us much about a star or other object in space.** The simplest kind of spectroscope can be made by a child for a couple of dollars' worth of materials. It consists of a box with a narrow slit in one end and a triangular piece of ordinary glass called a *prism* inside. The light that passes through the slit and, in turn, through the prism is spread out into the individual colors that make up the light (just like we see when sunlight passes through a piece of broken glass). Over the years, scientists have learned that a thorough analysis of those colors can yield a truly amazing amount of information about the *object* whose light is being analyzed—all from a piece of glass and a slit in a box. Sometimes nature is kind. Of course, the spectroscopes used on modern telescopes are somewhat more complex, but they work by the same principle.

556 The *spectra* of different-colored stars look different. In the late 1800s and early 1900s, astronomers began to examine the spectra of lots of stars. They immediately noticed that all these spectra were not alike. Some stars had a spectrum that was crossed with many fine dark lines, not unlike our Sun. The spectra of some other stars, however, had far fewer lines, while still other stars had spectra that displayed so many lines they seemed to almost blend together in the form of shaded bands.

557 In an effort to understand what was going on, astronomers looked for patterns in the spectra of stars. One of the great methods of science when you don't really un-

derstand what's going on in a particular situation is to look for patterns in the data and then try to figure out why such patterns exist. In short, when in doubt, catalog it, then figure out what it means. And so the astronomers did, arranging the spectra according to what looked like the pattern shown on page 186. Taking the process one step farther, the astronomers asked themselves if this pattern represented anything physical about the stars in question. As it turned out, it did. Stars near the top of the set of spectra were blue-white, those farther down were white, and those still farther down were yellow, orange, and red, respectively. Thus, the different kinds of spectra of different stars corresponded to differences in color and hence surface temperature. Hot stars had different-looking spectra than cool stars and both had different-looking spectra from stars with temperatures in between.

558 Astronomers applied a simple letter and number classification to stars with different spectra. Originally, the spectra of the very hottest blue-white stars were given the letter A and the spectra of successively cooler stars received the following letters of the alphabet down to about the middle of the alphabet, at which point astronomers ran out of different types of spectra. Over time, as they understood better what was actually going on with stars and their spectra, some of the letter designations got rearranged, other letters got dropped, and the scheme evolved into the current sequence: O, B, A, F, G, K, M, R, N, and S. Stars that have the highest surface temperatures are designated with an O, while the letters at the other end of the sequence generally represent the coolest stars.

559 **Remembering the spectral classification sequence is easy, but it can get you a slap in the face.** One way to remember the curious sequence O, B, A, F, G, K, M, R, N, S is to remember the mnemonic "Oh, Be

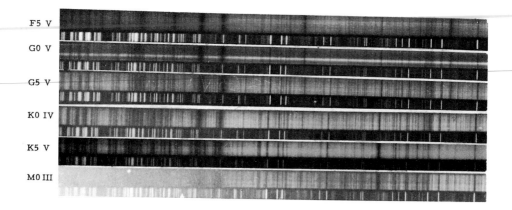

Astronomers classify the spectra of stars by using a series of letters with O stars the hottest and M stars the coolest. Dark lines due to helium (He) are seen in O stars while hydrogen (H) lines are most dominant in A stars. Numerous lines appear in the spectra of G stars like our Sun and are mostly due to the presence of metals, such as calcium and iron. Bands of lines in cool M stars are due to the presence of molecules, such as titanium oxide (TiO). (R,N, and S stars are also cool but have different abundances of certain carbon molecules.) The lines of each element are as unique as a fingerprint. They identify that element in the atmosphere of a particular star and also contribute to an understanding of that star's surface temperature. (University of Virginia)

A Fine Girl, Kiss Me Right Now, Sweetheart." Many an astronomy major has tried this out on a date. Some have met with success. Others have not. One astronomy professor used to hold a contest to see which of his Astronomy 101 students could come up with the cleverest alternative. The winning entry one year suggested a bit of strife back in the dorm and went: "Oh, Brutal And Fearsome Gorilla, Kill My Roommate Next Sunday!"

560 **Scientists learned to link spectral lines in different classes of stars to unique substances.** It was one thing to recognize the link between the different classes of stellar spectra and the surface temperatures of the stars involved and another to link these differences to the patterns of *specific* lines seen in the different stars. In time, however, by burning various gases in laboratories and observing the specific spectra produced by different gases at different temperatures, scientists were able to identify the specific lines in the spectra of different stars and figure out what substances they corresponded to. And so, unique lines in the very hottest O and B stars were seen to come from helium atoms, while the very strong lines in A stars were linked to hydrogen. Many of the lines in stars like our Sun were traced to a wide variety of metals from iron and magnesium to nickel and strontium. And the "banded spectra" of many red stars were found to be due to the presence of pairs and multiples of atoms bonded together, *molecules* containing such things as carbon, silicon, and oxygen. In

effect, these M, R, N, and S stars were so cool that molecules could actually exist in their atmospheres, whereas in hotter stars, such molecules would be torn apart.

561 **The lines in a spectrum are like a fingerprint.** Each substance in nature has a *spectrum* (a set of colored lines) and the spectrum of that substance is as unique to that substance as a person's fingerprints are to that person. Burn a substance in a flame in a laboratory and photograph its spectrum. Find those same lines in the spectrum of a star and you have discovered that the substance is found in that star. It's simple but wonderfully elegant detective work that has literally allowed astronomers to sit in their observatories on Earth and determine what the most distant stars are made of.

562 **A star may have lots of a certain kind of element, even though we don't see much evidence for it in the spectrum of the star.** At first glance, it might be tempting to conclude that A stars have more hydrogen in them than any other type of star because the hydrogen lines are strongest in the spectra of A stars. It turns out, however, that this isn't the case. The actual unraveling of this paradox took a thorough understanding of atomic physics, but in time astronomers realized that the hydrogen lines were most dominant in A stars because the temperature of their atmospheres was most conducive to the formation of these lines. Similarly, the higher temperatures of O and B stars were most conducive to the formation of helium lines. On average, each type of star had about the same amounts of hydrogen and helium.

563 **Sometimes there are differences that count between stars of the same class.** Sometimes, however, astronomers would find two stars of the same color and, therefore, the same temperature but whose spectra looked somewhat different. Both stars, for example, might have the same color as our Sun, but one might have lines as prominent as those in the Sun's spectrum, while the other might have somewhat weaker lines. Since many of these lines are due to the presence of *metal atoms,* the difference here is indeed the result of the one star having a greater abundance of metals in it than the other. Unraveling why some stars have more metals than others has, in turn, led to a better understanding of how stars evolve, as we shall see later.

564 **In time, the spectra of stars were further subdivided.** In classifying the spectra of stars, astronomers soon realized that simply using a handful of letters did not provide a fine enough distinction, so stars and their spectra were further subdivided by numbers. In this way, the hottest of the B-type stars were called B0 stars, while those that were a bit cooler were called B1 stars down to B9, at which point astronomers went to the designation spectral type A0, followed by A2, and so on.

565 **Soon other differences in stellar spectra were noted.** It soon became apparent that not all stars that had received the same spectral classification were the same and so even the increasingly complex system that had evolved thus far was inadequate. For example, astronomers had run across yellow G-type stars that had virtually identical lines in their spectra and hence were virtually the same temperature but whose spectra differed in one significant way, namely, in the *width* of the lines. One star might have rather wide lines, while another would have very narrow ones. In time, it was realized that the *width* of the lines in a spectrum was an indication of the *pressure* of the gas creating the spectrum. Gas under *low* pressure created *narrow* lines, while the same gas at the same temperature under *higher pressure* produced a spectrum with the *broader* lines. Astronomers reasoned correctly that the gas in the atmosphere of a small star would be under higher pressure than the gas in the atmosphere of a more distended star and so realized that the width of a star's spectral lines told how large the star was.

---

566 **Stars come in an amazing variety of sizes.** When it comes to size, our Sun is very average with a diameter of about 864,000 miles. Some stars are far larger and are appropriately called *giants*. If a giant suddenly replaced the Sun, all the planets out to the Earth would be inside the star. Stars called *supergiants* are even larger. Some measure nearly 2 billion miles across

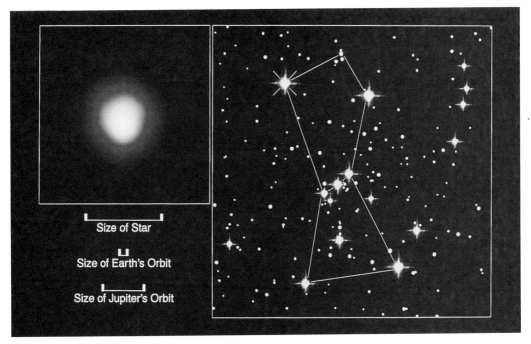

Using the HST, the star Betelgeuse in the right shoulder of Orion is resolved into a true disk (instead of merely a point, as seen in other telescopes). Betelgeuse is a supergiant star that has already moved off the Main Sequence and will someday explode as a supernova. The size of the orbits of the Earth and Jupiter are shown for comparison. (NASA/STScI)

and, if swapped for the Sun, would swallow up all the planets out to Saturn! In contrast, there are also *dwarf stars* no larger than Earth and tiny *neutron stars* that are barely as large as a city. Thus, while most stars are far larger than planets, others can be much smaller.

567 **To accommodate differences in size, more numbers were added to a star's spectral classification.** Using the width of the lines in a star's spectrum as a guide to the size of the star, astronomers differentiated between smaller stars, which generally came to be called dwarfs, and large stars, which came to be called giants. Dwarfs were given the Roman numeral V, while successively larger stars were given the designations IV, III, II, and I. Class III stars, it turned out, were actually pretty big and so wound up being called giants, which led astronomers to then apply the term supergiants to stars of Class I. Eventually, astronomers had to break Class I into Ib and finally into Ia for the really, really bright supergiants.

568 **What is the spectral classification for the Sun?** Given the temperature and size of the Sun, it is classified as a G2V star, that is, a rather small yellow star (as yellow stars go) with a surface temperature of about 11,000°F. Most other stars throughout the universe that also have G2V spectra would be very similar to our Sun.

569 The *shape* of a spectrum's lines also has another tale to tell. The shape of the lines can also lead to a determination of how fast the star is spinning on its axis. The faster the spin, the shallower the lines.

570 Some stars are regular whirling dervishes. Just like the Earth, most stars spin on an axis. Because the stars are made of gas, however, different latitudes rotate at different speeds. Regions near the poles of our Sun take about 31 days to make one rotation, while lower latitudes near its equator only take about 25 days to go around. While some stars spin slower than the Sun, others gyrate at incredible speeds. Some stars spin so fast they actually spin off their outer atmospheres into space like great cosmic lawn sprinklers. The fastest spinners, however, are tiny stars called *pulsars.* Some rotate hundreds of times each second!

571 Some stars have a real magnetic personality. The spectral lines of some stars are split so that each line actually appears double. This phenomenon is known as the *Zeeman effect* and is due to the presence of a magnetic field in and around the star. As we have seen, our Sun has a magnetic field that is particularly strong around its sunspots. But some stars have magnetic fields that are hundreds of thousands or even millions of times stronger than that of our Sun.

572 Many stars seem to have spotty faces. For centuries, we have known that the Sun's face is not uniformly bright but instead is marred here and there with dark spots that astronomers cleverly call *sunspots.* In recent years, telescope technology has allowed astronomers to begin to resolve some of the larger stars into tiny disks instead of infinitesimally small dots. One of the discoveries that has resulted is that some of these stars apparently also have blotchy faces. In some cases, the *star spots* seem to be quite large, being many times the size of the Earth. As on our Sun, the spots are likely caused by differences in temperature on the surface of the stars due to varying strengths in the star's magnetic field from place to place.

573 The spectra of stars can be used to help determine their distances. The distance to some stars can be determined by a measurement of their *parallax,* that is, how much a particular star appears to shift back and forth compared to much more distant stars. This technique results in what is actually termed a star's *trigonometric parallax* because it uses a branch of math called *trigonometry* and relies on direct measurements of the angle of shift of the star's position. Trigonometric parallaxes are very reliable but are only accurate out to a distance of several hundred light-years because the shift in the star's position beyond that distance becomes too small to measure. Astronomically speaking, several hundred light-years is still very close to home. To develop more than just a very local map, astronomers had to come up with another technique to determine stellar distances. By examining the spectra of the stars, astronomers quickly figured out they could extend distance measurements considerably farther. The technique in principle is very simple and based on the fact that two stars with virtually identical spectra are virtually identical stars. Say there's a star that we want to know the distance to, but it's too far away to allow us to measure its trigonometric parallax. Just take a picture of its spectrum and find a nearby star with the same kind of spectrum. If the spectra are the same, then the two stars are the same. If we know how bright the nearby star is and how far away it is, we can then easily figure out how far away the identical but more distant star would have to be to look as faint as it does. The technique is appropriately known as *spectroscopic parallax* because

it uses a star's spectrum to figure out its distance. Using this technique, astronomers can measure the distances of stars out to thousands or even millions of light-years.

574 **The universe is a big and complex place, but the chemistry of the universe is actually very simple.** There are billions upon billions of stars in the universe. Yet, using spectroscopes, astronomers have learned that in all the universe there are only 92 naturally occurring elements. And they are the same 92 naturally occurring elements that we find right here on Earth. To be sure, an enormous number of substances can be created by combining these elements in different ways. Hydrogen and oxygen bond in a simple way to make water, while, in contrast, hydrogen, oxygen, nitrogen, carbon, and phosphorus come together in complex unions to create DNA. But at the root of it all are the same basic 92 building blocks. Conceivably, the universe could have been a lot more complex . . . but it isn't.

CHAPTER 10

THE LIVES OF THE STARS

575 Scattered across space are the stellar maternity wards of the galaxy. Stars do not just pop into existence randomly in space. Instead, they are born in huge clouds called *nebulae* (simply Latin for "clouds"). The nebulae are made of gas and dust, out of which the stars are formed.

576 *Gravity* is the key player in the creation of stars. The gas and dust in nebulae is always in a state of motion. As a result of this motion, some regions in the cloud periodically have a higher concentration of gas and dust than others. The higher the concentration of gas and dust in a particular region of the cloud, the greater is the force of gravity in that region because gravity is generated by the presence of matter. If a region of space has more matter, the force of gravity in that region is stronger.

577 When the force of gravity is sufficiently strong in a particular region, a star is formed. If there is sufficient gravitational force present in a particular region of a nebula, the gas and dust in that region will begin to collapse upon itself and, in turn, draw more gas and dust into the collapsing region. (Astronomers call this gathering process *accretion*.) As the accretion process continues, temperatures at the center of the region climb higher and higher. Should enough matter be gathered together to allow temperatures to exceed about 18 million°F degrees at the center of the region, thermonuclear reactions will set in and a star will be born.

578 Some nebulae are hundreds of light-years across and can give birth to thousands of stars. Many nebulae can contain enormous quantities of gas and dust—enough for the formation of scores upon scores of stars. Sometimes several stars are created in a close grouping inside a nebula at about the same time. Such a grouping is called a *star cluster*. All the stars in the cluster have similar chemical compositions, evolve together, and usually travel together as a family through space. Later in time, the same nebula can give birth to other stars and other star clusters.

579 A large beautiful nebula can be seen in the winter sky with the unaided eye. A short distance below the three belt stars in the constellation Orion lies a little fuzzy patch of light just visible to the unaided eye on a clear dark winter night. Binoculars and telescopes, however, reveal it to be one of the true wonders of the heavens: the Great Nebula in Orion. The nebula's tiny apparent size is due to its great distance—about 1,200 light-years—from Earth. In reality, the Great Nebula is an immense and magnificent region of glowing multicolored gas and dust in which new stars are being born.

580 The Great Nebula is being lit up by young hot stars embedded within. The Great Nebula receives much of its glow from the light of a tiny grouping of stars within known as the Trapezium, which can be seen in even a small telescope. Their intense blue-white color reveals not only that they are very hot stars but also that they are very young, astronomically speaking—probably little more than 100,000 years old. This means that when our early humanoid ancestors looked at Orion in the sky, they saw it without the glow of the Great Nebula.

The Great Nebula in Orion—a giant stellar maternity ward about 1,200 light-years from Earth.
(©Anglo-Australian Observatory)

581 The Great Nebula is a giant three-dimensional "tapestry" of color and light. We see the Great Nebula in two dimensions from Earth, but, in reality, it may be thought of as a vast three-dimensional cavern, light sculpture, or "tapestry" of color and light—denser in places, more gossamer in others. Its apparent shape is determined by the varying densities of gas and dark dust, the location of the stars within, and the direction in space from which we view it.

582 The "tapestries" of color and light within the nebula are woven by some of its stars. Even a small to medium-sized

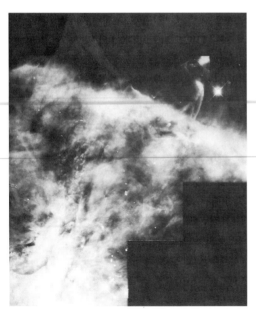

Complex detail in the inner regions of the Great Nebula in Orion are revealed in this mosaic of 15 images taken with the HST's Wide Field Planetary Camera. The region pictured here is about 2.5 light-years across (or almost 15 trillion miles), but its total area in our sky is only about 5 percent of the area of the full Moon. (NASA/STScI)

telescope reveals the Great Nebula's majestic beauty. Its intricate structures, consisting of strings, curved loops, and irregular walls of color, are created as light from its stars pass through the complex weavings of gas and dust that make up the nebula. In many instances, intricate details within these structures have been created by the pressure of the powerful radiation of the young stars, as well as jets of material that are blasted and spewed out of these stars at speeds of up to 100,000 miles per hour. Like colored streams of water shot into clear gently swirling pools, they create fanciful patterns that appear to the eye as much works of art as objects of science.

583 Recently, the Hubble Space Telescope revealed striking images of stars actually being born inside the Great Nebula. Astronomers had long said that stars were born in regions of gas and dust like the Great Nebula in Orion, but actually catching stars in the process of formation was another matter. Recently, they used the Hubble Space Telescope to peer deep within the Great Nebula and created a mosaic of incredible detail. Within the mosaic, they found images of over 700 stars in various stages of formation. Compared to the Trapezium stars, some are truly infants, possibly only a few tens of thousands of years old.

584 Probing deeper into the Great Nebula, Hubble has even glimpsed the "seeds" of stars that have not yet been born. Also sighted within the Great Nebula was a

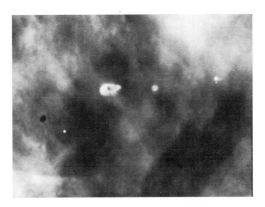

Deep within the previous image, the HST ferrets out three of many proplyds, or protoplanetary disks. Inside each "cocoon," an embryo star and possibly a family of planets is forming. (NASA/STScI)

breed of celestial object that proved even more exciting to astronomers—over 150 disk-shaped objects that are nothing less than embryonic stars still enveloped in "cocoons" of gas and obscuring dust. In time, the pressure of the radiation from these stars will push on the dust, dissipating the disks and revealing the stars themselves for the first time. For now, however, they appear dark against the brightly lit background of the nebula.

585 **The disks in the Great Nebula may harbor more than embryonic stars; some may also contain developing planets.** In some cases, the outer regions of dust within these disks may have begun to stick together in clumps large enough for gravity to continue the "gathering process." If so, some of these disks not only contain stars in the process of formation but systems of developing planets. In short, we may be

witnessing the creation of entire new solar systems, which, in turn, may someday give rise to life. For now, astronomers have named these objects *protoplanetary disks,* or *proplyds* for short. A few, studied in considerable detail, appear to range in size from about two to eight times the size of our solar system (out to Pluto) and harbor stars that are between about one-third and one and a half times the mass of our Sun.

586 **The first possible solar system in formation to be discovered is in orbit around the star Beta Pictoris.** Using an infrared telescope with a tiny opaque disk inside, astronomers managed to see a disk of material in orbit around the star Beta

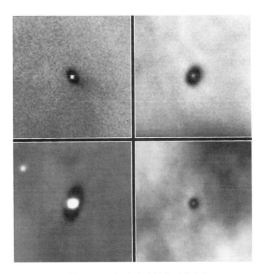

Close-ups of four proplyds in M42 with infant stars emerging from the obscuring disks of dust in which they were born. Each image is 167 billion miles across (thirty times the diameter of our solar system). Our own solar system probably looked much like this about 4.6 billion years ago. (NASA/STScI)

An image of the star Beta Pictoris reveals a disk of gas and dust in orbit around the star which may enshroud a system of planets in the process of formation. To help make the disk visible, the glare of the star itself has been blocked by a circular mask suspended by crosshairs. (NASA/JPL)

Pictoris in 1985. The disk is about ten times the diameter of our solar system (out to Pluto) and about 425 million miles thick. Because the disk happens to be turned nearly edge-on to the Earth, it appears as a line with the star at the center. Within, objects that range in size from grains of dust to large boulders may be coming together under the force of their mutual gravitational pull to form planetesimals and, in time, full-fledged planets.

587 **Chemical analyses of the Beta Pictoris disk reveal the presence of intriguing chemicals for building a solar system.** Studies reveal that the Beta Pictoris disk contains atoms of carbon and silicon and molecules of ice. Such materials are the stuff of comets, asteroids, and planets. The presence of these materials lends weight to the hope that material of sufficient total mass will come together in the disk and form comets, asteroids, and planets that resemble the ones we find in our own solar system.

588 **Other recent studies suggest that comets may have formed around Beta Pictoris and at least one other star.** Studying the disk of material around Beta Pictoris in the infrared, scientists have found that its spectrum suggests that the dust is similar in size to that released from comets in our own solar system. The spectrum of another star called HD 163296 reveals the presence of cold magnesium atoms. Again, the source could be a swarm of comets. The possible discovery of comets around other stars is exciting, for it again points toward the possibility of other solar systems being similar to ours. Furthermore, since comets may have helped seed the Earth with the "molecules of life," they could conceivably do so for other planets elsewhere.

589 **Still other objects appear to be forming within the Great Nebula that may become planets without stars.** Do planets exist that roam space without orbiting stars? Among the other recent Great Nebula finds are dark collapsing objects that are estimated to range in mass from about 0.1 to 100 times that of the Earth. Such objects would span the range from small planets to objects several times the mass of Jupiter—objects that would someday have to be termed planets yet would not encircle any nearby sun. Should such objects be discovered in other nebulae, it would suggest that billions of "orphan planets" may roam the cold blackness of space.

590 **What's the difference between a star and a planet?** A star is an object that, for much of its life, generates its own light and other radiation by means of nuclear reactions. By contrast, a planet may leak heat radiation that is being generated by radioactive decay or gradual collapse (like Jupiter). But a planet is not capable of generating its own light. Stars shine by their own self-generated light. Planets shine only by the light they reflect from nearby stars.

591 **Stars create their energy by a process called *thermonuclear fusion*.** Thermonuclear fusion is a fancy-sounding term, but it simply means taking certain kinds of atoms and making their centers, or nuclei, stick together, or fuse. The nuclei of atoms typically resist this sort of thing, so it usually takes a lot of pressure to get them together and a lot of heat to make them collide with sufficient force to fuse. The insides of stars are hot high-pressure places, so they are places where lots of thermonuclear fusion is taking place. And it is precisely the energy released in such processes that causes the stars to shine and creates the pressure pushing outward that stops a star from continuing to collapse inward.

592 **There are "missing links" between stars and planets that roam space.** Recently, astronomers confirmed the existence of objects that have been labeled *brown dwarfs*. Evidence for the existence of brown dwarfs had been sought for over three decades. Brown dwarfs represent a kind of missing link between objects that do not have

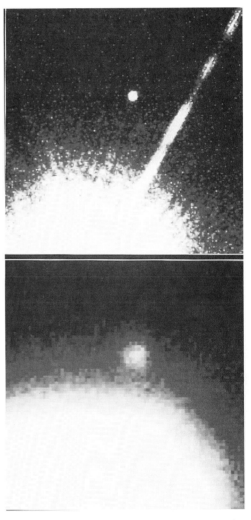

The brown dwarf GL229B is a small faint dot in these HST images. (NASA/STScI)

enough mass to ever become stars and yet are very massive compared to known planets. The first such object to be discovered is known as Glise 229B (or simply GL229B) and is in orbit around a small red star (known as GL229) about 19 light-years from Earth in the constellation Lepus, the Hare. It is estimated to be about twenty to fifty times as massive as Jupiter. Glise 229B has a surface temperature of about 1,300°F and gives off lots of heat—energy that is being leaked from its core, which is still slowly contracting under the force of gravity. The central temperatures in brown dwarfs, however, never reach levels high enough to allow thermonuclear reactions to begin.

593 **Brown dwarfs aren't really brown.** The name *brown dwarf* is somewhat misleading, for while such objects are indeed small compared to most stars, they would not appear brown to the eye but more likely a very, very dull red. The name was conjured up more as a joke than anything else at a meeting of astronomers some years ago and, for better or worse, it stuck. Since the first brown dwarf was found, several more have been discovered.

594 **How much mass can an object have and still not become a star?** When matter collapses out of a nebula, it will ultimately form a star if its central temperatures exceed about 18 million°F—hot enough for nuclear burning to begin. The more matter that goes into forming the star, the hotter the star will be. But how much mass can come together in a nebula and *just miss* becoming a star, that is, have an interior that just isn't quite hot enough to set off thermonuclear reactions? According to theory, the answer seems to be around 8 percent of the mass of the Sun. Recently, astronomers used the Hubble Space Telescope to photograph just such an object—a very tiny red star that is a companion to another star called GL 105A. GL 105A is very faint itself, but its companion is 25,000 times fainter still and qualifies as the intrinsically faintest star ever found—a star that almost didn't become a star.

595 **Astronomers also recently explored another region of star birth and discovered some truly remarkable structures.**

Another magnificent star-birthing region known as the Eagle Nebula, M16, about 7,000 light-years from Earth in the constellation of Serpens, the Serpent. (UCO/Lick Observatory photo/image)

An incredible image taken by the HST of serpentlike columns of gas and dust in the center of the Eagle Nebula. These "pillars of creation" contain new stars in the process of formation. (NASA/STScI)

Again using the Hubble Space Telescope, astronomers recently looked deep within another cloud of gas and dust called the Eagle Nebula, which is located about 7,000 light-years from Earth in the summer constellation of Serpens, the Serpent. There they found intricate fingerlike pillars of gas and dust up to 6 trillion miles in length. The "pillars of creation," which have been described as looking like the spires of enchanted castles or the heads and necks of sea serpents, are actually enormous towers sculpted by light and ultraviolet radiation. The radiation is coming from hot nearby stars and its pressure is literally eroding away less dense gas and dust to form the pillars. In this way, the pillars are similar to buttes in the desert where basalt and other dense rock have remained while surrounding softer stone has long ago been eroded away by wind and water.

596 Within the pillars of the Eagle Nebula lie "the Eagle's EGGs." As the pillars themselves are slowly eroded away, small globules of even denser gas and dust are uncovered within. These globules have been dubbed *EGGs* (short for Evaporating Gaseous Globules), but the term is appropriate for another reason as well because these EGGs may also be thought of as the Eagle Nebula's "eggs." In short, they are regions of gas and dust that are embryonic stars just like those also found in the Great Nebula in Orion. So when we look at these fanciful structures, we again are seeing the very incubators from which new stars, new planets, and, possibly, new life are now emerging.

597 There are dark nebulae as well as bright nebulae. Stars light up the gas and dust in nebulae, making them shine like multicolored tapestries in space. But if nebulous material lies far away from stars, it will

The Horsehead Nebula in Orion shows the presence of dark obscuring dust in space. The unilluminated dust is located between us and the bright illuminated gas and dust beyond, thus creating, in silhouette, the impression of a horse's head. (NOAO)

not be illuminated and therefore remain dark. In some places, unilluminated gas and dust happens to lie between us and more distant material that is illuminated. The result can be striking dark silhouettes, such as the Horsehead Nebula in Orion.

598 Unilluminated dust is also responsible for what appear to be "holes in the heavens." As seen from Earth, the Milky Way does not appear to be uniformly lit. Instead, it is crisscrossed and mottled with dark lanes and patches, which led some early astronomers to wonder if these "holes in the heavens" might be due to the appearance of fewer stars in these regions. In time, however, it was discovered that these regions appeared dark not because there were fewer stars but because clouds of unilluminated dust absorbed and hid the light of more distant stars beyond.

599 Striking examples of such obscuring dust clouds can be seen in both the northern and the southern skies. On a warm summer evening in the northern hemisphere, three bright stars can be found high in the sky. They are Vega, Altair, and Deneb, which together form an asterism known as the Summer Triangle. Arcing gracefully through the Triangle is the Milky Way. But at this point, the band of light is divided almost in half by a dark lane of dust just as it passes through the constellation Cygnus, the Swan. The feature, easily seen on clear moonless nights away from city lights, is known appropriately as the Dark Rift in Cygnus.

The constellation of Orion, the Hunter, seen in visible light, that is, as it appears to the human eye. The Great Nebula appears as a small milky patch of light below the three stars of Orion's belt. (Alan Dyer/Calgary Centennial Planetarium)

At 9 P.M. or so on a spring evening in the land "down under," the stars of the Southern Cross shine brightly overhead, along with some of the most striking star clouds of the southern Milky Way. In contrast, nearby the Cross lies an inky black hole in the sky about 5° across. This "hole" is another region of dark obscuring dust known as the Coal Sack.

600 This *interstellar dust* is made up of very tiny stuff. Studies of the spectrum of the interstellar dust reveal that it consists

This is an image of the same region in Orion but taken in the infrared by IRAS. The infrared rays cut through obscuring dust between us and Orion and reveal far more than the glow of the Orion Nebula. Indeed, we see that this entire region of the sky is a vast complex star-birthing nebulae. (NASA)

of very minute particles that range in size from about a thousandth of a millimeter down to things hardly larger than a couple of atoms stuck together. How the dust affects starlight gives clues to what it's made of. Studies of the spectrum of the interstellar dust and measurements of other ways that it affects the starlight shining faintly through it reveal that it contains atoms of silicon and carbon. Specifically, much of the carbon seems to be in the form of grains of graphite and simple soot. In short, parts of space are very sooty.

601 **Some of the dust also polarizes starlight the same way polarizing sunglasses do with sunlight.** Light may be thought of as traveling in waves like the waves you might make with a piece of rope if you tied one end to a post and moved the other end up and down. Normally, such light waves are oriented in all directions, just as you could make the waves in a rope move in different directions by moving your hand different ways. When light waves have no particular orientation, we say that they are *unpolarized*. But now imagine trying to make waves with a rope that passes through one of the openings in a picket fence on its way from the post to your hand. The opening in the fence would allow you to only make waves in one plane—the same plane as the opening itself. Similarly, a pair of polarizing sunglasses reduces light intensity by only allowing light waves in one plane to pass through. It can do this because the molecules in the lenses are all lined up in long rows like the slats in a picket fence. You can tell if a particular pair of sunglasses contains polarizing lenses by rotating a pair of polarizing sunglasses in front of them while looking at a source of light. If both pair are *polarized,* the intensity of the light will change. In the same way, astronomers rotate polarizing filters within their telescopes to see if any starlight is polarized. In so doing, they have discovered that starlight passing through interstellar dust is indeed slightly polarized.

602 **The discovery that interstellar dust has some polarizing properties tells us something about it and the space that it's floating in.** The fact that interstellar dust slightly polarizes the light that passes through it means that, like the molecules in polarizing sunglasses, the particles of dust are somewhat aligned. Carbon and silicon molecules alone don't do this, but substances with metallic properties do (as can be seen when iron filings line up in the field of a magnet). There is evidence that some grains of interstellar dust become coated with ice, attract atoms of iron floating in space, and eventually become aligned with weak magnetic fields that extend out from many stars (including our own Sun) and permeate the whole galaxy. In this way, the dust between the stars acts a bit like a giant pair of sunglasses.

603 **The source of the soot and other components in interstellar dust has been tracked to red giant stars.** Late in life, stars like the Sun become red giants and their atmospheres cool sufficiently to allow atoms to bond together to form compounds of carbon and silicon. The pressure of the star's light then pushes these compounds out into space. Thus, in effect, the dust clouds we see across the sky are the result of countless red giants "burning sooty" during certain periods in their lives. Iron and the components of water can also be found in the atmospheres of such cool stars, so they are also liberated into space at the same time.

604 **So why is all this dust important?** In order for new stars and planets to form, nature frequently needs sheltered environments in which gas and dust can gather under the force of gravity and begin to grow, or accrete, into larger and larger clumps. Open space is usually a difficult place for this to occur because the pressure from the light of nearby stars tends to disrupt the accretion process. Dusty regions serve to shield protostars and protoplanets from such disruptive forces and, therefore, serve as catalysts for the creation of new stars and planets.

605 **Such dark regions can be seen embedded in bright nebulae across the galaxy.** Known as *Bok globules* for Bart Bok, the Dutch-American astronomer who researched them, these dark knots are typically 5 to 10 light-years across and contain the mass of anywhere from 10 to 100 stars like our Sun. From such regions, the light of newborn stars will someday shine forth.

606 **In addition to dust, there is also gas between the stars.** Most of the material out of which nebulae are made consists of gas. The most common type of interstellar gas is hydrogen, the most abundant element in the universe. Since the early 1970s, however, about 100 additional types of sub-

Small dense patches of dark obscuring dust can be seen against the bright background of a nebula known as IC 2944. Some of the so-called Bok globules may someday condense to form stars and solar systems. (©Anglo-Australian Observatory)

stances have been found in interstellar space, including water, ammonia, methane, and such complex organic molecules as formaldehyde, acetylene, and ethyl alcohol.

607 **Until the 1970s, astronomers didn't believe that such complex molecules could exist in interstellar space.** The general belief was that the powerful ultraviolet radiation from young hot stars would tear complex molecules apart as soon as they formed. Clearly, however, this is not the case, as the few astronomers who first decided to look for these molecules happily discovered. The existence of the complex molecules is due to the accompanying interstellar dust, which sufficiently shields the molecules from the UV radiation, allowing the molecules to form in significant abundance.

608 **The discovery of complex organic molecules in space may be very significant.** To find such molecules abundant throughout space and in the very places where new stars and possibly new planets are forming is very exciting because it is believed that such molecules are essential to the creation of life. In short, if the molecules that form the building blocks of life are spread across the galaxy, so may life itself.

609 **There are many of certain types of stars in the universe and relatively few of other types of stars.** There are billions upon billions of stars very similar to our Sun

across the universe. There are even more stars that are smaller and redder than the Sun. But as we move in the opposite direction to stars that are progressively larger and hotter than our Sun, we find fewer and fewer.

610 **Like people, stars are born, live out their lives, grow old, and die.** Hey, it's just that simple.

611 **Trying to understand stars is a little like trying to understand people.** When we look out into space, we see a "snapshot" of lots of different stars at particular moments in their lives. The challenge is to try to piece the evidence together so that we can try to understand how any particular star evolves. It's rather like imagining that you are an alien on a very brief visit to Earth who is given the task of trying to understand how this creature called Homo sapiens lives out its life. A clever tactic might be to land your flying saucer on the 50-yard line during the Rose Bowl game and quickly take lots of snapshots of the crowd. Upon returning to your world and carefully examining your pictures, you would notice that there were different kinds of Earthlings: little smooth-skinned ones, bigger smooth-skinned ones, and ones who lacked smooth skin and had white hair. In trying to understand the creatures, you might try to arrange the images to see if you could find patterns and trace a logical path to understand how one type of human might, over time, evolve into the other kinds of humans you observed. The approach is good basic science. When you observe a phenomenon that you don't understand, gather data, classify it, try to find patterns and trends, and figure out why and what it all means. When it came to the stars, scientists did the very same thing.

612 **How long a star will live is frequently determined by how much it weighs at birth.** The stars that are the most massive at birth typically live the shortest lives, while the least massive stars usually live the longest.

613 **Stars live a lot longer than people, but some can live millions of times longer than others.** The most massive stars in space live for less than 3 million years as active stars, while the least massive ones can last for trillions of years—longer than the current age of the universe.

614 **In an early attempt to better understand the stars, astronomers sorted them according to temperature and true brightness.** Nature makes lots of different

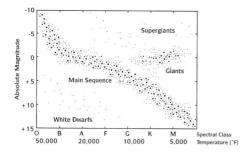

The H-R diagram categorizes stars according to temperature (or spectral type) and luminosity (or absolute magnitude).

colored stars, but nature doesn't make all colors of stars in *all* brightnesses. The illustration above shows a *Hertzsprung-Russell diagram* (H-R diagram), named after the two astronomers, Ejnar Hertzsprung and Henry N. Russell, who first constructed it. The H-R diagram has led scientists to some great insights into the workings of the universe, including an understanding of what makes stars shine and how different stars live out their lives.

615 **The H-R diagram takes us into a land of giants and dwarfs.** What the diagram basically says is that nature only makes certain kinds of stars. Pick a particular color of star, such as red, and we see that there are intrinsically faint red stars (down in the lower right corner of the diagram) and there are intrinsically bright red stars (in the upper right portion of the diagram). If two stars have the same color, it means that they also have the same temperature and that, in turn, means that they radiate the same amount of energy out of every square inch of their surfaces every second. If one is intrinsically *brighter* than the other, it means it is *bigger*. And so we see that stars across the *top* of the H-R diagram are *big* stars and those near the *bottom* are *smaller* stars. The smaller ones are called *dwarfs*. The big ones are called *giants*. Even bigger ones are called *supergiants*.

616 **The H-R diagram gives great insight into how stars live their lives.** The H-R diagram shows that there are lots of certain kinds of stars in space and very few of other kinds. There can be two reasons for this. Either nature, for some reason, just doesn't make certain kinds of stars or, during the course of their lives, stars move around on the H-R diagram and spend a lot of time in those parts where we see lots of stars and very little time in parts of the diagram that are virtually empty. As it turns out, both of these hypotheses are correct.

617 **The H-R diagram's main street is called the Main Sequence.** While stars may be found in many parts of the H-R diagram, most cluster in a line that runs from the upper left to the lower right. Astronomers refer to this as the Main Sequence and this is where most stars spend a good portion of their lives—from youth to just before middle age.

618 **The Main Sequence represents a region of stability.** Stable stars are stars that remain the way they are for long periods of time, growing neither larger or smaller, hotter or cooler. There are two forces present inside every star: the force of gravity pulling inward and the pressure of the star's radiation and hot gases pushing outward. To remain stable, however, a star has to successfully perform a delicate balancing act. At each point inside the star, the outward and inward forces must perfectly balance. Once stars reach the Main Sequence, they stop contracting because gravity is now counterbalanced at every layer within the star by the force of the slow steady nuclear reactions that

have begun within. The star has become a stable healthy Main Sequence star.

619 **The Main Sequence is where stars reside while they are undergoing their first stage of thermonuclear fusion.** The first and most common type of thermonuclear fusion that occurs inside stars takes place when nature transforms the most abundant element in the universe, namely hydrogen, into the second most abundant element in the universe, namely helium. In the process, the nuclei of 4 hydrogen atoms are fused to become a single nucleus of helium and tremendous amounts of energy are given off. Astronomers say that hydrogen is "burned" to make helium, even though the process is far from burning in the normal sense. Main Sequence stars are stars in this first stage of energy creation, undergoing "hydrogen burning." And the Main Sequence is a nice clean line on the H-R diagram because it simply marks the line along which stars of various masses finally stopped contracting due to gravity and became stable as they began burning hydrogen into helium.

620 **Just where a star comes to reside on the Main Sequence depends on the initial mass of the star.** As regions of gas and dust inside nebulae collapse to form stars, their central temperatures rise. The more gas and dust collapse to form a star, the higher the central temperature will become and, in turn, the higher its surface temperature will become before the object settles down to become a stable star. So stars that form from

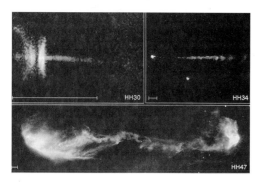

Jets of gas are seen being shot out of young stars. In the upper left, the star is hidden by an obscuring layer of dust. The jet in the upper right image is pulsed, showing that some young stars somehow produce repeated spurts in a stiletto fashion. The image at the bottom shows a 3-trillion-mile-long jet being shot out of a young star near the left edge of the frame. (C. Burrows [NASA/STScI and ESA], J. Hester [Arizona State University], and NASA, J. Morse, STScI, and NASA)

large clumps of gas and dust and start out life with a lot of mass (the "hefty babies" of the stellar world) come to rest toward the top of the H-R diagram. In other words, they start out their youth as white or blue-white giants and supergiants. Stars that happen to form from smaller clumps of gas and dust settle on to the lower end of the Main Sequence as orange and red dwarfs. And stars that start out with an average amount of mass wind up liv-

Additional jets from young stars. (NASA/STScI)

ing out their youth in the middle of the Main Sequence as rather average-sized, rather average-temperature yellow stars.

621 **From the positions of various stars along the Main Sequence, we can see that there is an important relationship between the mass and overall brightness of these many stars.** Stars along the upper portions of the Main Sequence are blue giants and supergiants. They are therefore very hot, very bright, and very massive. Stars along the lower portions of the Main Sequence are cooler, smaller, not very bright, and have less mass. Thus, for Main Sequence stars, the more massive a star is, the brighter the star is.

622 **Just how massive can stars get?** This is a matter of some debate. Some observations and theoretical calculations suggest that stars might not be able to get much more massive than about 40 to 50 times the mass of our Sun. Beyond this point, nature may create instabilities in the cloud out of which the star is trying to form or create such strong outward pressure in the star that it rapidly loses mass into space. Nevertheless, there is a very extraordinary star in the skies of the southern hemisphere known as Eta Carinae that may somehow be as much as 100 times the mass of our Sun.

623 **For Main Sequence stars, it doesn't take much of a difference in mass to create a huge difference in brightness.** It turns out that there is a fairly simple relationship between how much mass a Main Sequence star has and how bright it is. Astronomers cleverly call it the *mass-luminosity relationship* and it says that, on average, if one star has twice the mass of another star, it will be 10 times as bright, but if one star has 10 times the mass of another star, it will be 3,000 times as bright. Since Main Sequence stars have masses that range between about 0.08 times the mass of our Sun to perhaps 50 times the mass of our Sun, this means that the brightest Main Sequence stars are something like 10 billion times brighter than the faintest ones.

624 **The more massive Main Sequence stars are much more luminous than less massive ones because they burn their nuclear fuel at a faster rate and are much hotter inside.** The central temperatures of all stars climb as they form through accretion in nebulae. As the atoms and molecules of gas and dust fall toward the center of the knot that will ultimately become the star, they fall faster and faster under the pull of the knot's gravity (just like a ball dropped off a building falls faster and faster). The temperature of a gas is really just a measure of the average speed of the atoms, molecules, or ions making up that gas. The higher the average speed, the higher the temperature. Since larger, more massive stars are created by having more accreting gas fall farther into the knot, their final central temperatures are the highest of all Main Sequence stars.

625 **Our Sun's place in the H-R diagram tells us that it is a garden-variety star.** Our Sun is currently located just about smack-dab in the middle of the H-R diagram—a stable yellow Main Sequence star. As such, it is about as ordinary a star as you can find. There are stars that are bigger and stars that are smaller. Stars that are hotter and stars that are cooler. The Sun may in no way be extraordinary, but we should be thankful for this, for it is precisely this mediocrity that has made life possible on its third planet.

626 **Our Sun is currently somewhere between youth and middle age.** By our best reckoning, the Sun has been a stable star, living on the Main Sequence for about 4.6 billion years and it will remain on or near the Main Sequence for another 5 billion or so more before it leaves to squarely face middle age. In short, the Sun is no longer a teenager or even a young adult but a rather comfortable . . . thirtysomething.

627 **All stars enter the H-R diagram from stage right.** Since all stars are born in cool clouds of gas and dust and grow hotter as they form, we would expect infant stars to enter the H-R diagram from the right or cool side of the picture, and indeed they do.

628 **Before a star is a star, it's called a *protostar*.** Objects in nebulae that are in the process of becoming stars but aren't stars yet are called protostars. Protostars may have surface temperatures of hundreds to a few thousands of degrees and internal temperatures approaching 15 million°F. But all this heat comes directly from the fact that the protostars are still slowly collapsing.

---

629 **A big change occurs when a protostar finally gets hot enough inside.** A protostar finally becomes a full-fledged star when its central temperatures become high enough for the object to independently create new sources of energy within itself through thermonuclear fusion. This process also stops its own collapse by counteracting gravity.

---

630 **The amount of time a star remains on the Main Sequence also depends on its mass.** This connection is a simple one. The more massive a star is once it reaches the Main Sequence, the hotter it is inside. And the hotter it is inside, the faster it burns its nuclear fuel. And the faster it consumes its nuclear fuel, the sooner it leaves the Main Sequence.

631 **"Internal problems" ultimately force a star to leave its comfortable life on the Main Sequence.** Basically, it's a case of "too much lead in the gut" or, in this case, too much helium. As a star sits on the Main Sequence, hydrogen burning, which began at the star's center, gradually progresses outward. The result is a growing dense core of helium nuclei surrounded by the star's remaining hydrogen. Ultimately, the core becomes so heavy that it collapses under its own weight. Instantly, temperatures at the center of the star surge upward. As the excess energy finds its way to the surface of the star, it pushes the photosphere outward. The star grows and grows until equilibrium is again achieved and gravity balances outward pressure everywhere within the star. By now, however, the star has typically moved well off the Main Sequence—up and to the right in the H-R diagram—to become a redder *giant* star.

632 **We can tell from the H-R diagram that the transition from the Main Sequence to the realm of the giants is a quick one.** There are many stars along the Main Sequence and quite a few in the region where the giant stars dwell but very few in the space in between. This means that if stars go from one region to the other, they must do so quite quickly because we have caught few in the act of making the transition.

633 **Some stars have yet to make it to the giant branch of the diagram.** Many low-mass red dwarf stars burn their hydrogen so slowly that they are still far from the point when their cores will collapse. And so they are still on the Main Sequence. Some of these stars are young, but others are very old. Indeed, low-mass red dwarfs burn their hydrogen so slowly that they can quietly stay on the Main Sequence for hundreds of billions to trillions of years—longer than the universe has been alive.

634 **After leaving the Main Sequence, different stars suffer different fates,**

again depending on their mass. Many possible roads lie ahead for stars after they leave the Main Sequence, but some stars suffer rather gentle fates on the way from middle to old age, while others undergo catastrophic growing pains.

635 **The Sun and most other Main Sequence stars look pretty much the same day after day (and night after night), but for other stars this isn't the case.** The ancient Arab sky watchers knew of certain stars that did not behave like the rest. One, in the southern skies of autumn, would shine as brightly as many others at times, then, within weeks, would fade from view, only to return again within a year. They called the star Mira, meaning "the wonderful or amazing one." Many other *variable stars* have been discovered by astronomers over the years.

636 **How a star's light varies over time can be displayed in a *light curve*.** A star's light curve is a tracing or plot of how the star's light output or brightness varies with time. Different kinds of variable stars typically have different-looking light curves.

637 **Many variable stars are variable because they are unstable.** The Sun looks essentially the same in the sky every day. The same color, the same size, the same brilliance. (And it's a good thing, for any significant changes in the Sun would have significantly devastating effects on the Earth's climate.) The Sun currently remains the same in appearance because it is a *stable* star. That is, at each point inside the Sun, the force of gravity pulling inward is perfectly balanced by the pressure of the Sun's hot gases and radiation pushing outward. But at various times in the lives of all stars, these forces are not in balance. When this occurs, the stars are not stable but instead are *unstable,* which means they will change over time. Sometimes the time periods are short; sometimes they're long. Sometimes the changes are small; sometimes they are catastrophic.

638 **Specific imbalances inside a star usually lead to specific changes on the outside of a star.** If the core of a star should suddenly heat up, the extra energy spreading outward and the extra pressure pushing outward will eventually reach the surface of the star. The likely result will be that the star's photosphere will be pushed outward, expanding the size of the star. In effect, the outward pressures now exceed the inward force of gravity, so the star must expand. As the surface pushes outward, however, it will also begin to cool because the escaping radiation will have more surface area to escape through. And so the star will become both larger and redder than it was before.

639 **Sometimes this process is then reversed and a once-expanding star begins to collapse.** Once some stars expand and cool down, they stay that way for a long time. In other words, they achieve a new state of equilibrium as, once again, their internal forces come into perfect balance throughout the star. But when other stars expand and cool, they effectively "overshoot the mark" and expand too far. The result is that their photospheres become too thin and too transparent, so the stars leak more radiation into space than they should to be stable. This makes the star too cool to generate enough outward pressure to counteract gravity, so the star begins to collapse. As it does, it heats up again and again it overshoots the mark. And so the star oscillates, growing larger and smaller, hotter and cooler, time and time again. And because a star's size and temperature dictate its brightness and color, these stars appear to get brighter and dimmer and change color as well.

640 **Some stars vary in very regular ways.** Some stars are very regular in their cycles. Their changes in brightness and color happen like clockwork again and again. One such star is located in the northern constellation of Cepheus, the King, and is known as

Delta Cephei. Easily visible to the naked eye and well placed in the autumn for viewing, Delta Cephei goes from being a little brighter than fourth magnitude to a little fainter than fifth magnitude and back again every 5.37 days, month after month, year after year. There is a whole group of stars that behave similarly to Delta Cephei. Their periods of variability vary from as little as about 1 day to as long as about 50 days and the amount by which they vary also varies, but they all change in fundamentally the same way and so, as a class, are known as *Cepheid variables,* or simply *Cepheids.*

641 **Polaris is also a Cepheid variable.** You may have never noticed that Polaris, the North Star, is a variable star because it varies very little in brightness but does so like clockwork every 3.97 days.

642 **On average, the changes that Cepheid variables undergo are pretty substantial.** When a Cepheid variable star goes from smaller to larger, it changes its diameter by about 10 percent. Since these stars are somewhat larger than our Sun, this amounts to about a quarter of a million miles. In the process, the surface temperature of the average Cepheid goes from about 9,000°F to about 11,000°F. This, however, doesn't result in much of a change in color. Most Cepheids go from being almost white to yellow and back again.

643 **The changes in brightness and size are somewhat out of sync in**

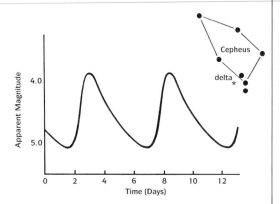

The light curve of the variable star Delta Cephei. (The insert shows the location of the star within the constellation.)

Cepheids. Ironically, Cepheids reach their maximum brightness when they are on their way from being smallest to being largest, not when they are largest. Similarly, they are faintest when they are in the process of contracting. This lack of synchronization is primarily due to the fact that it takes time for changes that are taking place inside a star to travel to the surface of a star for us to see.

644 **Some other stars are very similar to Cepheids.** These stars have periods of variability similar to Cepheids and physically function in about the same way, but, on average, they are about one and a half magnitudes fainter than classical Cepheids. As a group, they are known as *W Virginis stars,* named after the first of their kind to be discovered. The reason for the difference has to do with where they are found in the galaxy. Our galaxy consists of a large disk of stars surrounded by a spherical halo of stars. Classical Cepheids are typically found in the disk, while W Virginis stars are found in the halo. Beyond that, the primary difference between Cepheids and W Virginis stars is one of chemistry. Disk stars typically have higher concentrations of metals and other heavy elements than halo stars and this can sometimes affect how the different stars behave and evolve.

645 **The variable stars in another well-known class aren't as bright but pulsate more quickly.** These stars, known as *RR Lyrae variables,* are named after the first of their kind to be found. These stars are typically located in large clusters of stars in the galactic halo, but the number in any particular cluster can vary from none to a couple hundred. Their periods are usually less than a day and the amount by which they vary in brightness is also quite small.

646 **Cepheids, W Virginis stars, and RR Lyrae stars have become powerful tools for astronomers because of one very important thing they have in common.** Back in 1912, an astronomer at Harvard University named Henrietta Leavitt discovered a curious fact. She was studying Cepheid variables in a nearby galaxy known as the Small Magellanic Cloud (SMC). Leavitt noticed that the *brighter* a Cepheid variable appeared, the *longer* it took to vary from bright to faint to bright again. Since all the Cepheids in the Small Magellanic Cloud were at approximately the *same distance* from us, the *apparent* magnitudes of these stars were proportional to the *actual* brightnesses, so Leavitt knew that the relationship she was observing was real. In time, similar *Period-Luminosity Relationships* were found to be true for W Virginis and RR Lyrae stars as well. If you measured the period of the star, you could automatically deduce the absolute magnitude of the star. By then observing the star's average *apparent* brightness and combining this with its *absolute* brightness, you could calculate the *distance* to the star and hence to the cluster or galaxy the star was in. (If you know how bright a light looks and how bright it really is, you can calculate how far away it must be.) A new yardstick to the universe had been handed to astronomers—one that would allow them to measure distances out to tens of millions of light-years.

647 **Astronomers use a simple system to name variable stars.** In general, the first variable star discovered in a particular constellation is given the letter R followed by the Latin possessive of the name of the constellation. Thus, the first variable star to be discovered in the constellation Orion, the Hunter, was called R Orionis, while the first variable star to be discovered in Lyra, the Harp, was called R Lyrae. The second variable star to be discovered in a constellation is given the letter designation S, the third is given T, and so on to Z. After Z, the next variable star is called RR, as in RR Lyrae, followed by RS down to RZ, then SS down to SZ, and so on to ZZ. After ZZ, the sequence goes back to AA down to AZ, then BB down to BZ, and so on to QZ (skipping the letter J, just to be quirky). By now, if you've been counting, you need to get a life, but you also know we have gone through 334 letter combinations. If still *more* variable stars lurk within a constellation, they are simply designated V335, V336, and so on. (And wouldn't it have been nice if someone had just started with V1 in the first place!)

648 **In addition to Cepheids and their stellar cousins, there are other stars appropriately called *long-period* variables.** These stars typically have periods that range from about three months to two years. Mira ("the wonderful one") is part of this general class. While the changes in brightness of these stars is typically quite slow, they can also be very dramatic, going from being easily visible to the naked eye to being visible only with the help of binoculars or even a telescope. As with Cepheid-type variables in general, the longer the period of one of these stars, the greater is its average brightness. The relationship, however, isn't nearly as strict.

649 **Some stars vary in ways that aren't nearly as well behaved.** Such stars are known as *erratic, eruptive,* and *cataclysmic* variables. As the labels imply, they are prone to irregular and sometimes even violent outbursts.

650 **Among the erratic types we find the *semiregulars* and the downright *irregulars*.** As the names suggest, semiregulars are somewhat predictable in their behavior, while the irregulars seem to play by their own sets of rules. Of course, they don't—that's the really nice thing about physics—but in these cases, we either don't really understand what's going on yet or the phenomena that give rise to the star's changes in brightness are not themselves periodic.

651 **A phenomenon known as *mass loss* causes the variations in some erratic variables.** A good example of this can be found in a star known as R Coronae Borealis. Normally bright enough to see on a clear dark night, R Coronae Borealis will suddenly and dramatically fade to magnitude 12 or fainter (only a five-hundredth as bright) and then, slowly and erratically, work its way back up to its original brightness over the course of the next several months. In this case, the star's seemingly strange behavior is due to a process called mass loss. Simply put, R Coronae Borealis, a large cool star, is slowly shedding its outermost layers into space. R Coronae Borealis is so cool that much of this material is actually in the form of tiny grains of carbon, or soot. The soot builds up for a while around the star, thus dimming its light until the pressure of the radiation from the star finally pushes it out of the way, allowing the star's light to again shine forth.

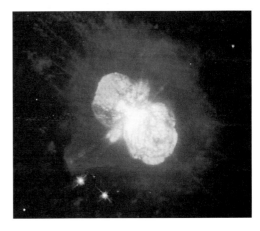

The massive supergiant star Eta Carinae is seen ejecting huge clouds of gas and dust from its atmosphere in a possible precursor stage to becoming a supernova. (NASA/STScI)

652 **Another erratic mass loss object is a spectacular star known as Eta Carinae.** Located in the southern constellation of Carina, the Keel (named for the keel on the *Argo,* the ship sailed by Jason and the Argonauts), lies the erratic object known as Eta Carinae. While Eta Carinae was the second-brightest star in the sky in 1848, it was too faint to see with the naked eye by 1880. The star is barely visible today. Recently, the Hubble Space Telescope zeroed in for a truly remarkable image of huge billowing clouds of gas and dust being blown from a supergiant star that's radiating 5 million times as much power as our Sun. With as much as 100 times the mass of our Sun, Eta Carinae is one of the most massive stars in the galaxy.

653 **Tiny stars can sometimes do tricks, too.** Some red dwarf stars occasionally undergo significant eruptions and, when they do, are known as *flare stars.* Flares also occur on the surface of our Sun, but unlike these localized eruptions, a flare star's entire surface may briefly flash brilliantly. The mechanism involved is still not understood.

654 **While the study of variable stars, especially erratic and long-period variables, is important to astronomy, professional astronomers frequently can't study them on a regular basis.** Observing time on large professional telescopes is always at a premium. Indeed, for every ten astronomers who want time on an instrument like the Hubble Space Telescope, only one actually receives it. Since the changes in long-period and erratic variables are so long-term and quirky, it is difficult to justify large amounts of "big telescope" time on them. Yet understanding such stars is important to the overall understanding of the cosmos.

655 **The study of variable stars is one area where amateur astronomers can make a real contribution.** Since there are many more amateur astronomers than professional ones and since many variable stars are well within the range of amateur telescopes and even binoculars, well-trained amateurs can and do monitor hundreds of variable stars. The data provided help fill in the gaps on what such stars are doing over long periods of time and help alert professionals to some stellar behavior that is sufficiently unusual to warrant a closer look from a large observatory telescope or the HST.

656 **Where can interested amateurs get information on making scientific studies of variable stars?** Persons interested in carefully monitoring variable stars should contact:

The American Association of Variable
Star Observers
25 Birch Street
Cambridge, MA 02138

You don't need a large telescope (for some stars, you only need binoculars), just a serious interest and a willingness to put in the time.

657 **Eruptive and cataclysmic variables are the real fireworks factories of the universe.** As the names imply, these stars typically change brightness by at least several magnitudes over very short periods of time.

The triggering mechanism is frequently a truly violent event.

658 *Novae:* **Something old is new again.** Many of the ancient Greek philosophers held that the starry realm of the heavens was a serene place where nothing ever changed. The ancient Chinese sky watchers knew better, however, for the royal astronomers would periodically chronicle the appearance of a star in the sky where no star had been seen before. In Western tradition, these stars were called *novae* (from the Latin, meaning "new"). In time, however, it was realized that these stars were not new at all but, on the contrary, were actually old stars that had suddenly flared to become much brighter than they had previously been and, in the process, were noticed for the first time. Today, astronomers understand what makes these novae flare up. It has to do with "close but troubled relationships." More in the next chapter, but first we have to understand . . .

659 **Not all stars lead solitary lives.** To the best of our knowledge, the Sun is a single star. But about 60 percent of all stars similar in age to the Sun are members of *double* or *multiple star systems:* systems of two or more stars that orbit around each other, bound by their mutual gravitational attraction. For stars that are younger than the Sun, the percentage that are members of double or multiple star systems appears to be even higher.

660 **Only recently have astronomers been able to understand how multiple stars**

are created. For years astronomers knew that most stars did not live solitary lives, but they needed the help of the latest generation of supercomputers to figure out why. These supercomputers allow the astronomers to create mathematical simulations of what happens in regions of star birth that, for the first time, are detailed enough to really be able to glimpse this process in detail. It seems that stars form in those regions of interstellar clouds of gas and dust that are the most turbulent, where different clouds or subclouds collide. When this happens, a shock wave of compressed gas is created. But the shock wave, which is long and thin, quickly becomes unstable and breaks up into knots. Disks of matter a billion times denser than the original colliding clouds form from the knots and these, in turn, collapse into stars.

661 **Once knots and disks form in colliding clouds, several circumstances favor the development of double and multiple stars over single stars.** Because the clouds of gas and dust that collide to form stars seldom collide exactly head-on, the filaments, knots, and disks that form usually do so tumbling end over end. Sometimes when this happens, adjacent knots that are moving slowly enough fall toward each other under their mutual gravity and form a double star system. Sometimes a single disk forms but is so massive that it continues to attract more matter from its parent filament. But because the filament is tumbling, it makes the disk spin faster and faster until it breaks up into two or more disks and so becomes two or more stars that

orbit around each other. In either case, we have a natural mechanism that leads to the formation of double or multiple stars.

662 **You can see a multiple star system on most clear nights.** If you find the Big Dipper and look carefully at the middle star in the Dipper's handle, you will see that it is not a single star at all but rather two stars that lie quite close together. Indeed, both ancient Arab sky watchers and some Native American tribes considered the ability to split this pair a test of good eyesight (a kind of natural eye chart in the sky). The Arabs gave names to the pair that survive on star charts to this day. They are called Mizar and Alcor, meaning "the horse" and "the rider."

663 **A closer look at Mizar and Alcor reveals even more.** A person with keen eyesight can see that the middle star in the handle of the Big Dipper is really two stars lying close together in the sky. But turn even a small telescope on the pair and you will see that the brighter of the two (Mizar) is, in reality, two stars itself. Thus, the handle of the Dipper contains a *triple star system:* three stars that orbit around each other, bound by gravity.

664 **High in the summer sky, you can find a *quadruple sun.*** High overhead on a late summer evening, you can easily spot the bright star Vega. Nearby is a faint little parallelogram of stars that forms the "frame" of the constellation Lyra, the Harp. The star in the parallelogram that is closest to Vega is known as zeta Lyrae. Zeta Lyrae, Vega, and another star form a neat little equilateral triangle (one whose sides have equal lengths). Look carefully at this third star and, if you have really good eyesight, you will see that it is really a double star (binoculars will help). Now look at this double star in a telescope and you will see that each of these stars is really two stars. You have found the quadruple sun, or *double double,* known as epsilon Lyrae.

665 **Nearby in the sky are other very colorful double stars well worth checking out.** Not far from the double double lie a couple of double stars known for their wonderful contrasts in color. The brightest star in the constellation Hercules is known as alpha Herculis, or Ras Algethi. In a modest telescope, you can see that it is a beautiful pair of stars—one is orange; the other is blue. And in Cygnus, the Swan, we find another. The star marking the swan's head is known as beta Cygni, or Albireo, and is, in reality, a stunning pair—one is a deep bluish-purple; the other is brilliant gold.

666 **In the winter sky, we find Castor, with more solar sisters than you can count on one hand.** In the winter sky lies the bright star Castor, which marks the head of the right twin in the pair known as Gemini. If we could take a flight to Castor, we would see that the light that enters our eyes on Earth really comes from a total of six suns, all in orbit around each other. Imagine living on a planet with six suns in the sky!

667 **Some stars look like doubles, but it's just an illusion.** Sometimes we see what looks like a close double star in the sky when there is really no double star there. In such cases, the two stars just happen to lie along almost the same line of sight as seen from Earth, so they look like they are close when, in reality, they can be many light-years apart. Astronomers refer to such tricksters as *optical pairs* or *optical doubles.*

668 **So how can you tell if a pair of stars is really a couple or just pretending?** You watch and see how they move. Stars that are members of a true *binary system* are gravitationally bound to each other and so, over time, will travel in curved paths around each other. Stars that are unattached and just lie along the same line of sight will move in more or less straight lines and eventually go their separate ways.

669 **Stars that move around each other obey the very same laws of motion that the planets do as they travel around our Sun.** This was one of the most important contributions made by Sir Isaac Newton to science and is of fundamental significance, for it showed that the same laws that govern the parts of the universe right in our own "backyard" also govern parts of the universe that are far away. The universe could have been more complex . . . but it isn't. As further defined by Johannes Kepler, the closer a planet or a star is to another, the faster it will orbit that star. Just as Mercury orbits the Sun faster than the Earth and the Earth orbits faster than Pluto, so two stars that are close together orbit each other in less time than they would if they were farther apart.

670 **Astronomers can track the movements of many binary stars over time.** To track the motions of particular stars, astronomers typically take a series of images of the appropriate part of the sky over the course of years or even decades. The positions of the stars' images are carefully measured relative to each other by using a device called a *measuring engine* that can detect differences in position down to a ten-thousandth of an inch or better. The positions are fed into computers and the apparent motions of the stars in the sky are calculated. The motions of several hundred binary star pairs have been extensively measured in this way.

671 **The time it takes double stars to orbit each other can vary considerably.** Many double stars orbit each other in periods of from 25 to 100 years, while some complete an orbit around their companions in less than a decade. Stars that are physically much farther apart, of course, take longer to complete their orbits, but even if astronomers are only able to observe and measure a small fraction of the total orbit, they can usually fill in the rest, since the physical laws that govern such motion are very well understood. The pair of stars that comprise alpha Herculis take about 3,600 years to orbit each other once, while two stars that make up a

pair known as sigma 2 Ursa Majoris take almost 11,000 years to do the same. The latter are separated by a distance approximately 500 times greater than the distance between the Earth and the Sun.

672 **Many binary stars are so close to each other and/or so far away from us that they cannot be seen as separate stars in even the world's largest telescopes.** This, of course, begs the obvious question: "How, then, do we know that they are indeed binary stars?" The answer is another example of the ingenious modern detective work that *is* astronomy. And once again, the *spectroscope* comes to the rescue. As noted earlier, spectroscopes break down the light of luminous objects into their spectra of colors. By analyzing these colors and the dark lines between them, astronomers have been able to determine an incredible amount of information about objects in space, including their temperatures, their chemical compositions, and their velocities toward or away from us.

673 **Shifting spectral lines from a spectrograph can reveal a binary pair of suns.** Imagine a binary pair of stars (A and B) so close to each other and/or so far away from us that they appear to be only one star in a telescope. Pointing a telescope equipped with a spectroscope toward the point of light would, of course, yield a spectrum. But the spectrum we would see in this case would actually be a *combination* of the spectra from the two individual stars. Now picture the stars moving around each other. First star A might approach us while star B moves away. Half an orbit later, star B would approach us while star A moves away. Then the process would repeat over and over again. What the astronomer would see with his spectroscope is the lines in the spectrum of each star alternately shifting toward the red and blue ends of the spectrum as the light from each star is alternately Doppler-shifted one way and then the other. (See entry 791.) The shifting spectral lines would thus magically reveal that two stars exist here instead of only one and, at the same time, give us information on how fast they are moving around each other, the size of the orbits, and how long it takes them to complete each orbit.

674 **Studying the shifting spectra of the stars can also tell us something about the masses of the two stars.** If two stars have equal masses, they will orbit each other around a point that lies halfway between them. This point is called the *center of mass* of the system and is where the whole thing would balance like a giant baton if you could attach the two stars on the ends of a giant stick. If one star is a lot more massive than the other, the two will still orbit around the system's center of mass, but that point will be somewhere along that invisible stick closer to the more massive star. In orbiting each other, the less massive star will have to cover a lot of the ground, while the more massive star completes a much smaller orbit around the center of mass, so the massive star's spectrum will show much less Doppler shifting than that of

its lightweight companion. And so, from seeing how much shifting each star's spectrum undergoes, we can get an idea of the masses of the stars involved.

675 **The ancient Arabs called attention to a star they referred to as "the demon."** In addition to the wondrous red star they called Mira, the ancient Arabs knew of another star that also changed its brightness. Every few nights its ghostly white color would dim for a period of five hours and then brighten again. They called it Algol, "the demon." Today, we know this star does not behave strangely because it is somehow possessed. And we also know this star does not vary in brightness as do Mira, Delta Cephei, or RR Lyrae, that is, by physically growing larger and smaller. Instead, Algol is really made up of two stars—stars that don't vary in brightness but instead *eclipse* each other every few nights as seen from Earth.

676 **Stars that eclipse each other reveal more stellar secrets.** Such stars, known appropriately as *eclipsing binaries,* can give us information on the *sizes* of stars, even though all stars just look like dimensionless points of light from Earth. For example, as a smaller star in an eclipsing binary pair begins to pass in front of a bigger star, the total light from the two doesn't instantly become less but instead diminishes *gradually.* This is because the small star gradually covers up more and more of the larger star as it crawls in front of the disk of the larger star as

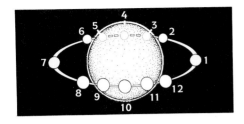

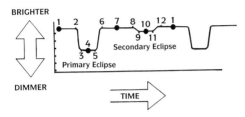

The orbit and corresponding light curve of an eclipsing binary. The surface of the smaller star is brighter, square inch for square inch, than the larger star.

seen from Earth. Then, for a while, the total light given off by the pair remains constant as the full disk of the smaller star passes in front of the disk of the larger star. Next, the total light we see from the pair will begin to increase again as the small star crawls off the face of the large star. Knowing how fast the small star is moving as it orbits the large star and knowing how long it takes to pass in front of it, we can figure out how big the large star actually is.

677 **Usually, star eclipses last for relatively short periods of time.** Such eclipses usually last from a few hours to a few days.

678 **In the northern sky, we find a mind-boggling exception to the short solar eclipse rule.** In the constellation Cepheus, the King, not far from Delta Cephei lies another variable, VV Cephei, which consists of a pair of red and white stars. As the white star passes behind the red one every couple of decades, it remains hidden in eclipse for a total of 1.2 years! From the speed of the two stars around each other, we can calculate that the red star is an astounding 2 billion miles in diameter. To put that into perspective, if you replaced the Sun with this red star, the planets Mercury, Venus, Earth, Mars, Jupiter, and Saturn would all orbit *inside* it. Appropriately, the red star in VV Cephei is known as a *supergiant*.

679 **A mystery object eclipses the supergiant Epsilon Aurigae.** Epsilon Aurigae is one of the most curious objects in the sky. It consists of a yellow-white supergiant star more than 1,000 times the size of our Sun. Yet every 27 years, this gigantic star is itself eclipsed by something so large that it takes 714 days to pass between the supergiant and us. The mysterious object is believed to be a gigantic rotating disk of gas and dust several times the size of our solar system orbiting around a pair of blue-white stars, each of which is several times the size of our Sun.

CHAPTER 11

STELLAR GERIATRICS: SUPERNOVAE, BLACK HOLES, AND MORE

680 **Some stars tell a tale by wobbling across the sky.** Astronomers have found stars that appear to do a "drunk walk" across the sky. Instead of traveling in straight lines, they weave back and forth across the heavens as they proceed on their way. It was soon suggested that the strange behavior was not due to inebriation but instead to the fact that these stars were orbiting around other objects that tugged on them gravitationally as the two traveled through space together— objects that were too faint to be easily seen— objects that, at first, were termed "dark stars."

---

681 **Early searches for such "unseen companions" led to the discovery of a whole new type of star.** In 1844, an astronomer named F. W. Bessel discovered that two bright stars in the winter sky, Sirius and Procyon, both wobbled instead of moving in a straight line and suggested the cause was unseen companions or "dark stars" in orbit around them. As it turned out, both stars did indeed have companions, but they weren't "dark stars," only stars that were so faint they had been lost in the glare of their brilliant companions. The first was discovered when a new lens for a telescope at Northwestern University was being tested in 1862. In orbit around Sirius, it was only one four-thousandths the brightness of its companion. Thirty-three years later, the companion to Procyon was finally found and it proved to be even fainter.

---

682 **Combining information on the brightness and distance to the companion of Sirius led to an amazing discovery.** Since it was known that Sirius was one of the closest stars to Earth, it was realized that its faint companion not only looked faint but that, intrinsically, it must be *very faint* indeed. Adding to the suspicion that astronomers were dealing with a whole new breed of star was the fact that this faint companion was also *white* in color. If a star is white, that means it's literally *white-hot* with a surface temperature that is far higher than that of our Sun. If this star was that hot and yet that faint, it meant only one thing—that it must also be *tiny*—only about the size of the Earth. Astronomers had discovered what came to be appropriately called a *white dwarf star*—a star no larger than a small planet.

683 When astronomers combined information on the tiny star's size with its mass, they were in for another surprise. Because the companion to Sirius was part of a binary star system, astronomers were able to calculate its mass. What they found was a star the size of Earth with a mass approximately equal to that of the Sun. Having that much mass packed into that small a space could only mean one thing: this little star was the densest thing yet known. A teaspoon of the star, if brought to Earth, would weigh a couple of tons!

684 Astronomers use a simple scheme to differentiate between the various members of such double and multiple star systems. To differentiate between the two stars that made up Sirius, astronomers simply began referring to the much brighter of the pair as Sirius A and to its faint companion as Sirius B. And the same scheme has carried over to other star systems as well, whether or not they contain white dwarf stars. Thus, the nearest star to Earth, Alpha Centauri, is really composed of three stars known as Alpha Centauri A, B, and C, with C happening to be the closest star to us at this time.

685 White dwarfs are dense because they represent the final stages in the lives of many stars. Young stars occupy real estate along the Main Sequence, while older stars move off the Main Sequence and into the Giant Branch of the H-R diagram as they begin to burn helium to create carbon in addition to burning hydrogen to create helium. In the process, the star continues to generate more energy, thus keeping itself stable against the force of gravity. But since no star has an infinite supply of nuclear fuel, eventually it must run out. When this occurs, the star is no longer *creating* energy, so the star is no longer generating the necessary pressure that has kept it stable against the force of gravity for its whole life. Gravity, of course, is still there, so the star has no alternative but to collapse. In the case of stars in the mass range of our Sun, the process is reasonably painless, but the shrinkage doesn't stop until the star is about 8,000 miles across and has a density of several tons per cubic inch. Put another way, a teaspoonful of a white dwarf, if brought to Earth, would weigh as much as a family minivan with the entire family and a lot of camping gear thrown in. For stars considerably more massive than the Sun, the collapse can be more spectacular and the end result even more mind-boggling.

686 Before entering old age, most stars have to lose weight. Different stars do this in different ways, depending on their mass in middle age and how much weight (mass) they have to lose. In some stars, this weight loss is a painless affair. In others, it's downright catastrophic.

687 Missing links between red giant stars and white dwarf stars are beautiful objects known as *planetary nebulae.* Across the galaxy a variety of beautiful objects can be found that go by the misnomer *planetary nebulae* because many look rather round, like planets, in telescopes. In reality, these objects, which look like delicate puffs and smoke rings of light, represent the evolutionary stage many stars go through between being red giants and white dwarfs. Red giants go through a "sooty" period in which they lose significant portions of their upper atmospheres, slowly leaking carbon-containing molecules into space and contributing to the interstellar dust. Over time, this dust forms a kind of halo around the star that continues to be pushed outward by the pressure of the radiation of the star at about 6 miles per second. Eventually, the star's entire outer hydrogen atmosphere escapes into this dusty halo, leaving the hot inner helium core of the star exposed. The radiation from this white-hot core creates a stronger *stellar wind* that begins to move outward from the star at over 1,000 miles per second. As the faster wind encounters the dusty halo, it acts like a giant snowplow and sets up a shock wave that we see as an expanding bubble of luminescent gas surrounding the stellar core.

688 The planetary nebula stage of a star's life is very short. In only a few tens of thousands of years (the blink of an eye in the lifetime of a star), the bubble dissipates, the nebula fades, and the core of the remaining star settles down to be a white dwarf. Despite the brevity of the planetary nebula stage, as-

tronomers have discovered many such stars. In all, over 1,000 planetary nebulae have been found thus far in our galaxy and many more have been identified in other galaxies.

689 Planetary nebulae are the psychedelic paintings of the universe. The central stars inside planetary nebulae (known as *Wolf-Rayet stars*) are very hot with surface temperatures ranging from about 60,000°F to as much as 400,000°F. Such very hot stars give off huge quantities of high-energy ultraviolet radiation that is not visible to the eye. But as this ultraviolet energy encounters the gas in the nebula, it is absorbed and reradiated as visible light (rather like a "1960s black light painting" absorbs ultraviolet rays and glows in the dark in visible light).

690 A beautiful planetary nebula can be seen in even a small telescope in the

The Ring Nebula, M57, in the constellation Lyrae, the Harp, is an example of a planetary nebula. It is visible in even modest telescopes on a summer night. (NOAO)

summer sky. High in the summer sky lies Vega, or Alpha Lyrae, the brightest star in the little constellation Lyra, the Harp. Nearby a parallelogram of faint stars forms the frame of the harp. Along the side of the harp farthest from Vega lies a little gem known as the Ring Nebula, M57. It indeed looks just like a tiny smoke ring. In large telescopes its central star can be seen, but in small telescopes it is too faint a target.

691 **Scientists recently made a striking discovery in the nearest planetary nebula to Earth.** Using the HST, astronomers recently took the most detailed images yet of a section of the Helix Nebula, which is located about 450 light-years from Earth in the constellation Aquarius. The images show thousands of tadpolelike features called *cometary knots* (because of their cometlike appearance). The curious objects are far from comets, however, for each has a head twice the size of our solar system and a tail that stretches 100 billion miles or about 1,000

A section of the Helix Nebula close up. (NASA/STScI)

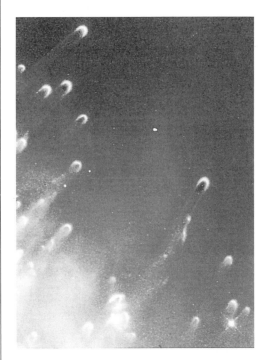

Within the Helix Nebula, the HST reveals cometary knots billions of miles long. They are the result of the luminous gas of the nebula colliding with cooler gas released by the star 10,000 years earlier. (NASA/STScI)

The Helix Nebula in Aquarius is another excellent example of a planetary nebula and the closest such nebula to Earth. (NOAO)

times the distance from the Earth to the Sun. The objects are the result of the collision between the faster-moving gas that is now escaping from the star and the slower-moving material that left the star some 10,000 years ago.

692 **A bizarre object called the Egg Nebula may be a missing link between a red giant star and a planetary nebula.** This unusual object has been dubbed a *protoplanetary nebula* (as opposed to a *protoplanetary disk* in which a protostar and a possible family of planets is developing). The Egg Nebula is one of several recently discovered that may be what planetary nebulae look like before they take on their more classic spherical shape. In addition to the odd X-shaped beams of light, however, the nebula does exhibit

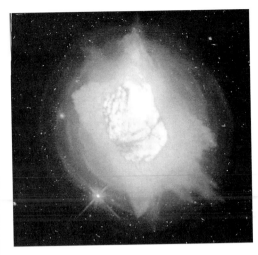

Planetary nebula NGC 7027. (NASA/STScI)

faint concentric circles of light. These may well show periodic sheddings of material by the central star (hidden behind a band of dust).

693 **If the truth be known, all white dwarfs aren't white.** Since these stars have no continuing supply of new energy, they have no choice but to slowly cool off over time and thus go from being white to being yellow and then orange and red before ultimately fading from view. The first white dwarfs that were discovered just happened to be white because their surfaces were white-hot. In time, cooler dwarfs of other colors were found, but the misnomer of *white* dwarf is still usually applied to them all.

694 **White dwarfs have their own special neighborhood on the H-R diagram.** White dwarfs are found on a line well below the Main Sequence in the H-R diagram.

The Egg Nebula is an example of what a star looks like before it develops into a full-fledged planetary nebula. (NASA/STScI)

Thus, they share the temperature range of Main Sequence stars but are much fainter because they are much smaller.

695 **Just to really confuse things, the Sun is sometimes referred to as a *dwarf star.*** When stars first started getting put on the H-R diagram and astronomers realized that some stars were a lot bigger than others, they decided to call the big ones *giants* and the little ones *dwarfs.* Hey, it seemed cute at the time. In this sense, the Sun was a dwarf because it sure wasn't a giant. But then the astronomers realized that they also had to accommodate white dwarfs like Sirius B. So, to differentiate, the Sun is sometimes referred to as a *Main Sequence dwarf* and what are usually called *white dwarfs* (which, of course, can be yellow like the Sun) are sometimes referred to as *degenerate dwarfs . . .* but not in all books. Understandably, this can all be a source of some confusion.

696 **The use of the word *degenerate* in the term *degenerate dwarfs* does not refer to any alleged deviant behavior on the part of these stars.** The term *degeneracy,* as used above, actually refers to a particular physical state of matter. As a star like the Sun slowly collapses to become a white dwarf, its atoms get squeezed closer and closer together. In effect, the nuclei of the atoms become arrayed in a regular pattern similar to a crystalline solid, while the electrons "swim" somewhat freely in a great "electron sea" within the star. As the collapse continues, however, the electrons themselves put a final stop to how small the star can get. According to a tenet known as the *Pauli Exclusion Principle,* no two identical particles, like electrons, moving around at a specific speed can occupy less than a certain minimum amount of space. This means that there is a definite limit to how close these little guys are going to get to each other before the minimum space is filled and the star's collapse must stop. When this point is reached, we say that the gas is in a *degenerate state* or has entered a *state of degeneracy.* Since the star has become a white dwarf by this point, these stars are sometimes referred to as degenerate stars or degenerate dwarfs.

697 **Some books attempt to explain the situation in a somewhat different way.** Another way of looking at this state of affairs is to think of the electrons in the white dwarf as particles that all have negative charges and remember that particles having the same kind of charge repel each other. Furthermore, the closer they get to each other, the stronger the repulsion force gets. So as the star collapses and the electrons inside get squeezed closer and closer, the force with which they repel each other gets greater and greater until it gets strong enough to counterbalance the inward force of gravity. At this point, the inward and outward forces within the star are again balanced, so the star stops collapsing. Under these conditions, the electrons act like incompressible hard spheres.

698 **Some stars are so close to others that their shapes are distorted.** Normally, stars are shaped like spheres or, if they are spinning rapidly, like somewhat "squashed" spheres (known as *oblate spheroids)* that are larger in diameter across their equators than they are from pole to pole. But if two stars in a binary pair are reasonably close, their gravitational fields can actually distort one or both stars, raising bulges on their surfaces at the closest and farthest points between them. These bulges, in effect, are tides—just like those raised by the Moon in the oceans of Earth.

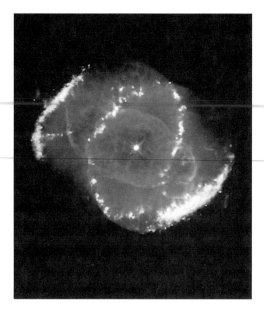

NGC 6543, also known as the Cat's Eye Nebula, is the result of a pair of stars orbiting around each other and spewing their atmospheres into space. Like sprays of water creating patterns on sand, the jets of ejected material create patterns of light as they interact with interstellar gas surrounding the star. (NASA/STScI)

699 **Some binary stars are so close that they virtually touch.** Normally, two stars may be in orbit around each other, but they remain distant enough so that they each function as separate and distinct stars. At a certain point in the lifetimes of very close binary star systems, however, this does not remain the case. As one of the two stars enters its red giant phase, its atmosphere blossoms outward until it becomes so close to its companion star, the companion star is able to begin pulling the atmosphere off the giant star and onto itself by using its own gravity. Such stellar pairs are known as *contact binaries* because they are literally in contact with each other.

700 **The passing of matter from one star to another leads to some interesting twists and turns in the lives of these stars.** The rate at which a star evolves and the type of star it becomes along the way depends on the star's mass. Contact binaries, however, break the rules on normal aging because their masses can change significantly as they grow older. If the two stars are of different initial masses, for example, the more massive star will leave the Main Sequence and become a red giant star first. But as it swells to become a red giant, its close companion star will begin to swipe its atmosphere for itself. This means that the more massive star will get less massive and the less massive star will get more massive. This, in turn, will mean that the evolution of the more massive star will slow down while the less massive star will begin to age faster. If you're a star and you want to age more slowly, you just give some of your mass to someone else. Ultimately, many such systems can evolve into a situation where you have a red giant star in close orbit around a white dwarf . . . and then the trouble really begins.

701 **Stealing your solar companion's atmosphere can lead to some nasty consequences.** As mentioned earlier, some binaries can develop "close but troubled relationships" and here's where the *troubled* part comes in. As the hydrogen from the atmosphere of a red giant star is stolen by its white dwarf companion, it collects on the companion's surface and becomes enormously compressed under the white dwarf's strong force of gravity. This accumulation process can go on for only so long, however, before the hydrogen is heated and compressed to the point of detonation. At that moment, the outer layers of the dwarf are instantly turned into one giant hydrogen bomb, which blasts them into space at enormous speed. Instantly, the tiny star lights up to become tens of thousands of times brighter than it was before. From Earth, the tiny star may become bright enough to be visible for the first time and hence look like a new star in the sky. To the ancients, such a star did indeed appear new. Today, we know better but still refer to such an event as a *nova* (from the Latin, meaning "new").

702 **Stars that "go nova" always belong to contact binary systems.** The mass transfer process described above appears to always be necessary to ultimately trigger a nova. Thus, single stars or double or multiple star systems whose members are far apart can never experience a nova.

703 **Many stars that go nova do repeat performances.** When a nova explosion is detonated, it blows the outer layers of the white dwarf into space. Indeed, in the years that follow such explosions, telescopes frequently reveal clouds of expanding debris. But the inner layers of the dwarf star are so compact they aren't really affected. As a result, the dwarf can soon set about cannibalizing its larger companion once again and, after a while, will have accumulated enough hydrogen to explode as a nova again.

704 **How often a nova explodes seems to depend on the star system involved.** Some star systems harbor what are called *dwarf novae.* These brighten unexpectedly and irregularly by a few magnitudes every few weeks or months. Here the release of energy is relatively small and may somehow occur in a disk around the dwarf star, rather than on its surface. Other novae brighten more regularly once every 50 to 100 years and are called *recurrent novae* because we have had occasion to see two or more outbursts from the same systems over the course of modern science. A *full-fledged nova,* as described above, can brighten by about twelve magnitudes or more in a day or two and may suffer repeat detonations but not more often than perhaps every few hundred thousand years.

705 **Novae are named in a pretty simple way.** When a star goes nova, it is simply referred to by the constellation in which it occurred followed by the year we saw it explode. The star V 1500 Cygni went nova in 1975 and for a brief time rivaled Deneb as the brightest star in the constellation Cygnus, the Swan. At this point, it received the new designation Nova Cygni 1975.

706 **Many more nova explosions occur in our galaxy than we see.** We have recorded about half a dozen bright full-fledged novae during this century, but we know that many more have undoubtedly occurred. They would have gone unnoticed because they lie too far away and their brightness would have been obscured by dark clouds of interstellar dust that lie along the plane of our galaxy.

707 **There are stars that make novae look like cap pistols.** In 1572, the famous Danish astronomer Tycho Brahe saw a "new" star light the heavens in the constellation Cassiopeia, the Queen. It reached magnitude −4 and rivaled Venus in brilliance. In 1604, the German astronomer Johannes Kepler witnessed a similar stellar spectacle in the constellation Ophiuchus, the Serpent Bearer. It reached magnitude −2.5, outshining Jupiter. For many years, it was believed that these two events were also examples of what were termed *novae*. In time, however, astronomers developed telescopes big enough to see stars flare in other galaxies millions of light-years away. They soon realized that these stars, as well as those seen by Tycho and Kepler, could not have been ordinary novae. Instead, on such occasions, astronomers had been witness to stellar explosions that were orders of magnitude more grand and destructive than an ordinary nova—a phenomenon that has since come to be known as a *supernova*.

708 **All supernovae are not created equal.** Different supernovae rise to different levels of brightness and stay bright for different amounts of time because the mechanism that triggers the explosion as well as the type of star involved can vary from supernova to supernova. In general, astronomers classify supernovae as *Type Ia, Type Ib, Type Ic,* and *Type II.*

709 **Type II supernovae: big stars with even bigger appetites.** A star destined to become a Type II supernova begins life as an intensely hot blue-white giant on the upper end of the Main Sequence. Such stars are among the most brilliant stars in all the universe, but they achieve their brilliance by burning their nuclear fuel at a prodigious rate. A star like our Sun has been a nice stable Main Sequence star for around 4.6 billion years and will remain one for another 5 billion years, all the while slowly burning its hydrogen to make helium and generating enough outward pressure to keep itself poised against the force of gravity. By comparison, to counteract the gravitational pull of its enormous mass, our blue-white giant has to run through most of its hydrogen supply in barely 10 million years. At this point, still converting some hydrogen to helium, the star also needs to start fusing its newly created helium supply into carbon to keep shining and poised against its own gravity. But the star finds that this ploy only works for about another million years before it has to take the next step and begin converting its carbon into magnesium, neon, and sodium to generate still more energy. But as internal temperatures soar and make each new fusion process possible, they also drive the reactions faster and faster. While hydrogen burning had gotten the job done for 10 million years and helium burning had carried the show for a million more, the star wipes out its supply of carbon in a scant 12,000 years, its neon in less than a human lifetime, and its silicon in less than a week. The star is rapidly running out of options and the consequences will be dire indeed.

710 **The consequences of being a stellar spendthrift are ugly.** A blue-white giant star continues to burn its nuclear fuel to keep itself poised against gravity. And up to a point, the scheme works, for each time the star's soaring internal temperatures transform a chemical element into a more complex one, energy is released—energy that keeps the star shining and keeps the star stable. Once the star has created the element iron, however, its fate is doomed, for the next set of reactions don't *release*

energy, they *absorb* it. Instantly, it's like turning on a giant fire extinguisher inside the star. The outward pressure suddenly drops and the outer layers of the giant star collapse inward in seconds, accelerating to nearly the speed of light. Simultaneously, the star's iron core collapses in on itself, driving central temperatures to nearly 18 billion °F, and then explodes, blasting the outer layers of the star into space at 6,000 miles per second. The star has become a supernova.

711 **The real culprits in blowing such a giant star to bits are tiny things called neutrinos.** The word *neutrino* is Italian and means "little nothing." It was coined by the nuclear physicist Enrico Fermi to describe a class of subatomic particles that have little or no mass and zip around the universe at or near the speed of light. They are created in stars as a by-product of the nuclear reactions raging within, but once created, neutrinos are also among the most "antisocial" things known. Antisocial because they very rarely interact with matter, passing right through lots and lots of it as if it wasn't even there. Our Sun generates oodles of neutrinos that constantly race out and across the solar system and right at and through you. Indeed, about 500 billion neutrinos produced by the Sun are passing through every square inch of your body every second! And it doesn't matter whether it is day or night. The subatomic assault goes on just as fiercely at night because neutrinos have no more trouble tunneling through the entire Earth and zapping you from below at night than they have raining on you from above during the day.

712 **A Type II supernova really knows how to manufacture neutrinos.** Considering the number of solar neutrinos cruising through your body in the time it takes to read this sentence, one would be inclined to say that the Sun is a pretty good neutrino factory. But in the moments leading up to the explosion of a Type II supernova, a blue-white star has our Sun beat by a very comfortable margin. Indeed, it is because of the mad exodus of neutrinos from the star—and the energy they take with them—that the star tears itself asunder. During the course of its life, as each new nuclear reaction is ignited inside the star, more and more neutrinos are created and escape. But in the final ten seconds before the star explodes, the energy loss from the star due to the loss of neutrinos grows to equal fully 30,000 times as much energy as the brilliant star is pouring into space from all its visible light, radio waves, infrared, ultraviolet, and X rays combined! It is, in effect, this monumental energy loss that acts as the star's fire extinguisher and leads to its deadly collapse.

713 **A Type II supernova is definitely over the top.** Within hours, the exploding star brightens to shine as brilliantly as several million other stars put together. During the moments of the cataclysmic detonation itself, this one star gives off more total energy than all of the other objects in the universe combined!

714 **Type I supernovae come in three varieties: Ia, Ib, and Ic.** Unlike Type II supernovae, which are the fate of very massive single stars, Type I supernovae occur in binary star systems in which either a white dwarf star or a giant companion is blown to bits. In the case of Type Ib, it is the giant star that explodes. Ironically, Type Ia supernovae, in which the dwarf explodes, are the real dazzlers of the universe. Brightening as much as thirty magnitudes over the course of a day or two, they can briefly outshine all the other 100,000 million stars in their galaxy put together! If such a star exploded 32 light-years from Earth, it would light the sky with the combined brilliance of more than 500 full moons! The supernovae seen by Tycho and Kepler were both probably Type Ia but were fortunately much farther away. Type Ic supernovae are similar to Type Ib in that the larger star in the pair is the one that detonates. In this case, however, detonation typically occurs later in life, after the dwarf has relieved its companion of almost all of its atmosphere, and the large star's iron core has grown massive enough to collapse.

715 **Different types of supernovae are found in different parts of the galaxy.** Type Ib and Type II supernovae occur in relatively young stars that have not had much time to wander far from their birthplaces and hence are confined to the disk of the galaxy. Type Ia supernovae, however, occur in much older binary stars and so are distributed in both the galactic disk and the halo.

716 **The remains of a Type II supernova can be found in the autumn sky.** In the constellation Taurus, the Bull, just above the tip of the bull's lower horn, even modest telescopes reveal the twisted wreckage of a once-great star. In 1844, its tendrillike appearance led the English astronomer Thomas Parsons (also known as the Earl of Rosse or Lord Rosse) to dub it the Crab Nebula because these features reminded him of the legs of a crustacean. The name has stuck to this day. It is also the very first object that Charles Messier put on his famous "avoidance list" while looking for comets in 1758, so it carries the codesignation M1. Messier's telescope just wasn't good enough to clue him into the fact that he had stumbled upon one of the most scientifically intriguing objects in the sky—the exploded atmosphere and core of a Type II supernova.

717 **Ancient sky watchers actually recorded the appearance of the supernova that created the Crab Nebula.** In the summer of 1054, Chinese astronomers chronicled the appearance of what they called a "guest star" in the constellation we know today as Taurus. Their records say the object became as bright as the full Moon—bright enough to be seen in broad daylight for almost a month before gradually fading. Curiously, while the dazzling sight would

have also been visible from Europe, no records of its appearance have been found there. In the American southwest, however, in the states of New Mexico and Arizona, pictographs that date to the same period have been found that depict a bright star near the crescent Moon. Using computers, modern astronomers have turned back the clock and found that the "guest star" seen by the Chinese would have been in just this position shortly before dawn on the morning of July 4, 1054.

718 **Modern telescopes reveal the Crab Nebula to be an amazing object.** What Chinese and Native American sky watchers saw nearly a thousand years ago was the explosion of a star 7,000 light-years out in space that, for a brief time, blazed with the light of 400 million stars the size of our Sun.

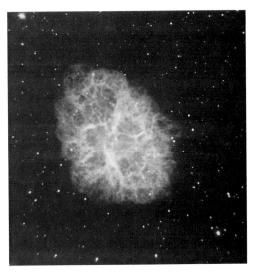

The Crab Nebula, M1, in the constellation Taurus, the Bull, is the remnant of a supernova seen in 1054. (J. Hester/Palomar Observatory)

When twentieth-century telescopes take aim at this point in the sky, they reveal an amazing twisted mass of gas and dust—the shredded atmosphere of a once-great star still racing outward at several miles per second.

719 **The twisted tendrils of the Crab Nebula may be an intriguing package, but the real prize lies deep inside.** Did the star that exploded to form the Crab Nebula totally disintegrate or did part of it somehow survive? In 1942, the German-born American astronomer Walter Baade suggested that the core of the massive star might have somehow remained intact. He pointed to a star near the center of the nebula as a possible candidate, but except for the star's location, there was little other evidence to support Baade's suggestion. Then, in the late 1960s, a remarkable series of discoveries were made . . . which takes us to Merrie Olde England.

In 1967, a British astronomy graduate student named Jocelyn Bell was working on her Ph.D. dissertation. She was using an array of radio telescopes to study how moving interstellar dust particles might make objects in space appear to twinkle in the radio spectrum the way moving currents of air in the Earth's atmosphere cause stars to appear to twinkle to the eye. In the course of her sky survey, Bell came upon a number of objects that mysteriously pulsed on and off, emitting sharp bursts of radio waves. Astronomers had found that lots of objects in space, including the Sun and the planet Jupiter, emitted radio waves in magnetic disturbances or lightning storms, but the newly discovered radio *pulses* were repeating with the accuracy of an atomic clock! No natural phenomenon in the universe was known to exist that could perform such a feat and so, for a little while, astronomers only half-jokingly began to catalog these objects as LGM 1, LGM 2, and so on, with LGM standing for "Little Green Men." Serendipitously, Jocelyn Bell had apparently discovered a whole new breed of object in space—something that promised to be truly strange indeed. And while it was quickly assumed that lots of little green men all over the sky were probably *not* trying to signal us, the mystery of the pulsing radio beams with the accuracy of atomic clocks remained. Observational astronomers had stumbled on something that the theoreticians were now challenged to explain.

720 **The pulsing sources of radio waves didn't turn out to be little green men signaling Earth but something almost as mind-boggling: *neutron stars.*** Astronomers were well aware of variable stars that slowly got brighter and dimmer over the course of weeks, days, or even hours. Jocelyn Bell's objects were a different matter. They appeared to flash on and off and did so in a few seconds or less. Soon a theory evolved that has held and been confirmed by observation. To understand, we must return to the fiery moment when a supernova detonates. At this moment, the star has a core of pure iron about the size of the Earth but with a mass greater than that of the Sun. As neutrinos flood outward, robbing the core of energy and thus cooling it, the core collapses. It is already the size of a white dwarf star but far more massive. And so, its own gravity continues the collapse past the white dwarf stage, past the point where the star would be held up by the repulsive force of its own electrons. Instead, the core continues to collapse until its electrons and protons are squeezed together and the entire star becomes a ball of neutrons less than 10 miles across—an object astronomers appropriately decided to call a neutron star.

721 **To come to grips with a neutron star, we have to visit a baseball stadium.** Most things are made of *atoms.* Atoms consist of tiny things called *protons, neutrons,* and *electrons.* The protons and neutrons are crowded together in the nucleus of the atom and the electrons are orbiting around them. If you went to a baseball stadium and put a small bunch of peas on second base to represent the protons and neutrons in an atom, then the electrons would be a little swarm of gnats circling around second base at about the distance of the bleachers. In a white dwarf star, you have atoms packed so close that they are arrayed "shoulder to shoulder"—like a bunch of Yankee Stadiums placed side to side. But in each Yankee Stadium, notice that virtually all the space between second base and the bleachers is empty. (Most of an atom is empty space!) In a neutron star, the atoms are crammed so tightly together that the electrons are squeezed into the protons to make neutrons and all the neutrons, acting like incompressible hard spheres, are sitting on top of each other. In our model, this means going from a density where your Yankee Stadiums are virtually empty to a situation where the gnats in the bleachers are squeezed into some of the peas and each stadium is packed to the brim with peas. What's the result when this happens with stars in outer space? A mass greater than the mass of the Sun gets squeezed down into a ball only 10 miles across and the densities go through the roof. How high do the densities go? While a teaspoonful of a white dwarf, if brought to Earth, would weigh a couple of tons, a teaspoonful of a neutron star, if brought to Earth, would tip the scales at over 500 million tons!

722 **But what would make a neutron star pulse and why do they pulse so fast?** As the core of a massive star collapses to form a neutron star, the rate at which the core spins dramatically increases (just like the rate of spin of an ice skater increases as she pulls in her arms). By the time the core has turned into a neutron star (a matter of seconds), the 10-mile-in-diameter ball of neutrons may be spinning at the incredible rate of a dozen or more times a second! Indeed, only an object as dense as a neutron star can rotate that rapidly and not break up. At the same time, the magnetic field of the star also increases dramatically in strength. It traps charged particles from the exploding atmosphere and forces them to spiral frantically about at nearly the speed of light as they are dragged around with the spinning star itself. In so doing, these particles emit a flood of light and other forms of radiation in two narrow beams that shoot out from opposite points on the star. If the star happens to be oriented in the right way, these narrow beams of energy sweep rapidly across our line of sight here on Earth, causing the star to appear to pulse like a giant cosmic lighthouse. Just as the light-house light is always on, so too are the neutron star's beams, but we only detect them as sudden bursts of radiation at the moments when the beams sweep past the Earth. Neutron stars that appeared to pulse in this way were soon dubbed *pulsars.*

723 **A pulsar was soon found inside the Crab Nebula and it held a surprise**

of its own. In 1968, inspired by Jocelyn Bell's discovery and the models of pulsars (a.k.a. neutron stars) conjured up by the theoreticians, observational astronomers had a closer look at that little star Walter Baade had pointed out back in 1942. Baade's star was indeed pulsing away at the remarkable rate of thirty times a second. And surprisingly, the star not only appeared to be flashing on and off in radio waves but was doing so in *visible light* as well like some kind of hyperactive disco strobe light. The reason that no one had noticed the flashing before was that its rate of blinking was simply too fast for the human eye to catch, so to the eye, as well as

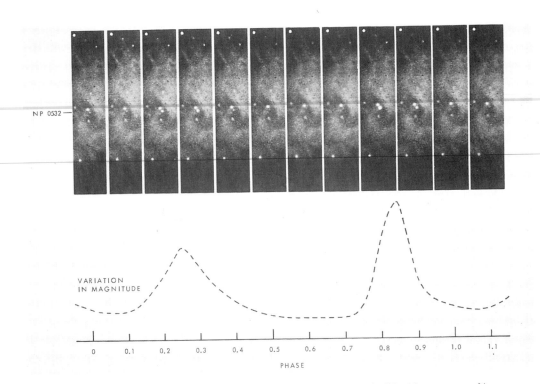

The pulsar at the center of the Crab Nebula is actually seen flashing on and off in this sequence of images. This neutron star is barely 10 miles across and spins at the rate of 30 times a second. (NOAO)

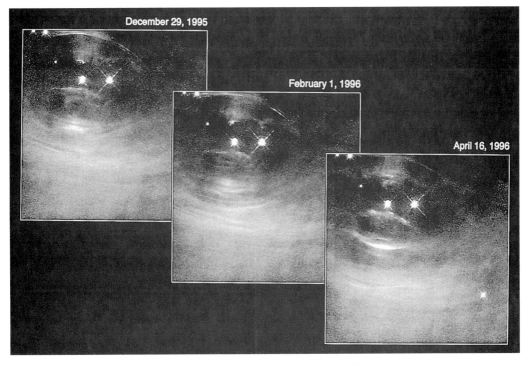

December 29, 1995

February 1, 1996

April 16, 1996

The central area of the Crab Nebula (shown in the rectangle) was repeatedly scanned by the Hubble Space Telescope in late 1995 and early 1996 to produce the sequence of images shown at the right. The right star of the bright pair in each image is the pulsar at the center of the nebula. Below it, bright wisps of light can be seen propagating out through the nebula. (NASA/STScI)

in pictures taken with time exposures, the star appeared to be "always on."

---

724 **The pulsar in the Crab Nebula is the nebula's powerhouse.** As the hulking mass of the pulsar spins at a furious rate, the star carries its intense magnetic field around with it and, in so doing, acts as nothing less than a monstrous electrical generator. Electrons and other charged particles are trapped in these magnetic fields and shot out through the nebula at half the speed of light. The power thus released keeps the nebula expanding (it's already larger than 600 solar sys-

tems) and lights it up with the combined energy of 75,000 stars the size of our Sun! In radio waves alone, the Crab is a remarkable source of energy. Indeed, if our eyes were sensitive to radio waves instead of visible, the Crab Nebula would be the brightest object in the nighttime sky.

---

725 **Recently, astronomers have discovered remarkable rapid changes in the Crab that further reveal its dynamic nature.** In 1996, using the Hubble Space Telescope, astronomers recorded dramatic changes in the nebula that occur in time spans of days or even hours—bright wisplike features that are seen to propagate outward through the nebula at half the speed of light and look somewhat like an endless series of waves rolling on to a beach. The cause is again the powerful pulsar lying deep in the nebula's heart. As it wildly whips trillions upon trillions of charged particles out through the nebula, these particles crash into the escaping tendrils of gas and cause them to light up like neon signs.

726 **Like many of us, the pulsar in the Crab Nebula is slowing down.** Its pulsar is the powerhouse of the Crab Nebula, but what is recharging the powerhouse? The answer? Nothing. Day by day, second by second, the pulsar is losing energy as it taps into its reserves to expand and light up the gigantic nebula. The energy comes from the star actually draining the power stored in its own rapid rate of spin. And so, over time, just as a spinning top slows down as it loses energy, the pulsar also slows down. We know it happens because we can simply watch as the flashing rate of the Crab Nebula pulsar (and that of other pulsars) slows down little by little, year after year. The change isn't much (about 0.0001 percent per day), but it's measurable with modern instruments.

727 **Not all neutron stars appear to pulse.** A neutron star's beams of energy can only be detected on Earth if these beams of energy happen to sweep past the Earth as the star rotates. Since these beams are very narrow, there are many pulsars that go undetected because they happen to be oriented in directions that cause their beams to sweep past other stars but not our Sun.

728 **Over the years, many pulsars have been discovered.** Over 500 pulsars have been discovered to date and more continue to be discovered all the time.

729 **The Crab Nebula pulsar is the youngest pulsar known.** Because this pulsar is so young, its rate of spin is still very fast. Indeed, it is the high energy of this pulsar that allows it to radiate X rays and visible light as well as radio waves. Older, slower spinning pulsars only tend to emit radio waves because they have less energy. In time, this will also become true for the Crab pulsar and millions of years from now, drained of energy, it will fade from view.

730 **White dwarfs are like crystals. Neutron stars are like drops.** Calculations indicate that the interior of a white dwarf star resembles the way atoms or parts thereof are assembled in a crystal, that is, in a very regular geometrical pattern. Ironically, while the interiors of neutron stars are hundreds of millions of times denser, their interiors act like a frictionless fluid known as a *superfluid*. Thus, except for a hard exterior crust, the entire star behaves as if it is one giant drop.

731 Occasionally, "glitches" occur in the pulse rates of some pulsars. These occur when the hard crust of the neutron star occasionally cracks and reseals as waves randomly slosh around the interior.

732 Gravity on a neutron star is a very pressing matter. Because a neutron star has so much mass squeezed down into so small a space, the resultant gravity at its surface can be more than 300,000 times stronger than it is on the surface of the Earth. Indeed, if you could somehow withstand the temperature, your sojourn on a neutron star would nevertheless be very short because you would be instantly spread out into a thin film only an atom thick—a veritable paint spot of your former self.

733 In addition to the Crab Nebula, the remnants of other supernovae can also be found in the sky. In the summer constellation of Cygnus, the Swan, lies the remains of an ancient supernova. Noting its long graceful form, astronomers have named it the Veil Nebula. In the southern hemisphere, the constellation Vela, the Sail, serves as the backdrop for another supernova remnant and, in Cassiopeia, radio telescopes have ferreted out two more (one of which is what remains of the star that Tycho saw explode back in 1572). Indeed, the entire sky is literally strewn with the debris of old supernovae and our galaxy is catacombed with spaces cleared by the blast waves of these exploding stars.

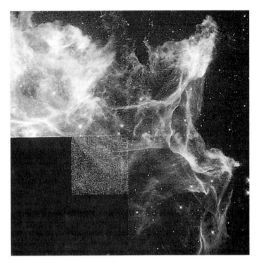

The twisted tapestry of gas and dust known as the Cygnus Loop shows a region of space where supersonic blast waves from a supernova explosion race through the interstellar medium and act as catalysts for the creation of new stars and possibly new planets. (NASA/STScI)

734 Not all supernova remnants appear to have pulsars lurking within. Type Ia and Type Ic supernovae may actually result in the total—or almost total—annihilation of the exploding star and hence no mass great enough to become a neutron star remains. Type Ib and Type II supernovae can result in the formation of pulsars, but if they happen to be oriented in such a way that their radiation beams don't sweep past Earth, they can't signal us, so we don't notice them, even though they're out there. Still other clouds of stellar debris may have "outlived" their pulsars, which are now so old they have stopped pulsing.

735 In 1987, the brightest, closest supernova since the invention of the telescope blew its top. About midnight on

Wispy clouds of stellar debris mark the remains of a star that exploded as a supernova in the southern constellation of Vela about 12,000 years ago. Near the center of the debris, which spans 6° of the sky, lies a pulsar. (©Anglo-Australian Observatory/Royal Observatory, Edinburgh. Photo made from U.K. Schmidt plates by David Malin.)

February 23, 1987, Oscar Duhalde, a night assistant at the Las Campanas Observatory in Chile, was taking a break. Stepping outside the observatory, he casually looked up into the sky and noticed a star near the Tarantula Nebula that he had never seen before. Returning to work, he forgot to mention it to the astronomers on duty. A few hours later, astronomer Ian Shelton was developing a picture of the same region of the sky that he had taken with a nearby telescope at the same time Duhalde was on his break. As he pulled the picture from the developing tank, Shelton looked down and saw the image of the star. Thinking it might just be a defect in the film, he ran outside to convince himself the star was real and then hurried to the neighboring observatory to tell Duhalde and the others on duty. Both men thought they had likely discovered a nova. But the star was in the same area of the sky occupied by the Large Magellanic Cloud (LMC), a small galaxy in orbit around the Milky Way, 160,000 light-years away. The astronomers on duty quickly realized that if the newly sighted star was in the Large Magellanic Cloud, it was much too far away to be as bright as it was and still be an ordinary nova. Soon the distance was confirmed. Indeed, Duhalde and Shelton had stumbled upon the closest, brightest supernova in nearly four hundred years—the first since the invention of the telescope.

---

736 Naming supernovae is very straightforward. When supernovae are discovered, they are simply given the designation SN (for "SuperNova") followed by the year of their outburst and by successive letters of the alphabet. Thus, the super nova of February 23, 1987, being the first one spotted that year, is known as SN 1987A.

---

737 SN 1987A drew immediate attention from the worldwide astronomical community and allowed astronomers to confirm that supernovae are indeed the alchemists of the universe. Until the outburst of SN 1987A, many of the "facts" frequently quoted about such exploding stars were really little more than theory because astronomers lacked the necessary data to confirm them as well-established science. SN 1987A, however, was sufficiently bright and so closely studied it allowed astronomers to change much of what really had been speculation into accepted fact. Among the most significant findings was the confirmation of a

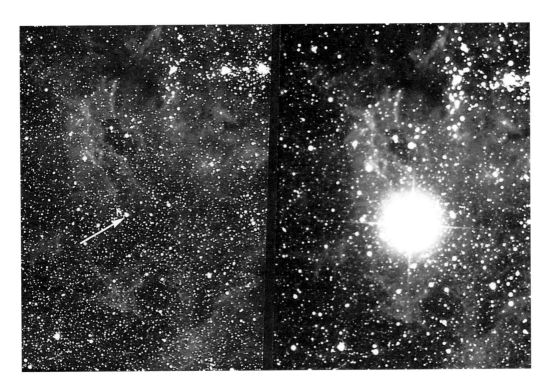

"Before" and "during" images of the region in the Large Magellanic Cloud where the supernova SN 1987A exploded. SN 1987A became the brightest, closest supernova since the invention of the telescope. (©Anglo-Australian Observatory. Photography by David Malin.)

STELLAR GERIATRICS:
SUPERNOVAE, BLACK HOLES, AND MORE

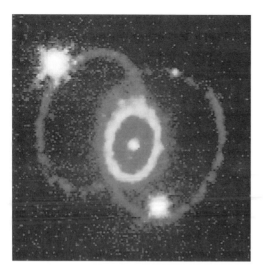

Mysterious ring structures around SN 1987A.
(NASA/STScI)

certain isotope of the element cobalt in the exploding star's spectrum—an isotope that has a half-life of only a few months. This observation proved conclusively that heavy elements are indeed created during a supernova explosion and hurled into space.

738 **SN 1987A also confirmed a theory about the role of neutrinos in supernova explosions.** Another theory had suggested that it was a flood of neutrinos from deep inside the star that led directly to the collapse of the star's iron core, which, in turn, led to the supernova explosion. Using special *neutrino detectors,* scientists detected a blast of neutrinos passing through the Earth at the very moment of the supernova's first appearance, thus confirming the invisible but highly significant activity in the very heart of the distant star. In so doing, scientists literally witnessed the dawn of the science of *extragalactic neutrino astronomy,* that is, the study of neutrinos from beyond our galaxy.

739 **Some pulsars are part of double star systems.** At first you might think a star that goes supernova would utterly destroy any other nearby star, but in recent years astronomers have found a number of neutron stars and pulsars that are members of close binary systems. In many cases, they call attention to themselves not because their beams of radiation sweep past Earth but because of a violent interchange between the two stars. In the process, the neutron star continues to draw matter from its companion. The matter first spirals around the neutron star in a disk and finally crashes down onto the star's surface, where it accumulates until it periodically explodes in a brilliant burst of X rays. Astronomers appropriately refer to such objects as *X-ray bursters.*

740 **Pulsars in binary systems provide a unique opportunity to test some of the predictions of Einstein's General Theory of Relativity.** One prediction of the theory is that the way the orbits of the stars are pointed relative to the Earth should change with time by a specific amount. Another is that the two stars should gradually be drawing closer to each other because they are losing energy. Both predictions seem to be verified by the observation of the pulse rates and orbital motion of binary pulsars under study. Another verification of the theory may come someday when detectors become sensitive enough to record distortions in space called *gravity waves* that such stars are supposed to be generating. Indeed, it is this energy, being radiated away by gravity waves, that would account for the stars gradually drawing closer together.

741 **Ultimately, such stars will merge and, when they do, they should give off some real "cosmic vibes."** As these super-dense stars rush to collide and merge with each other, computer simulations suggest that tremendous energy should be given off. Some would be in the form of high powered X rays and gamma rays, while more would come from huge pulses of very energetic gravity waves. Within a few years, *gravity wave detectors* on Earth should be sensitive enough to record such catastrophic events out to distances of 70 million light-years. The rate of neutron star collisions is higher than you might think. With an estimated 30,000 neutron star binaries in our galaxy alone, collisions may average one every twenty minutes! If so, future gravity wave detectors should be kept pretty busy as, one after another, the resulting distortions ripple through the Earth.

742 **Some fast-spinning pulsars have found a fountain of youth by cannibalizing their companion.** When it was first discovered, the Crab Nebula pulsar was the fastest-spinning pulsar known. But a few years ago, a pulsar was found that was rotating an incredible 885 times per second (that's once every 0.00113 seconds). And since, other such *millisecond pulsars* have

been discovered. It was soon realized that all of these really fast spinners had one thing in common: they were part of close binary star systems. From their understanding of stellar evolution, astronomers knew that such stars in binary systems all had to be relatively *old,* but a rapidly spinning pulsar (such as the one in the Crab) is a *young* pulsar. Hence, there seemed to be a paradox. In time, it was realized that the pulsars in binary systems were indeed old but were managing to masquerade as youngsters by slowly eating up their close companions. Simply put, as matter from the companion star continued to accumulate on the pulsar, it not only added mass to the pulsar but also added to the pulsar's rate of spin, thus allowing it to "dance faster" than its single contemporaries. What price deception? Some stars have no shame.

743 **An extreme example of such cannibalizing stars may be found in the case of the Black Widow pulsar.** In this case, a fast-spinning pulsar has been found orbiting a tiny object with little more than two-tenths the diameter and two-hundredths the mass of our Sun. Once this object was a formidable star, but the voracious pulsar has managed to eat away almost all of it and will likely continue to do so until there's little if anything left.

744 **Planets have been found around a pulsar.** A few years ago, astronomers announced the rather extraordinary discovery of two planets in orbit around a millisecond pulsar in the constellation Virgo, the Virgin. At first, such a star seemed an unlikely candidate for a solar system, but the discovery of the Black Widow pulsar has led some to suggest that these planets, each with about 3 times the mass of the Earth, may once have been companion stars that another Merry Widow had her way with.

745 **The existence of pulsars in binary systems allows us to determine their mass.** From studying their motions, astronomers have been able to determine that the masses of most neutron stars are in the range of 1.4 to 2 times the mass of our Sun and, in no cases so far encountered, exceed about 3 times the mass of the Sun.

746 **Supernovae are the true alchemists of the universe.** Many a medieval alchemist tried to turn base elements into gold. Alas, they were all doomed to failure for the simple reason that the flames they used to heat their concoctions were several million degrees shy of the mark. Only in the high-temperature high-pressure interiors of the stars do nuclei collide with sufficient force to be fused into more complex ones. Stars like our Sun turn hydrogen into helium and eventually even process helium into carbon. But only high-mass stars eventually muster the heat to go farther. In the final hours before detonation, a high-mass star has created all of the elements up to iron. Then, in the unimaginable heat of the supernova blast itself, all of the other elements in nature—the entire Periodic Table of Elements, in fact—is manufactured right up to uranium, the most complex element found in the universe.

747 **In dying, stars that explode as supernovae can serve as catalysts for the birth of new stars and planets.** When stars explode as supernovae, they tear themselves apart. But in the process, they also do several important things that are very constructive in terms of the future of the galaxy. First, they create all of the different elements found in the universe. Second, they blast the atoms of these elements at very high speeds out across space, where they mix with and enrich the surrounding gas and dust of interstellar space. Third, they create shock waves as their atmospheric debris slams into this surrounding, slower-moving gas and dust. The result? The shock waves compress the gas and dust in their paths and squeeze filaments of it together to the point that new stars and planets begin to form. Hence, without the explosive death of one generation of stars, the birth of the next might not take place.

748 **Without the explosion of supernovae in the ancient past, neither we nor our planet would ever have come to be.** When the universe was very young, it consisted of only three elements: hydrogen, helium, and a little lithium. Today, our world (and the rest of the universe) contains 92 elements, from hydrogen all the way up to uranium. The reason for the change, in a word, is *supernovae*—supernovae that created all of the other elements and subsequently blasted them into space. Every carbon atom in our bones, every iron atom in our blood, every oxygen atom that we draw into our lungs was forged inside the fiery heart of a star that went supernova billions of years ago and ultimately found its way into the cloud of gas and dust that formed our solar system. It might sound like the wildest science fiction, but it is literally true: We are all made of "star dust."

749 **If a supernova leaves behind a core that has more than about three times the mass of the Sun, it becomes a *black hole*.** A black hole is a region in space where gravity is so strong that nothing, not even light, can escape. You can understand black holes better by understanding that the force of gravity on the surface of any object in space depends on just two things: the size of the object and the mass of the object. Change an object's mass or its size and you change the amount of gravitational force it generates. Increasing its mass or decreasing its size will increase the gravitational force. Decreasing its mass or increasing its size will have the opposite effect. Nature usually creates black holes by crunching down an object into a sufficiently small space to create a very strong gravitational force—one so strong that nothing, not even light, can escape.

750 **How to turn the Earth into a black hole.** The good news is that this isn't really possible, but running this little "thought experiment" will help you understand black holes better. Imagine standing in an open field and throwing a ball up into the air. The faster you throw it, the higher the ball will go before ultimately returning to the Earth. Is there a speed at which you could theoretically throw the ball so that it would never return to Earth? The answer is yes, about 7 miles per second. While you will never have such power, many rockets do and indeed this is how they hurl satellites into orbit or even farther out into space. This speed is known as the Earth's *escape velocity.*

For the sake of our little "thought experiment," however, we are going to give you the strength to fling a ball upward at 7 miles a second and even faster. But before we ask you to put your pitching arm back in gear, we also are going to engage the services of a very large and powerful giant who is going to squeeze the Earth down a bit in size before you do your super-speedball trick. The giant does his job and crunches the Earth down a bit in size and again you toss a ball skyward. But this time you find that giving the ball a speed of 7 miles per second just isn't enough and the ball falls back to Earth. Thanks to the giant's action, the escape velocity of the Earth has now become about 10 miles per second. Next, the giant comes in and again squeezes the Earth down some more and this time you find that you have to hurl the ball upward at 100 miles per second to allow it to escape the Earth's gravitational pull. And so the experiment continues as the giant repeatedly squeezes the Earth's atoms into a smaller and smaller space. As he does, the Earth's escape velocity goes higher and higher—to 1,000 miles per second; 100,000 miles per second; 185,000 miles per second. Finally, you are standing on an Earth that is less than an inch across and whose escape velocity is nearly 186,000 miles per second, the speed of light. The giant enters the scene one last time and gives the Earth a final squeeze—enough to make the Earth's escape velocity greater than 186,000 miles per second, greater than the speed of light. *The Earth has been turned into a black hole.*

751 **Very massive stars *can* sometimes become black holes.** When a very massive star explodes as a supernova, its core continues to collapse. Should the mass of the core be greater than about 1.4 times the mass of the Sun but less than about 3 times the mass of the Sun, the Pauli Exclusion Principle will ultimately halt the core's collapse at the point where it becomes a neutron star. But should the mass of the core exceed about 3 times the mass of the Sun, the force of gravity, as the collapse continues, will reach a point where it exceeds the force of the outward pressure that maintains a neutron star. Once this occurs, there is no force in the universe that can halt the further collapse of the core. Within a fraction of a second, the core proceeds past the state of a neutron star and becomes a black hole.

752 **Black holes consist of two and only two things: *singularities* and *event**

horizons. All the mass that goes into forming a black hole is scrunched down into a point. Not a real small space. Not an object the size of the head of a pin. Not an object a million times smaller than the head of a pin. A point. A point with no dimensions. A point that scientists simply refer to as a singularity. A black hole's singularity is surrounded by a region in space that is totally dark. The boundary between this region and the rest of the universe is known as the black hole's event horizon.

753 **As you approach a black hole, gravity gets stronger and stronger.** Just as with any other object in space, including the Earth, the strength of a black hole's gravitational grip on you depends on how close to it you happen to be. Gravity obeys what is called an *inverse square law.* This means that if you halve the distance between you and an object, its gravitational force on you becomes four times as great. Chop the distance between you and an object in space to a third of the original distance and the gravitational grip goes up nine times. This is true for planets, stars, and black holes. Thus, as you approach a black hole, the speed you need to escape the grip keeps getting higher and higher. Finally, at the point where you reach the hole's event horizon, gravity has become so strong that the escape velocity equals the speed of light. This is why the region interior to a black hole's event horizon appears black.

754 **The event horizon of a black hole should have a sign that reads: ABAN-**

DON ALL HOPE, YE WHO ENTER HERE. The event horizon of a black hole marks the distance from the black hole's singularity where the escape velocity has reached the speed of light. Get closer to a black hole than its event horizon and you will be pulled kicking and screaming into the hole no matter what because, from here on in, you would have to travel away from the singularity at a speed greater than the speed of light in order to escape. You can't do it. Rockets can't do it. Not even light can do it. So from here on in, there's no escape.

755 **The distance between a black hole's singularity and its event horizon depends on how much mass is buried in the singularity.** The more mass that is buried in a black hole, the larger the black hole will appear; that is, the larger will be its cloak of darkness. Stars that collapse to become black holes typically have event horizons that are several miles across. Should far more mass create a black hole, the associated regions of darkness can be far larger, as we will see later.

756 **How can we detect black holes when they're so darn dark?** Outer space is black and so are black holes. So how can we hope to find them if they exist? The answer lies in not looking for the black holes themselves but instead looking for their *effects* on *other* objects. Because the gravitational pull around a black hole is so strong, it can tug on nearby objects and wrench matter from them, heating it up in the process. These are the telltale signs that astronomers have been on the lookout for in their search for black holes.

757 **There is evidence that *stellar* black holes actually exist.** In the summer constellation Cygnus, the Swan, there is a powerful source of X rays. When astronomers turn their telescopes to this point in the sky (known as Cygnus X-1, since it was the first source of celestial X rays coming from this constellation), they see only a single star. It is a blue-white supergiant, but it is in orbit around something which cannot be seen. From the motion of the supergiant star, astronomers are able to calculate that the star's unseen companion is an object that has a mass between four and sixteen times the mass of the Sun. Scientists surmise that the blue-white star is in orbit around a black hole. The object that is now the black hole was once a star in orbit around its blue-white neighbor but late in life collapsed to form the black hole. The X rays are the result of the intense gravitational grip of the black hole, which is literally tearing the atmosphere off the nearby supergiant. The matter, caught in the grip of the hole, heats to intense temperatures and emits the X rays as it spirals faster and faster, closer and closer, before finally disappearing down the hole's throat.

758 **There are other candidates for stellar black holes as well, including one called the Great Annihilator.** With a name that sounds like the title of an Arnold Schwarzenegger movie (and that probably belongs in one), the Great Annihilator lies in the constellation Sagittarius and is a powerful source of *gamma rays,* which are even more powerful than X rays. Matter seems to be streaming into the Great Annihilator from a binary star system where it is heated to something like 2 billion °F. In the process, antimatter particles called *positrons* (equivalent to electrons but with a positive charge) are produced. The positrons are then shot out of the vicinity of the black hole and, upon colliding with electrons in the surrounding gas, undergo mutual annihilation in incredible bursts of gamma rays. Other places where stars may have collapsed to form black holes exist in Cygnus; in the southern constellation Monocerous, the Unicorn; and in the Large Magellanic Cloud (LMC), a nearby galaxy. Unseen objects in each of these regions emit powerful X rays and appear to have masses in the range of three to ten times the mass of the Sun.

759 **The existence of black holes was first conceived two centuries ago.** The first scientific discussion of black holes was actually published by a French mathematician and astronomer named Pierre Laplace way back in 1796. He said that he saw no reason why nature would prevent such an object from forming. However, Laplace also recognized that his idea might never become more than a mental exercise, since the science of his day held out little hope of actually being able to discover such an object. And so the notion gathered dust until early in the twen-

tieth century when it was rekindled by a German astronomer named Karl Schwarzschild who did further work on the theory. The actual term *black hole* was coined in 1968 by an American physicist named Harold Weaver at the University of Texas. The name stuck.

760 **Gravity bends light.** One of the more fascinating predictions of Einstein's General Theory of Relativity (published in 1915) was that gravity should actually bend beams of light because any object that has mass distorts the space around itself. While this was a truly far-fetched idea at the time, a total eclipse of the Sun in 1919 provided just the right circumstances to allow other scientists to see if the theory might be true. The apparent positions of stars around the Sun at the time of the eclipse were compared to the positions of these same stars when the Sun was not in that part of the sky. Amazingly, the apparent positions of the stars were different by exactly the amount Einstein suggested would be the case if the Sun's gravitational field had indeed bent the beams of starlight as they passed close to the Sun on their way to Earth.

761 **Black holes *really* bend light.** Black holes have far stronger gravitational fields than the Sun, so they're even better at this light-bending trick. Indeed, if you were inside the event horizon of a black hole and shined a flashlight up and away from the singularity, the beam would curve back around toward the flashlight (just like a ball thrown up into the air on Earth) because nothing, not even light, can escape from such a region.

762 **If the truth be known, black holes really aren't black.** In the 1970s, Cambridge University professor Stephen Hawking showed mathematically that black holes can "leak." The concept is pretty abstract, but it is based on the fact that parts of matter and antimatter particles can randomly pop into existence and then quickly collide, annihilating each other and thus pop back out of existence. Should this happen, as it inevitably will, near the event horizon of a black hole, we could have a situation where one of the particles could cross the event horizon and be lost to the hole before it could collide with its counterpart and disappear. If this occurred, the particle that was left beyond the event horizon could escape into space and thus be "leaked from the hole." Hawking calculated that in this way, low-mass black holes could actually lose mass faster than they could gobble up new matter and so eventually evaporate out of existence in a brilliant burst of gamma radiation.

The black holes that would be most likely to do this might be termed *mini-black holes.* Such an object would have something like the mass of a mountain buried in its singularity and would not be the result of a stellar collapse. Instead, such objects would have to have somehow been created as a consequence of the fiery birth of the universe itself. Ironically, calculations indicate that they would be reaching the end of their lives

about now, but no radiation has been detected that has the right characteristics to be attributed to such evaporating mini-black holes. Thus, astronomers conclude that they probably don't exist.

763 **The Cold War gave rise to a whole new branch of astronomy: the study of gamma rays from space.** In the early 1960s, at the height of the Cold War, the United States and the Soviet Union signed the Limited Test Ban Treaty, which banned aboveground nuclear testing. Just to make sure the other side was being honest, the United States launched a series of satellites into space that were able to detect the extremely high-energy radiation called gamma rays that are released in a nuclear explosion. On July 2, 1967, one of the satellites detected an intense burst of gamma radiation. Others were to follow. But the source was not the Soviet Union. The gamma rays were somehow coming from deep space. The discovery was first made public in 1973 and thus, serendipitously, the science of *gamma-ray astronomy* was born— the study of the shortest wavelength and highest energy waves in the entire electromagnetic spectrum.

The real age of gamma-ray astronomy began in the early 1990s with the launch of the *Compton Gamma Ray Observatory (CGRO)* by the Space Shuttle. The *CGRO* is the heaviest spacecraft ever deployed by a space shuttle and joins the Hubble Space Telescope as the second in a series of Great (Space) Observatories planned by NASA. Two more—

one for the infrared and one for X-ray astronomy—will be launched.

764 **The universe as seen in gamma-ray is a chaotic, incredibly violent, and mysterious place.** The early gamma-ray spy satellites had hinted at what the gamma-ray universe was like, but the *CGRO* painted the picture in far greater detail. It has mapped hundreds of sources of gamma rays from all over the sky and, in particular, confirmed and recorded far more detail on the mysterious sources of explosive gamma rays first detected in the 1960s. These objects, called *gamma-ray bursters,* flash into existence (where, seconds before, nothing could be seen), radiate with incredible energy for anywhere from a fraction of a second to several minutes, and then, just as suddenly, disappear without a trace, never to be seen again. They occur at random moments in time and from random directions all over the sky and are not detectable with any other telescope in any other part of the spectrum.

765 **There are also *gamma-ray repeaters.*** In addition to gamma-ray bursters that seem to scream once and then forever hold their peace, satellites such as the *CGRO* have also identified gamma-ray sources that erupt over and over for days or even years and hence are called repeaters.

The nature of the gamma-ray bursters and repeaters is still a matter of considerable debate. Some astronomers believe that these sources are all quite close to us, that is, distributed in a halo around our Milky Way galaxy. Others believe that the sources lie at much greater distances out across the universe. Both sides may be partially correct. The *repeatability* of some of these objects suggests that they may be a variety of neutron star—in effect, a pulsar that can somehow switch on and off. If so, the ones we are detecting could be in and around our own galaxy. However, because of the very *random* distribution of the bursters all over the sky (a geometry that seems very independent of our galaxy), these objects may well be millions to billions of light-years away (astronomers refer to such distances as *cosmological*). If so, the amount of energy being released in some of these bursts must be mind-boggling. Potential sources could be colliding neutron stars or even colliding black holes.

766 **It's a burster. It's a pulsar. It's a *bursting pulsar.*** In December 1995, the *CGRO* discovered a new beast in the cosmic zoo. Known as GROJ1744-28, it is a neutron star that is part of a binary star system and emits both bursts *and* pulses of high-energy X rays. One explanation is that this pulsar doesn't have quite as strong a magnetic field as most. As a result, some of the matter (mostly hydrogen) from the companion star is drawn toward the pulsar and goes into creating the beams of energy that we see as pulses. But because of the weaker field, some of the hydrogen also accumulates on the neutron star's surface. It collects for a time until densities get so high that the hydrogen instantly fuses to create helium in a runaway thermonuclear explosion that races across the star's surface and gives off an intense burst of X rays. It's like coating the star with shoulder-to-shoulder hydrogen bombs. After a time, the hydrogen can again build up and another X-ray burst will ensue. The amount of energy emitted in such bursts can be formidable. In a single hour of bursting, GROJ1744-28 released 18 bursts, each of which was more powerful than the radiation from 100,000 stars the size of our Sun.

CHAPTER 12

OUR HOME STAR CITY:
THE MILKY WAY

767 We live in a vast city of stars. The Sun and all the stars we see on a clear dark night are part of a vast conglomerate of stars that are bound together by gravity. Such systems also can include planets and smaller objects—plus gas and dust—and are known as *galaxies.*

768 Piecing together a picture of our own galaxy has been a formidable task. Because our view of our galaxy is obscured in many directions by dust, putting together an overall picture of our home star city has been something like trying to construct a three-dimensional map of a huge metropolitan area from a vantage point in a single building located somewhere within the city on a dark and foggy night. The job has required the work of thousands of astronomers over several centuries who have compiled information on the distances, spacing, and motions of countless stars and the interstellar medium.

769 To confuse matters, the name of our galaxy and the name of the milky band of light that stretches across the sky on summer and winter evenings are both the same. They are both referred to as the Milky Way. But to help differentiate things a bit, the milky band of light is simply called the Milky Way, whereas our galaxy is correctly referred to as the Milky Way galaxy.

770 Until the invention of the telescope, the true nature of the Milky Way remained a mystery. To ancient cultures, the nature of the Milky Way was a matter of fanciful speculation and mythology, not science. To the ancient Greeks, it was milk from the breasts of Hera, while to the Romans, it was a trail of wheat scattered across the sky by the goddess of the harvest. Both the Vikings and the ancient Maya saw the Milky Way as the path followed by departed souls on their way to the afterlife.

In the early 1600s, the Italian astronomer Galileo became the first person in history to point a telescope toward the heavens. Scanning along the Milky Way, Galileo saw that its glow was due to countless stars, too faint to be seen as individuals by the unaided eye. With one gaze upward, Galileo had solved a centuries'-old mystery.

771 **The general shape of our galaxy can be deduced from simple observations with the naked eye.** Look casually around the sky on a summer or winter evening and you will see the Milky Way arcing across the sky as a narrow band of light that stretches from horizon to horizon. And since we see a different portion of the Milky Way in summer than we do in winter, we can correctly conclude the band of the Milky Way extends around the entire sky. And this, in turn, is precisely what we *should* expect to see, since the Milky Way galaxy is a relatively flat system of stars with us embedded inside.

772 **To better picture our place in the galaxy, imagine yourself to be a bug in a bicycle wheel.** Imagine yourself to be a bug on one of the spokes of a bicycle wheel. As you look around, you see the rim of the wheel as a narrow band that makes a circle that completely surrounds you. But if you look away from the rim, you are able to look out in both directions far beyond. In the same way, when you look along the band of the Milky Way in the sky, you see an impenetrable band of countless stars. But when you look away from this band, you see far fewer stars as you look out into intergalactic space. Looking along the plane of the Milky Way and then far away from it in the sky with a pair of binoculars will make this very apparent.

773 **Until early in the twentieth century, most astronomers mistakenly placed**

The band of the Milky Way as seen from Earth in visible light. Note the intervening dark areas within the Milky Way due to obscuring dust. (Lund Observatory)

OUR HOME STAR CITY:
THE MILKY WAY

our solar system close to the center of the **Milky Way.** The astronomer who discovered the planet Uranus also was the first to try to make a map of our galaxy. By actually counting stars in many different directions in the sky and trying to estimate their distances, the English astronomer William Herschel made an early map of the Milky Way back in the late 1700s. It placed us and our Sun near the center of a flat, roughly disk-shaped grouping of stars.

The idea that we were located near the center of a large, roughly lens- or grindstone-shaped assembly of stars remained in vogue for quite a long time. In 1918, however, Harvard University astronomer Harlow Shapley took special notice of the positions in the sky of dozens of large star clusters called *globulars,* each of which contained tens of thousands to hundreds of thousands of stars. While the band of the Milky Way extended completely and quite uniformly around the sky like a belt dividing two great hemispheres, the globular clusters had anything but a uniform distribution. Instead, most of them were clustered around the summer sky in the general direction of the constellation Sagittarius, the Archer. This meant that either we were in the center of the galaxy and all the globular clusters were down on one end or that this vast system of enormous star clusters orbited around our galaxy's center and we— and our garden-variety Sun—were in fact the ones that weren't in the center of the galaxy. Shapely argued and ultimately proved that the latter was indeed the case.

774 Careful measurements revealed just how far from the center of our galaxy we were. It was one thing to note that the distribution of the globular star clusters implied that we weren't at the center of the galaxy and another to determine just how far off-center we were. Fortunately, the globulars contained many RR Lyrae variable stars. From measuring their periods of oscillation and their apparent magnitudes, astronomers were able to measure the distances to the globulars. The results were a bit of a shocker. In one stroke, we were catapulted from the very center of galactic real estate to a place far removed from the galactic core. No longer could we smugly think of ourselves as being in the center of the action. Instead, it was determined that our Sun and its planets were about 30,000 light-years from the center of the galaxy. Far from center city, we actually had been living in the galactic boondocks all along.

775 Even with big optical telescopes, we don't come close to seeing all the way down to the center of our galaxy. While we can gaze in the direction of the center of our galaxy with our unaided eyes and see stars with binoculars that only appear as a glow to the eye, even powerful optical telescopes only allow us to penetrate a relatively short distance toward the galactic core. The reason is the same interstellar dust that creates the dark patches and rifts along the Milky Way. Indeed, within the plane of the galaxy in any direction we choose to look, we can typically see only a few thousand light-years.

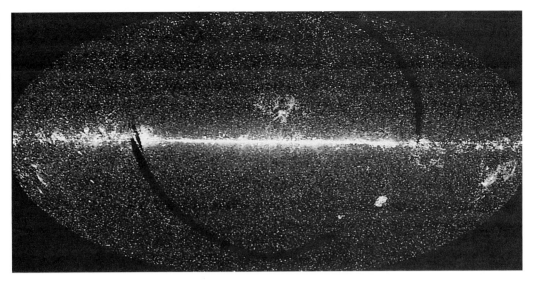

The band of the Milky Way seen in infrared. Note that the dark patches seen in visible light in the previous image are missing. This is because either the clouds of obscuring dust are heated in the infrared by more distant stars or we are seeing through these obscuring clouds to the stars beyond. (NASA/JPL/IPAC)

For this reason, mapping the galaxy using an optical telescope was a very difficult task. Indeed, it was very much like trying to map your location in an unfamiliar city on a foggy night. You may be near the edge of the city, but if you look around and see tall buildings in all directions, you might easily conclude that you are near the center of the city, even though you are not.

776 **Using infrared and radio telescopes, astronomers were able to cut through the dust as if it wasn't there.** The dust between the stars dims starlight but does so in a peculiar way. The dust lets more of the light from the red end of a star's spectrum pass through than from the blue end. The reason has to do with the size of the grains that make up the dust, which, on average, are closer in size to the wavelength

of blue light. Since infrared and radio radiation have wavelengths even longer than red light, astronomers reasoned that they should pass through the interstellar dust even more easily. When infrared and radio telescopes

The *IRAS* (*Infrared Astronomy Satellite*) spacecraft. (NASA/JPL)

were developed, astronomers were proved right. Such instruments were able to receive radiation from much greater distances away and thus allowed astronomers to effectively see through the intervening dust as if it wasn't there.

777 **Astronomers have used radio telescopes to map the size and structure of our Milky Way galaxy.** Using radio telescopes, astronomers have been able to cut through virtually all of the obscuring dust in the plane of our galaxy. In so doing, they recorded radio waves from across the galaxy that are naturally emitted by hydrogen and other gases between the stars. An image created from tracing hydrogen atoms in this way clearly shows that our galaxy has a series of arms that spiral out from its center.

Spiral galaxy M81 in the constellation Ursa Major, the Great Bear, looks very much like what astronomers think our Milky Way would look like if viewed from afar. (NOAO)

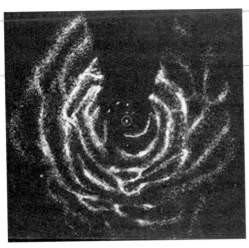

A map of our Milky Way galaxy made from radio waves emitted by hydrogen gas between the stars. These radio waves penetrate interstellar dust over great distances and allow astronomers to detect the spiral structure of our galaxy. (Gart Westerhout)

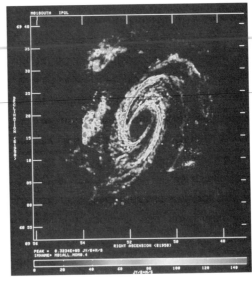

An image of M81 in radio waves clearly reveals its spiral arms. It is in such regions that the rich supply of hydrogen gas leads to the formation of new stars and, in places, new families of planets. (NRAO/Adler/Westpfahl)

If seen edge-on from a distance, our Milky Way galaxy would probably look like this spiral galaxy known as NGC 4565. Notice its thin disk, its small central bulge, and the dark lane of obscuring dust that runs along its disk. (NOAO)

Indeed, if seen from outside, our Milky Way galaxy would look like a giant pinwheel.

778 **The Milky Way galaxy is flat but very large.** How big is our "Milky Way pinwheel"? If you could travel at the speed of light (fast enough to orbit the Earth seven times in one second) it would take you approximately 100,000 years to fly clear across it from one side to the other. In other words, our galaxy is about 100,000 light-years across. Yet if seen edge-on, our galaxy would appear to be very flat (like a pancake) but with a little bulge near its center. This flat *disk* of our galaxy is a few thousand light-years' thick, while the central bulge is perhaps 10,000 light-years in diameter.

779 **The Earth appears to be located on an offshoot of one of our galaxy's spiral arms.** The Milky Way galaxy appears to have several rather complete spiral arms plus a few offshoots or pieces of arms called

spurs. Our Sun and its family of planets appears to be nestled on the inner edge of one of these spurs usually referred to as the Orion Spur or the Spur off the Orion Arm of the galaxy.

780 **When you look along the Milky Way in the sky, you are actually looking out in the direction of different spiral arms of our galaxy.** Gaze up into the winter sky at the stars in and around the constellation of Orion, the Hunter, and you are looking into the nearby Orion Arm of our galaxy. Higher along the Milky Way in the direction of the constellation Perseus, the Hero, you are looking across space toward the spiral arm that lies farther from the center of the galaxy than our own—an arm appropriately called the Perseus Arm. And half a year later when you look toward the Milky Way in the direction of Sagittarius, you are looking toward a spiral arm that lies between us and the galactic core—the Sagittarius Arm.

781 **Astronomers also find certain types of stars somewhat useful in trying to figure out what our galaxy would look like from the outside.** While dust greatly diminishes the light of stars and prevents us from seeing them over great distances, the very brightest O and B stars in the galaxy would naturally stand the best chance of being seen over the greatest distances. These blue-white supergiant stars, hundreds of thousands of times as luminous as our Sun, are the real "beacons of the universe." So astronomers have used these stars to also get a handle on the size and shape of our galaxy. Because such stars evolve very quickly, the ones we see are still very young and delineate the current positions of a galaxy's spiral arms very well. Indeed, it is the blue-white supergiants, strung out like pearls, that light up the spiral arms.

782 **Because of all the dust in our galaxy, you can see a lot farther in some directions at night than you can in others.** As we look around the plane of our galaxy, that is, along the band of the Milky Way in the sky, we can typically see out only a few thousand light-years in any direction. Put into perspective, this means that if the Milky Way were a couple of feet across, most of the stars we see with the eye on a clear dark night could easily be covered by your thumb. Because the galaxy is very thin, however, there is little obscuring dust in those directions in the sky that lie far away from the band of the Milky Way (for example, up in the general direction of the Big Dipper). So when we look in these directions, we can actually see much farther out into space.

783 **Why don't we see more stars in the direction of the Big Dipper than we do along the Milky Way?** The reason is because the galaxy *is* very thin, so when we look away from the plane of the Milky Way in the sky, we quickly look out to distances where we simply run out of stars to see. But space in these directions is far from empty, for what we do see instead are billions of other galaxies far beyond our own. Within the plane of the Milky Way, we may barely be able to see a few thousand light-years, but away from it, we can turn our gaze and look out across millions and millions of light-years. What a difference a turn of the head can make in terms of how far we can see.

784 **Radio telescopes have allowed us to peer all the way down to the galactic core.** In recent years, powerful radio telescopes have enabled astronomers to cut through almost 30,000 light-years of obscuring dust and obtain the first images of what lies near the very heart of our galaxy. They reveal huge clouds of gas moving at tremendous speeds around a strange object that appears to lie at the very center of the galaxy. Known as Sagittarius A* (pronounced "Sagittarius A star"), the object is an intense source of radiation no larger than the orbit of Jupiter. From the speed of the gas, the mass

This is an image taken in the direction of the constellations Scorpius and Sagittarius. The center of our Milky Way galaxy lies 30,000 light-years beyond. Dark clouds of interstellar dust prevent us from seeing anything near the galactic nucleus.
(©Anglo-Australian Observatory)

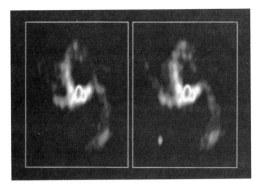

Radio telescopes such as the VLA cut through the obscuring dust in our galaxy and produce images like this of an object known as Sagittarius A*, believed to be located at the very center of our galaxy. The image shows spiraling bands of gas being drawn toward a very intense source of radiation that probably surrounds a massive black hole. (NRAO)

of the central object can be calculated and the result is nothing short of staggering, for buried within this relatively small amount of space lies a mass equivalent to 5 million times the mass of our Sun! Conclusion: Sagittarius A* is radiating tremendous amounts of en-

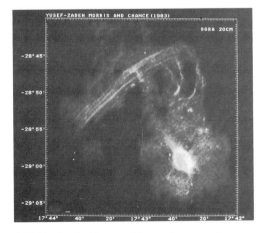

A highly detailed image of the region around Sagittarius A*. (NRAO)

ergy because the energy is coming from stars being torn asunder just before plunging down the throat of an enormous black hole.

785 **On warm summer evenings, you can gaze in the direction of the center of our galaxy.** If you look low in the southern sky around 10 P.M. on a summer evening, you will find a bright red star named Antares. Antares marks the heart of Scorpius, the scorpion whose body extends down and toward the left along a gentle curve of fainter stars. Just above the scorpion's tail lies a patch of sky deep within the Milky Way. Look in this direction and you are gazing toward the very heart of our galaxy some 30,000 light-years away.

786 **Our galaxy has a magnetic field.** Our galaxy has a weak magnetic field that is probably an extension of the individual magnetic fields of all of its stars, some of which, like neutron stars, can be millions of times stronger than the Earth's. Charged particles blown into space on stellar winds or ejected at high speeds by supernovae are directed in their motion by this galactic magnetic field, which also slightly polarizes the interstellar dust.

787 **The constellations always appear to have the same shape, but the stars all have individual motions.** Although the stars move around the sky during the course of the night, they always remain in the same positions relative to each other and hence are sometimes referred to as "fixed" stars. In reality, however, all of the stars actually have their own individual motions in three-dimensional space and some of these motions are as high as many miles per second. The reason we are not aware of these motions is simply because the stars are all very far away and so the apparent motions of individual stars in our sky are typically very, very tiny, even over the course of several human lifetimes.

Over truly long periods of time, the constellations do change their shapes. If we could magically move through time very rapidly, we would gradually see the constellations change their shapes due to the individual motions of the stars, but in most cases, it would take tens of thousands of years to see significant changes. In time, for example, the Big Dipper's bowl will become more pointed and its handle will become more bent.

788 **The apparent motion of stars in the sky is known as their *proper* motion.** The proper motion of a star is simply how much it moves in the sky in a year, that is, its motion perpendicular to our line of sight. Astronomers use *angular* measurement to express this motion in terms of *degrees* (°) in the sky. A circle has 360°, so a line in the sky that goes from the horizon, straight overhead, and down to the opposite horizon goes through one-half of a circle, or 180°. Since the apparent motion of the stars

is very small, stars have proper motions that are much smaller than a degree per year. For this reason, we talk in terms of tiny *fractions* of degrees. One degree is divided into 60 equal parts called *minutes of arc* and each minute of arc is divided, in turn, into 60 equal parts called *seconds of arc.*

789 **The "speediest" star around is a little red guy in the summer sky.** The prize for the star having the greatest proper motion of all the stars seen from Earth goes to a little red star called Barnard's Star. Its proper motion is 10.27 seconds of arc per year. To put that in perspective, the Moon is about 30 minutes of arc, or 1,800 seconds of arc, so it takes over 175 years for Barnard's Star to move a distance in our sky equal to the diameter of the Moon. The proper motions of most of the other stars we see are far, far less. And it must be remembered that Barnard's Star isn't so speedy in absolute terms, it just looks speedy to us primarily because it's so close.

790 **The proper motion of a star can have little to do with the actual motion of the star in space.** A star could be heading directly at us or directly away from us at a tremendous speed and its proper motion would be *zero* because we wouldn't see it move in the sky. Second, just like birds flying across our line of sight, closer stars, on average, would have greater apparent motions simply because they are closer. To get a handle on how stars *really* move through space, we also need to know each star's distance and how fast it's moving toward or away from us, that is, its motion in the line of sight. Combining all three bits of information, we can determine how fast the star is really moving through space.

791 **A star's spectrum can tell us whether that star is moving toward us or away from us and how fast.** By carefully watching a star from year to year, astronomers can determine its proper motion—that is, its motion perpendicular to our line of sight—by measuring how much it moves relative to much more distant stars. But how could we ever hope to determine how fast such a star is moving toward us or away from us? Again, an analysis of the spectrum comes to the rescue. But to understand this feat, we must understand a phenomenon known as the *Doppler effect.*

792 **What do moving trains and moving stars have in common?** Have you ever heard a train blow its horn as it sped past you? If you listened carefully, you heard a distinct change in pitch from higher to lower as the train went past. This change in pitch is known as the Doppler effect and is due to the fact that the train's sound waves were pushed together as the train approached and pulled apart as it moved away. This change in wavelength or frequency of the waves is heard by the ear as a

change in the pitch of the sound. Light may also be thought of as a wave. If an object that emits light (such as a star) approaches us, its light waves are pushed together, whereas if it is moving away from us, its light waves are pulled apart. Just as a change in the frequency of a *sound* wave is perceived as a change in *pitch,* so a change in the frequency of a *light* wave is perceived as a change in *color.* And so, by carefully measuring any change in the colored spectrum of a star compared to a similar spectrum of a stationary object in a lab on Earth, astronomers can actually determine if the star is moving toward or away from us and how fast. Objects moving *toward* us have their spectra shifted toward shorter wavelengths, that is, toward the *blue* end of the spectrum, while objects moving *away* from us have spectra that are shifted toward the *red* end of the spectrum. A star's velocity toward or away from us is referred to as its *radial velocity.* The greater a star's radial velocity, the more its spectrum is shifted toward the blue or the red.

793 **We can combine this data to figure out how a star is really moving through space.** If we combine information on a star's proper motion with its distance from us, we can figure out how fast a star is really moving across our line of sight. If we then combine this information with the star's radial velocity—that is, how fast it's moving toward or away from us—we know how fast and which way the star is really moving through space. By combining such data on many stars, we can get a general notion of how the stars around us move. Such studies have shown that, while many stars have individual motions, large swarms of stars can move together in streams. A general plot of such "star streaming," in turn, leads to an overall understanding of the general motion of stars out to several thousand light-years from Earth.

794 Our solar system has a general or average motion through space relative to the stars around us. If you look high up in the eastern sky on a warm early summer evening, you will see where our Sun and its family of planets are headed. The point is in the general direction of the constellation Hercules not far from the bright star Vega in the neighboring constellation of Lyra. We're moving that way at about 14 miles per second. But don't worry about bumping into Vega someday. It's moving too and *not* in our direction.

795 All the stars in our galaxy are orbiting around its center, but stars at different distances from the center orbit at different speeds. In general, stars closest to the center of our galaxy are revolving around it the fastest, but these velocities begin to drop off about 6,000 light-years from the core. Then, from about 16,000 light-years out to the galaxy's edge, they increase very slowly. Our Sun takes a long time to orbit the center of the galaxy: about 200 million years. In other

words, the last time the Sun and Earth were in the same place they are now, dinosaurs ruled the Earth.

796 The rate at which different stars orbit around the center of the galaxy presents an interesting puzzle. Our galaxy is estimated to be about 10 billion years old—a little over twice as old as our Sun and its family of planets. But if you look at how fast different stars at different distances from the center of the galaxy are orbiting around it, you discover a curious fact. The stars closer to the galactic core are moving so much faster than those farther out that the spiral arms of our galaxy should have wound up and disappeared in less time than the Sun has been around. Yet the Milky Way still has spiral arms. Why?

797 In solving the mystery, astronomers learned a lesson from traffic jams. We've all had the experience of being on a local highway where an accident has occurred. As we approach the area of the accident, the traffic slows down to a crawl. Once the accident has been passed, the traffic picks up speed again. Looking at the situation from a traffic helicopter high above, we would see that the spaces between the cars on the highway are at first quite large. Then, as the cars approach the scene of the accident, the density of cars increases dramatically and the amount of space between them decreases. Finally, after passing the accident, the spaces between the cars again increase as the density of cars decreases. From above we see individual cars flowing through a pattern of low density, followed by high density, followed by low density once again. The pattern itself doesn't move, just the cars through the pattern. Such a pattern might be called a *density wave.* In the same way, astronomers imagine a series of density waves existing in our galaxy. In this way, the stars that we see in the Milky Way's spiral arms today are like the cars that are passing the scene of the accident. These regions of maximum density in the galaxy are the places where gas and dust are being squeezed together to form new stars. In time, just as other cars will flow through the point of maximum density on the highway, so other gas and dust will flow through the various density waves that will continue to define the locations of our galaxy's spiral arms. By then, the brilliant O and B stars that delineate the spiral arms today will have gone supernova and been replaced by the next crop of O and B stars, which will then delineate the spiral arms. Just what sets up and maintains such density waves in the galaxy is still a matter of considerable debate. But observations of the way stars orbit the center of our galaxy infer the existence of considerable amounts of mysterious material in and surrounding the Milky Way that must play a role. Thus far, this material remains unidentified.

798 **Some stars are members of *associations.*** The hottest and most massive stars in the galaxy, namely the O and B su-pergiants, are typically members of loose aggregates of stars that astronomers refer to as associations. Dozens of associations are known across the spiral arms of the galaxy. Many of the bright blue-white stars in Orion are part of one such association. All of the stars in the same association were likely born in the same general region of space at about the same time but are not typically bound to each other gravitationally and so move out through space in separate directions at rather high speed.

799 **Some associations aren't limited in their membership to just blue-white supergiants.** In the summer sky, we find the constellation Scorpius, the Scorpion, and within it, an association of blue-white supergiants. But the brightest star in the constellation, Antares, is a brilliant red star. Antares's different color is due to its different mass. Simply put, Antares is more massive than its fellow association members and so it has already evolved off the Main Sequence and over into the red supergiant branch of the H-R diagram.

800 **Many stars are members of tighter knit groups called *star clusters.*** While double and multiple stars shun the solitary life, some stars are truly social beings and belong to large families called star clusters. There are two kinds: *open clusters* and *globular clusters.*

801 **Open clusters are the "jewel boxes" of the galaxy.** These are aggregates of anywhere from a few dozen to a few thousand stars. Such clusters range in size from a few light-years to about 25 light-years across. Many of the stars within can be relatively bright and come in a variety of colors, making them look like wonderful collections of jewels sprinkled against the velvety black backdrop of space.

Open clusters have their own special place in the galaxy. Because the stars in most open clusters are relatively young, these clusters are generally found within the plane of the Milky Way strung along its spiral arms. Thus, you can find many of them by simply scanning along the Milky Way using a simple pair of binoculars or a small telescope set at very low power.

802 **Some beautiful examples of open star clusters await your easy discovery.** In the skies of winter and spring lie a couple of open clusters worth finding. The first lies at the foot of the right twin in Gemini and is known as M35. A short distance in the sky to the right lies the bright star Capella. Between

The pair of open star clusters known as the Double Cluster, h & chi Persei. (NOAO)

Capella and M35 lie three additional clusters known as M36, M37, and M38. The finest is M37. In the opposite direction from Gemini lies the faint constellation of Cancer, the Crab, and another beautiful cluster known as the Praesepe. It's also called the Beehive, because it looks like a swarm of bees all buzzing around a hive. Almost overhead on a mid-December evening lie a couple of additional gems. Between the tip of the constellation Perseus, the Hero, and a W-shaped grouping of stars known as Cassiopeia, the Queen, lies a close-knit pair of open clusters appropriately called the Double Cluster, or h & chi Persei. Most of these clusters are between one and three times the diameter of the full Moon in the sky, making them spectacular in a small pair of binoculars.

803 **The most stunning open cluster in the sky can easily be seen with the naked eye.** In the autumn sky lies a beautiful open star cluster known as M45, or the Pleiades. It is located in the constellation of Taurus, the Bull, at a distance of about 450 light-years from Earth and looks a bit like a little swarm of gnats hovering over the bull's back. Just look for the three stars in Orion's belt and follow them upward and to the right to the bright star Aldebaran. The cluster is above Aldebaran and to the right. The Pleiades is also known as the Seven Sisters and with 20/20 vision you should be able to count six or seven stars in this part of the sky. With binoculars or a telescope at very low power, you will be able to see dozens more. The

Pleiades is sometimes called the "false dipper" because the stars within it that are visible to the naked eye form a tiny figure not unlike the Big Dipper.

804 Virtually every culture has taken special notice of the Pleiades. Because of its unique appearance and striking brightness, the Pleiades has become part of virtually every culture's sky mythology. To some Native North Americans, the Pleiades were seven girls who danced in the sky, while to others, they were warriors on the hunt. In Greek mythology, the Pleiades were consorts of Zeus, the king of the gods. Recognizing that the cluster returned to the same place in the sky at the same time each year, other cultures used it as a calendar. To this day, some Indonesian farmers plant and harvest their rice based on the position of the Pleiades in the sky, while Native South Americans of the high Andes use its rising at dusk as a warning that it is the time of year when killer frosts may harm the crops.

805 Another open cluster forms the face of Taurus, the Bull. Below the two stars that mark the tips of his horns, a V-shaped grouping of stars forms the face of Taurus. These stars are also part of a cluster or family known as the Hyades that is only about 150 light-years away. The stars of the Hyades are closer and older than those in the Pleiades. This accounts for the Hyades Cluster being more spread out in our sky. Also, since the Hyades stars are older, this cluster no longer contains the brilliant blue-white stars of the Pleiades. Thus, while closer, the Hyades also appears fainter.

806 Although the stars in open clusters are indeed physically related, just like people, some of them eventually leave home. While the individual stars in an open cluster each have their own motion, they move as a group through space together. This is because all the stars in an open cluster were born out of the same large cloud of gas and dust and share its original motion through space.

Eventually, some members of open cluster families move away from home. The member stars of an open cluster are all gravitationally bound to each other, but, over the course of time, random close encounters between various members can result in some of the stars winding up with a greater speed than they had before, while others wind up with less. Just as with a planet like Earth, a star cluster as a whole has a particular escape velocity and should a star's speed exceed it, the star will escape from the cluster. And so most clusters get sparser as time passes.

807 All the stars in the same cluster are about the same age. It appears that when a cloud of gas and dust in space gets ripe enough to develop stars, the birthing process happens over a relatively short period of time. Hence, while the ages of all the stars in a particular cluster are not identical, they may be thought of as being pretty much the same age in astronomical terms.

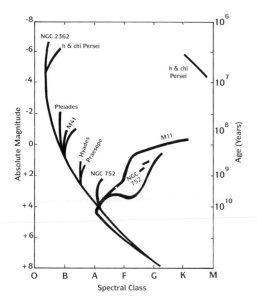

Combining the H-R diagrams of different open star clusters allows us to tell their relative ages. The older the cluster, the fewer of its stars remain on the Main Sequence. Match the point where each cluster's stars turn off the Main Sequence with the appropriate number on the right to tell that cluster's age.

808 The relative ages of different star clusters can be easily determined using H-R diagrams. Earlier in this century, astronomers found that superimposing the H-R diagrams of various open clusters was an easy way to determine their relative ages. The H-R diagram of the Pleiades, for example, shows that virtually all of its stars lie along the familiar line known as the Main Sequence. The H-R diagram of the Hyades, however, shows that a few stars at the top of the diagram have left the Main Sequence and are in the process of moving over toward the red giant branch. The combined H-R diagrams of h & chi Persei have even more stars over in the red giant region, while still another clus-

ter called M67 is missing most of the stars from the upper branch of its Main Sequence. Since the length of time a star can remain on the Main Sequence depends on its mass and since the stars at the top of the Main Sequence are the most massive, astronomers correctly concluded that the Pleiades was the youngest of all these clusters, while M67 was the oldest with the age of the other clusters listed above falling in between. In effect, the Pleiades has been around only long enough for a few of its most massive and, therefore, fastest-evolving stars to have left the Main Sequence, while in each successive cluster shown, more and more of the stars have already peeled off the Main Sequence because they have had more time to do so.

809 Combining H-R diagrams also allows us to determine the distances to various star clusters. Since the Hyades Cluster is quite close to Earth, the parallax of its stars can be measured and used to determine the distance to the cluster. But what about clusters that are too far away for the parallax technique to work? A little clever handiwork and the problem is solved. Simply put the H-R diagram of another cluster on top of that of the Hyades and match up the red stars at the lower end of the Main Sequence of both clusters. Since you know the distance to these stars in the Hyades Cluster and hence can calculate their absolute magnitudes, you automatically know that the identical red stars in the other cluster have the same absolute magnitudes. And since you can measure the apparent magnitudes of these stars, you can determine the distance to the cluster they are in. In this way, the distances to dozens and dozens of clusters can be determined based on the known distance to one.

810 **In addition to a large pinwheel-shaped disk of stars, gas, and dust, our galaxy also has a halo that opens a window into its past.** Our galaxy's halo isn't a luminous ring that you might envision encircling the head of a saintly person but rather a large number of stars that surround the galaxy in a roughly spherical distribution. Thus, if you imagine the disk of our galaxy as a pinwheel, you might imagine the halo stars as being a large swarm of bees slowly flying around it.

The halo stars of our galaxy point the way to its past. The spectra of stars in our galaxy's halo reveal that they have much fewer metal atoms in their atmospheres than do stars in our galaxy's disk, such as our Sun. Stars like our Sun got their metal atoms from earlier generations of massive stars that created metals inside themselves and then blasted these atoms into space during supernova explosions. Since the halo stars don't have many metal atoms, it must mean that they came into existence before many supernovas had occurred. Thus, we conclude that the halo stars are all very old stars—among the oldest in our galaxy. As time has gone on, succeeding generations of stars in our galaxy have had higher and higher abundances of metal atoms because the interstellar material out of which new stars are continuously forming is forever being enriched with more and more metal atoms from supernova explosions.

The fact that the stars in our galaxy's halo are arranged in a spherical distribution and are also very old tells us that the infant Milky Way was a spherically shaped mass that rather quickly flattened into a disk but left some stars behind in the form of a halo.

811 **Many of the stars in our galaxy's halo are quite faint and belong to large star clusters.** Since the stars in our galaxy's halo are all old, we would not expect to find massive young blue-white stars or even yellow stars like our Sun among them. Indeed, the stars of the halo are mostly smaller red stars. Many are actually members of rich clusters called *globulars.* Over 100 such clusters help delineate the halo itself. Globular clusters are great spherical or nearly spherical assemblies of between 50,000 and several hundred thousand stars each. The stars in each cluster are typically all packed into a space that is less than 200 light-years across. (The globulars

The magnificent globular star cluster known as NGC 104 located in the southern constellation Tucana, the Tucan. (NOAO)

were used by Harlow Shapley in 1918 as proof that we were not at the center of our galaxy [see page 258].)

812 **Down near the center of globular clusters, stars are packed pretty closely together.** In the largest telescopes on Earth, the stars near the centers of globular clusters seem to blaze together to create a single mass. Even the Hubble Space Telescope has had difficulty resolving such regions into individual stars. In reality, the single blaze of light that we see at the centers of these clusters is mostly due to the great distances over which we are viewing them. Nevertheless, it's been estimated that the average density of stars per cubic light-year of space in these centers *is* about a million times greater than in the region of the galaxy in the vicinity of the Sun! Imagine what the night sky would look like from such a place.

813 **Although most globular clusters are pretty far away, several are visible to the naked eye or in simple binoculars.** Because of their wide distribution around the galaxy, most globular clusters are far from Earth. The nearest is over 9,000 light-years from Earth and many are over 50,000 light-years away. Nevertheless, because such clusters contain so many stars, several of the nearer ones make rather easy targets for stargazers. In the skies of the southern hemisphere lies the brightest globular cluster seen from Earth. Known as Omega Centauri, it is easily visible to the naked eye. In the north-

ern hemisphere, several clusters are just visible to the unaided eye on very clear dark nights. The most striking of these is known as M13 and lies along one side of the "keystone" of the constellation Hercules in the spring and summer sky. (See the illustration on page 180.) To the eye, it appears as a faint, slightly out-of-focus star. In binoculars, it is easily seen as a tiny "cotton ball" and in a telescope of even modest size, its outer stars begin to be resolved, looking like uncountable bees all in frozen flight around a hive. Nearby M13 in the sky, is M92, another globular cluster best seen in a telescope.

814 **The H-R diagrams of globular clusters look different from those of open clusters.** The H-R diagrams of many open clusters typically have a rather well-defined Main Sequence with stars extending all the way from small red dwarfs up to white or even blue-white stars. The H-R diagrams of most globular clusters, however, look strikingly different in that virtually all of the upper branch stars are missing from the Main Sequence. The reason is simply that the globular clusters are much older than the open clusters, so their upper branch Main Sequence stars have all evolved off the Main Sequence and have gone on to become red giants, variable stars, and even white dwarfs.

815 **The centers of some globular clusters are sources of powerful X rays.** Images taken of globular clusters by X-ray satellites show the centers of many to be brightly lit.

Since X rays are a possible sign of the existence of black holes, many astronomers believe that massive black holes may lurk at the centers of many globulars. In such regions, the high density of stars, in effect, may have led a large number of stars to collide and coalesce into a black hole. The masses buried in some of these black holes may be of the order of dozens or even hundreds of stars the mass of our Sun.

816 Astronomers refer to the stars in the disk and the halo of our galaxy as being from different *populations.* Stars in the galactic disk are referred to as *Population I stars,* while those in the galactic halo are called *Population II stars.* Population I stars, in general, are young and rich in metals, while Population II stars are typically old and "metal-poor." In reality, of course, our galaxy has been a birthing place for new stars throughout its existence, so there is no clear-cut break between the time when Population II stars stopped being created and the creation of Population I stars began. Some astronomers even speak in terms of old and younger Population I stars. And while Population II stars are "metal-poor," they are certainly not entirely without metals. And so the term *Population III stars* is sometimes invoked to refer to the very first stars to form in our galaxy—stars that would have been virtually all just hydrogen and helium at birth.

817 Different types of stars and clusters typically belong to different populations. Most Main Sequence stars, from our Sun up through the O and B stars and the stars in open clusters, are all located in the galactic disk and hence are Population I stars. Stars in the globular clusters are Population II stars. Red dwarf stars may be either Population I or Population II stars. Some are young low-mass stars and others, being of low mass and evolving very slowly, still lie near the Main Sequence after billions of years.

818 The very best way to enjoy the Milky Way is with a pair of binoculars or a telescope at very low power. While some of the biggest telescopes in the world have been needed to unravel the complex structure of our galaxy, when it comes to single stargazing, little can take the place of leisurely scanning along the Milky Way on a warm summer night with a pair of simple binoculars or a telescope at very low power. This is a good example of "less is more," for this way you can see more of the overall beauty that the Milky Way has to offer as countless stars, nebulae, and star clusters slowly drift before you. It's just one of those "best things in life are free" things most people never take the time to enjoy. I hope you will some night soon.

CHAPTER 13

To Galaxies Beyond:
Islands in Time and Space

819 A major discovery in the early part of the twentieth century was the realization that the Milky Way wasn't the entire universe. By the turn of the century, astronomers knew the Milky Way was a vast system of stars. But was every star in the universe in the Milky Way? Was the Milky Way all that exists? Many thought the answer was yes. But for centuries, astronomers had seen small wispy patches of light scattered across the sky. Some, like the Great Nebula in Orion or the Eagle Nebula in Serpens, were known to lie within the Milky Way and be the birthing places of stars. It was suspected, however, that some other nebulae were different in nature. Particularly intriguing were some that had spiral shapes. In the early 1920s, an American astronomer, Edwin Hubble (for whom the Hubble Space Telescope is named), used the newly commissioned 100-inch telescope on Mount Wilson in California to have a closer look. The new telescope, then the largest in the world, revealed individual stars in several of the spiral nebulae. Even more importantly, it revealed some of these stars to be Cepheid variables, which can be used to determine distance. When Hubble calculated the distance to these stars, and hence the spiral nebulae they were in, he was astounded. While the Great Nebula in Orion and the Eagle Nebula in Serpens were less than 10,000 light-years away, the nearest of the spiral nebulae was over 2 million light-years from Earth. Indeed, far from being relatively small star-forming regions in our own galaxy, these spiral nebulae turned out to be full-fledged galaxies. It was instantly realized that our universe included far more than the Milky Way galaxy and was far larger and more complex than had been imagined. Between this study and the one by Harlow Shapley on globular clusters, mankind had to come to grips, in just a few short years, with the realization that we were not only not in the center of our own galaxy but that our galaxy was just one of many galaxies in the universe.

820 You can see the nearest major spiral galaxy to our own with your naked eye. In the eastern sky on a crisp autumn evening, look for four bright stars that make a large square. It is known as the Great Square of Pegasus and forms the wings and

Images of the same object taken in different wavelengths of radiation can reveal radically different looks and can be used to trace different objects and processes from image to image. The Andromeda galaxy, the closest major galaxy to our own, is seen in visible light along with its two tiny satellite galaxies, M32 and NGC 205. Most of the light in this image comes from stars and illuminated gas between the stars. (Palomar Observatory/Caltech)

part of the body of the fabled flying horse from Greek mythology. Branching off the Great Square toward the northeast is a curve of stars that leads from Pegasus into the constellation of Andromeda, the Princess. Just above the second star in this chain lies a fuzzy oval patch of light just visible to the naked eye on moonless nights. When you gaze at it, you are gazing on the combined light of over 100,000 million stars, for what you are seeing is the Andromeda galaxy (M31), a vast spiral system of stars that lies over 2 million light-years away! (On a clear night, you might not be able to see forever, but you can see over 2 million light-years with your eyes alone.) The Andromeda galaxy is so much farther away than all the stars we see in the sky that it's like pressing your nose to a windowpane to look at a distant mountain where the dis-

tant mountain is the Andromeda galaxy and the specks of dust on the windowpane are the stars in the sky.

821 While the Andromeda galaxy and our own Milky Way are similar in size and shape, not all spiral galaxies are created equal. Like the Milky Way, the Andromeda galaxy is a great spiral of stars, gas, and dust over 100,000 light-years across. Like the Milky Way, it also has a central bulge and a halo of old stars that include many globular star clusters. Some spirals are a tiny fraction of the size of M31 and the Milky Way. Whereas the Milky Way is about 100,000 light-years across, some *dwarf galaxies* are barely a few thousand light-years in diameter.

Galaxies that have spiral arms are sometimes just called *spirals* and are designated by the letter S. Thus, when astronomers talk about an S-type galaxy, they mean a spiral galaxy. But not all spiral galaxies have exactly the same shape. Some have spiral arms that are very tightly wound around the nucleus,

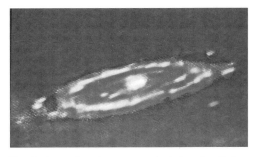

The Andromeda galaxy as seen in infrared radiation by the *IRAS (Infrared Astronomy Satellite)*. Here the central regions of the galaxy and spiraling bands of dust can be seen shining brightly. (NASA/JPL)

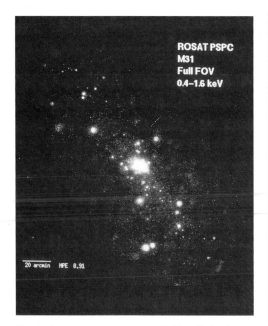

The Andromeda galaxy as seen in X-radiation by the *ROSAT* X-ray satellite. The overall spiral structure of the galaxy disappears and instead we see individual sources of X-radiation scattered throughout the galaxy. Most are the remnants of supernova explosions or X-ray-emitting binary stars. Intense X-radiation at the galaxy's center may be due in part to the presence of a large black hole. (Max-Planck-Institut)

The German X-ray satellite *ROSAT* launched in 1990. X rays are emitted by many objects in space but cannot penetrate the Earth's atmosphere and reach the ground. Therefore, X-ray astronomy must be carried out by satellites above the atmosphere. (Max-Planck-Institut)

others have somewhat looser spiral arms, and still others have spiral arms that are even looser. To differentiate, astronomers break these different types of spiral galaxies down into subclasses known as Sa, Sb, and Sc galaxies, respectively. Our own Milky Way seems to be an Sb galaxy.

822 **Some galaxies come with their own built-in bar.** While S-type galaxies have arms that gracefully spiral out from their centers, other spiral galaxies have arms that branch off the ends of a straight bar-line feature that runs through their centers. Appropriately, these galaxies are known as *barred spiral galaxies* or *barred spirals* and carry the designation SB. Just as with normal spirals, we find examples of barred spirals whose arms are wound to varying degrees of tightness, so these galaxies also have subclasses SBa, SBb, and SBc.

From the analysis of the motion of stars and hydrogen gas clouds in the Milky Way, a few astronomers suggested in the early 1990s that the Milky Way might actually have a very slight bar structure in its central regions. If so, we may have to reclassify the Milky Way someday, but whether its classification goes from Sb to SBb remains to be seen. For now, it is still officially listed as an Sb galaxy.

823 **Some galaxies have no spiral arms at all and just look like big elliptical blobs of stars.** Known appropriately as *ellip-*

tical galaxies or *ellipticals,* they are quite symmetrical and range in shape from almost spherical to elongated ellipsoids and, as such, are classified as E-type galaxies with subclasses E0 through E7. Assuming that overall, the elliptical galaxies across the universe are oriented randomly, there would appear to be approximately equal numbers of these galaxies in each of the above subclasses. As with spirals and barred spirals, elliptical galaxies also come in giant and dwarf varieties. Giant ellipticals can be fully 50 times the diameter of dwarf ellipticals and more than 20,000 times as bright.

NCG 1530 in the constellation Camelopardalis, the Giraffe, is an example of a barred spiral galaxy (Class SBb). Notice that the arms don't come directly out of the nucleus as with an Sb galaxy but instead extend from the ends of a straight bar that passes through the nucleus. (NOAO)

824 **Astronomers also recognize what they call S0 and SB0 galaxies.** These are galaxies that at first glance might appear to be ellipticals but have very slight tendencies to show spiral or barred spiral characteristics.

825 **There are also *irregular galaxies* and downright *peculiar galaxies.*** Irregular galaxies (which carry the designator I or Irr) typically have chaotic shapes or, in effect, no definitive shape at all. Instead, they are usually just a big jumble of stars, gas, and dust. Peculiar galaxies frequently have rather well-defined shapes but typically have a curious twist, such as an appendage or asymmetry, in part of their structure.

826 **Our Milky Way galaxy has two little companions.** They are known as the Large and Small Magellanic Clouds because western Europeans first heard of them from the crew of Ferdinand Magellan when the crew returned to Europe after circumnavigating the globe. Visible only from southern latitudes, they are both *dwarf irregular galaxies.* The Large Magellanic Cloud (LMC) is about 21,000 light-years across and 170,000 light-years away, while the Small Magellanic Cloud (SMC) is about 9,500 light-years in diameter and 190,000 light-years distant. Both are easily visible to the naked eye under dark skies, with the LMC being about 8° (or 16 times the diameter of the full Moon) across at its widest point and the SMC appearing about half that size. The Milky Way and its smaller

The Small Magellanic Cloud. (©Anglo-Australian Observatory/Royal Observatory, Edinburgh. Photo made from U.K. Schmidt plates by David Malin.)

companions are all gravitationally bound together, meaning that they will continue to orbit around each other.

827 **The Andromeda galaxy also has a pair of satellite galaxies.** Both are *dwarf elliptical galaxies* and are known as M32 and NGC 205. M32 is the thirty-second object in *Messier's Catalog* and NGC 205 is the two hundred and fifth entry in another astronomical listing known as the *New General Cata-*

The Large Magellanic Cloud. (©Anglo-Australian Observatory/Royal Observatory, Edinburgh. Photo made from U.K. Schmidt plates by David Malin.)

log. A little over 3,000 and 6,000 light-years in diameter, respectively, they are gravitationally bound to M31.

828 **Some galaxies have both young and old stars, while others only seem to have old ones.** Like our Milky Way, other spiral galaxies as well as the barred spirals seem to have mostly young (Population I) stars in their disks, surrounded by a spherical halo of old (Population II) stars. Elliptical galaxies, however, are made up of virtually all old (Population II) stars: lots of red giants (and presumably even more red dwarfs and white dwarfs) but no upper Main Sequence blue-white stars because the O and B stars passed from the scene as supernovae long ago.

829 **Different types of galaxies are typically different colors.** Spiral and barred spiral galaxies primarily get their luster from the hot young blue-white stars that are strung like jewels along their spiral arms. Indeed, the light from these O and B giants and supergiants delineates the spiral arms and makes them visible. Hence, the arms of spirals and barred spirals tend to take on a bluish hue in telescopes. The central regions of spirals and barred spirals have some older, cooler stars, so these areas can appear more yellow than the spiral arms. Since ellipticals have no O and B stars and derive most of their light from their large population of red giants, they appear somewhat redder in color.

The Sombrero Galaxy. Its "hatband" is a dark lane of dust that appears to cut the galaxy in two. (NOAO)

830 **Some galaxies have lots of gas and dust, while others are mostly just stars.** Spirals, barred spirals, and many irregular galaxies have prominent dark lanes caused by the presence of dust. These galaxies also exhibit areas of O and B stars and have abundant quantities of gas from which many new generations of stars will continue to be created. In contrast, most elliptical galaxies seem relatively devoid of interstellar gas and dust, so star formation in these galaxies has all but ceased.

When seen edge-on, dust in the central regions of spirals and barred spirals can appear to cut such galaxies in two, running like a dark belt or hatband completely around the galaxy's circumference. The Sombrero Galaxy is a good example. In reality, the dust is merely obscuring the light of millions of stars that lie beyond, within the galaxy's disk.

831 **Some galaxies go through periods of intense and rapid star birth.** While most spirals and barred spirals are lit up by brilliant young stars along their spiral arms, some galaxies or regions of galaxies appear to have undergone a recent period of almost explosive large-scale star birth—what has come to be called a *starburst* phenomenon. In some cases, such regions exist near the cores of the galaxies where, for some reason, a large number of stars may have gone supernova and their expanding shock waves have, in turn, led to the creation of many new stars. In other cases, the starburst region lies in or near a region of irregularity or peculiarity. Such large-scale distortions also suggest massive and/or violent agents at work, including the fact that . . .

832 **Sometimes galaxies collide.** The distances between most stars are large compared to the sizes of the stars themselves. (For example, the separation between the Sun and the stars of the Alpha Centauri system is about 4.2 light-years—almost 6 trillion miles—or almost 7 million times the diameter of the Sun.) For this reason, except in the case of very close binary star pairs or at the very centers of globular star clusters, collisions between stars are very, very rare. With galaxies, however, it's a different story. The distance between the Milky Way and the Andromeda galaxy is about 2.2 million light-years, but that's only about 20 times the diameter of each galaxy. And many galaxies are closer to each other than that. Thus, galaxies *do* collide with significant frequency.

833 **What happens when galaxies collide?** For one thing, a lot of stars *don't* go crashing into each other. As noted above, the distance between the Sun and the Alpha Centauri system is almost 7 million times the diameter of the Sun. Halve this distance, which is what you would get on average if another galaxy collided with the Milky Way, and you would still have lots of room between such stars. What does happen when galaxies collide is typically a massive distortion of the shapes of one or both galaxies as their immense gravitational fields wreak havoc with each other. In effect, each galaxy raises enormous tides in the other. Furthermore, while there's typically lots of room between the stars of each galaxy, the same cannot be said for the interstellar clouds of gas and dust that can permeate much of the space between the stars. Massive collisions of these clouds do occur and, like supernova shock waves, they can "squeeze" new stars out of the fray in great starburst episodes.

834 **Many of the *peculiar galaxies* we see in space show strong evidence of galactic collisions.** As we look out across space, we see many peculiar galaxies that are part of close galactic pairs. For years, astronomers believed these distortions were due to huge tidal disturbances that would occur during collisions. The latest supercomputers confirmed this by allowing astronomers to create simulated time-lapse movies of galaxies in collision. In many cases, the resulting computer models rather accurately reproduce many of the distorted galactic shapes actually observed in nature.

835 **Of ringtails and cartwheels.** Depending on the types of galaxies that collide, their individual masses, and their trajectories, a great variety of fancifully shaped galaxies can result. Two galaxies in the act of collision in the constellation of Corvus, the Crow, have dislodged long streamers of gas and dust from each other that have been made luminous by bursts of new stars. Because of their appearance, the galaxies have been nicknamed the Ringtail Galaxies or the Antennae. Elsewhere we find another grouping (see below). The small galaxy on the right has collided with the large one on the left and, in fact, has passed almost directly through its center. The result is a tidal wave that is radiating out through the large galaxy, creating an expanding circular burst of star formation as it goes. The large member of the pair is appropriately known as the Cartwheel Galaxy.

836 **Why do different galaxies have different shapes?** The answer is still far from clear but, in many cases, it may be a matter of how much spin the cloud of material that went into forming a particular galaxy had in the first place. We know from a basic understanding of physics, borne out in computer simulations, that a spherical cloud of material with a reasonably high degree of spin will quickly flatten out into a disk. We see this occur with clouds of gas and dust that evolve into stars and surrounding families of planets and we believe the same basic process, on a much larger scale, leads to the formation of spiral galaxies. In contrast, clouds of gas and dust with very little initial rotation never flat-

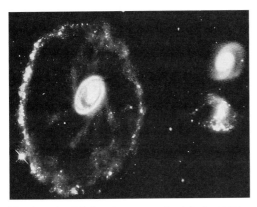

The Cartwheel Galaxy is seen on the left. Its name comes from the bright circle of young hot stars around its nucleus. They are an example of a starburst phenomenon induced in the galaxy when one of the small galaxies to the right passed directly through the center of the Cartwheel at some point in the past. (NASA/STScI)

ten out, so they become elliptical galaxies and possibly some of the irregular galaxies as well. Ellipticals were able to get down to the process of star formation right away, so they have little gas and dust left. Spirals took time to flatten out and started most of their star formation later, so they still have gas left to make more stars. In addition, collisions or close encounters between galaxies can produce other irregular and peculiar shapes and induce starburst phenomena.

837 **The Milky Way and the Andromeda galaxy are currently moving toward each other.** The two galaxies are actually drawing closer by about 172 miles each second or over 600,000 miles each hour. But don't sell the farm quite yet. At this rate, it would take them more than another 2 billion years to collide, even if they were headed right for each other, which they're not.

838 **Some peculiar and irregular galaxies show signs of massive disturbances from deep within.** Some of these galaxies may not owe their odd shapes to collisions but instead appear to exhibit regions that look as though they have been torn asunder or shredded by unimaginable force. Only explosions of massive proportions could account for such things—perhaps a rash of supernovae somehow triggered at almost the same point in time in the same region or perhaps some hitherto unknown phenomenon or combination of phenomena.

839 **Some galaxies are known as active galaxies and have small, dazzlingly brilliant centers.** In the 1940s, an American astronomer named Carl Seyfert completed a survey of spiral galaxies and realized that about 1 percent of them had brilliant, almost starlike centers. The spectra of these *Seyfert galaxies* reveal gas and dust in rapid motion around their central regions at speeds of over 6,000 miles per second. Seyferts are but one of several types of unusual galaxies that exhibit rapid motion in their cores as well as the outpouring of enormous quantities of energy from these regions. In some cases, such objects radiate a trillion times the energy of our Sun from a region that is far smaller in diameter than the distance from our Sun to the nearest star! Such amounts of energy cannot be accounted for through stars alone because you simply cannot pack that many stars that close together, even in the very heart of a galaxy. Because of these remarkable characteristics, all such objects are referred to as *active galaxies*. They are said to be galaxies

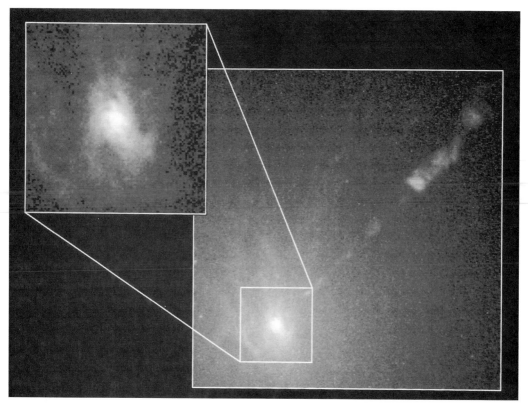

This image of the central regions of the giant elliptical galaxy M87 shows its 8,000-light-year-long jet and small brilliant nucleus. The insert shows clouds of gas that are spiraling rapidly around this nucleus. From an analysis of the motions of the gas, astronomers deduce that a black hole with 1.2 billion times the mass of the Sun lies within. (NASA/STScI)

that have *active cores,* or *active galactic nuclei* (AGNs for short).

840 **Some active galaxies have enormous jets.** In the constellation of Virgo lies a remarkable object—a giant elliptical galaxy known as M87. Shooting out of its very heart is a jet of material over 8,000 light-years long. The material in the jet is fleeing the center of the galaxy at nearly half the speed of light.

Radio telescopes yield even more wondrous revelations. Shooting out of a galaxy known as NGC 6251 is a jet that measures 300,000 light-years long. Another galaxy, known as Hercules A in the constellation Hercules, has a pair of jets each 450,000 light-years long and yet another galaxy known as NGC 315 has a jet fully three-quarters of a billion light-years long—more than seven times as long as the Milky Way galaxy is wide. The radio radiation in these enormous jets comes from electrons and other charged particles being shot out of the centers of these galaxies and then forced to spiral at nearly the speed of light along the galaxy's gigantic magnetic fields.

These objects are known as *radio galaxies* because they give off far more energy in the radio part of the spectrum than they do in the form of visible light. Put another way, if our eyes were sensitive to radio waves instead of visible light, these objects would all look much brighter to us than they do. Also, in all such cases, the image of the galaxy in visible light is tiny compared to what it looks like in radio waves—a great example of how examining celestial objects with radio telescopes can produce complementary and vastly different images and insights into the nature of these objects.

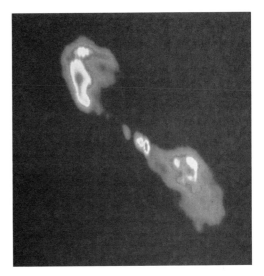

This is an image of Centaurus A made in radio waves. The entire object that we see as Centaurus A in visible light (see previous image) would be a mere dot at the center of these two enormous jets of radio energy. (NRAO)

This is a visible light image of an unusual object known as Centaurus A. Centaurus A may be a giant elliptical galaxy in collision with a spiral or irregular galaxy whose dust lanes we see in silhouette. Stretching several hundred light-years across, Centaurus A is an example of a radio galaxy and is indeed one of the brightest objects in the radio sky. (NOAO)

841 **Our Milky Way is not classified as a radio galaxy.** Our Sun and many other stars in our galaxy (as well as the gas between the stars) all give off radio waves. Thus, our galaxy as a whole emits radio waves. It is not, however, classified as a radio galaxy because it radiates more energy in the form of light than radio waves. Indeed, radio galaxies tend to be ellipticals rather than spirals or irregulars.

842 **Strange celestial beasts called *BL Lacertae objects* may be radio galaxies whose jets are pointed right at us.** The curious objects in this class are all named after the first of their ilk to be discovered—a point of light called BL Lacertae (in the constellation Lacerta, the Lizard) originally believed to be a variable star. When radio telescopes were pointed at BL Lacertae and similar objects, however, it was soon realized that they were powerful emitters of radio waves. Probing further with optical telescopes, astronomers confirmed that these were not variable stars in our own galaxy but actually the brilliant centers of giant elliptical galaxies far away. No great jets of material are seen coming from the centers of these objects and so, at first, astronomers thought they might be dealing with a totally new phenomenon. Astronomers now believe the jets are there, however. It's just that these radio galaxies happen to be oriented in such a way that we are looking right down the nozzle of one of the jets as its material is being shot directly at us.

843 **The single most powerful radiators of energy in the universe are known as** *quasars.* Quasars were first discovered in the early 1960s and baffled astronomers for much of the next three decades. The first quasar was picked up in a radio telescope as a very "bright" source. Radio astronomers gave the coordinates of the object to optical astronomers, who took a picture. What they found looked like nothing more than a faint bluish starlike object, so they called it a "quasistellar radio source," or quasar for short. This and other subsequently discovered quasars were found to be among the most distant objects in the universe. Thus, to be visible over such great distances, quasars also had to be intrinsically very, very bright. In fact, quasars were discovered to be the intrinsically brightest objects in the universe. Some quasars radiate as much energy as

1,000 galaxies the size of our Milky Way from a space not much larger than our solar system. In addition, like other "active" objects, many quasars were found to have jets shooting out of their centers at very high speed. But what were these strange powerful objects and what could be responsible for their awesome outpouring of energy?

844 **In the mid-1990s, telescopes finally discovered the truth about quasars.** For many years, quasars appeared to be truly unique objects that, although very bright, were much smaller than galaxies and so were thought to be mysteriously unrelated to anything else we saw in the universe. In time, however, telescopes (including the HST) were developed with sufficient sensitivity to allow astronomers to see that quasars were actually the incredibly brilliant centers of very active galaxies. (Up until that point, telescopes weren't good enough to make out the rest of the surrounding galaxy against the glare of its quasar core.) In the meantime, other objects (such as BL Lacertae objects) with active galactic nuclei had been discov-

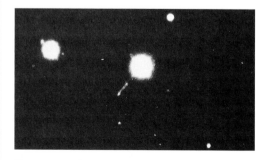

The quasar 3C 273 with its jet of material. (NOAO)

ered, so quasars were finally envisioned as possibly just the most active in a whole spectrum of such objects from Seyferts through BL Lacertae objects and on to quasars.

845 **Seyferts, radio galaxies, BL Lacertae objects, and quasars are probably all powered by the same awesome mechanism.** They were all discovered separately, so Seyferts, radio galaxies, BL Lacertae objects, and quasars were named and categorized differently. Yet they all share remarkably compact and brilliant centers that can vary in brightness over the course of weeks or even days—regions that can shine with the brilliance of a trillion times the energy of our Sun or more. Coming up with an explanation for such enormous amounts of energy in such a small space—as well as for the enormous jets hundreds of thousands of light-years long—at first proved difficult. In time, however, the speed with which gas was seen to whirl about the centers of these galaxies and the incredible compactness of their cores eventually led to only one plausible conclusion: These objects are all probably related manifestations of the same general type of object and all probably have massive black holes buried deep within them!

846 **Only massive black holes could create such power and energy.** The enormous energy released from active galactic nuclei is now seen as the direct result of matter being pulled toward a massive black hole at or near the galaxy's center. This matter, torn from nearby stars as well as from clouds of gas and dust, would be drawn into an accretion disk

surrounding the hole. Spiraling ever nearer the hole's event horizon, the material would accelerate and heat up until it gave off enormous amounts of energy across the spectrum. The high-powered jets also seen in some of these objects come from another characteristic of black holes, namely intense magnetic fields that lie twisted along the axis on which the black hole is spinning. These powerful force fields prevent some of the infalling matter in the accretion disk from entering the hole. Instead, they channel it, like gushing water is channeled in a fire hose, and shoot it at very high speeds out of the region of the hole in opposite directions along its axis of rotation.

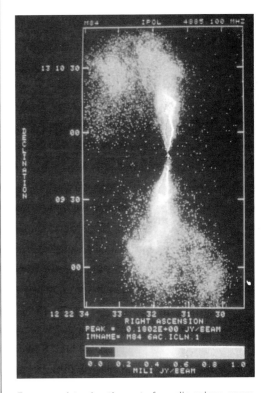

Enormous jets shooting out of a radio galaxy. (NRAO)

847 Gigantic galactic black holes have now been found in the hearts of several galaxies. In 1994, using the Hubble Space Telescope, astronomers were able to carefully analyze the speed with which clouds of gas were orbiting around the center of M87. As a consequence of this study, they were able to announce that a black hole with a mass equivalent to 2.4 billion times the mass of our Sun lies buried in the galaxy's core. Later that same year, radio astronomers analyzed the motion of another disk of gas near the center of a spiral galaxy known as NGC 4258. The results inferred the presence of a black hole at the center of the disk with a mass equal to 40 million times the mass of the Sun. In 1995, astronomers again used the incredible vision of the HST to obtain an image of a dark disk of dust near the center of elliptical galaxy NGC 4261. The disk, which is 800 light-years in diameter, encircles a black hole with a mass equivalent to 1.2 bil-

lion Suns buried within an event horizon larger than our solar system. In early 1997, astronomers announced the discovery of black holes in three additional galaxies with masses up to 100 million Suns.

These latter galaxies were all normal, that is, nonactive galaxies, and thus raise the likelihood that many normal galaxies harbor massive central black holes.

848 The black hole in NGC 4261 is off kilter. A surprising find about the massive black hole in NGC 4261 is that it is displaced about 20 light-years from the galaxy's center. One theory is that the NGC 4261 is really the result of the past collision of two galaxies and that the tidal forces from the interaction have knocked the hole off-center. Another interesting idea suggests that the two jets coming out of the hole may not be equally as powerful, so the huge black hole is being slowly rocket-propelled across its host galaxy by the unbalanced force of its more powerful jet.

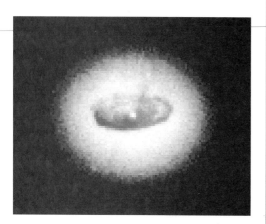

This HST image shows a dark disk of obscuring dust 800 light-years in diameter near the center of the galaxy NGC 4261. Within lies an enormous black hole (NASA/STScI)

849 Stellar and galactic black holes are horses of the same color. There is no intrinsic difference between stellar and galactic black holes. They are really the same phenomenon, produced, in both cases, by gravitational collapse. Galactic black holes, however, are much more massive and have much larger event horizons than stellar black holes.

850 Could there be an evolutionary link between active and nonactive galaxies? Indeed, there might be. Some astronomers suggest that quasars may be the centers of very young, very active galaxies and, in time, such objects may evolve into BL Lacertae objects or Seyfert galaxies, which, in turn, might evolve into nonactive elliptical galaxies and normal spiral galaxies, respectively. If this is true, many nonactive galaxies, including our Milky Way, may have black holes in their centers as well. Indeed, there is strong evidence that this is the case.

851 **The difference between active and inactive galaxies may be a matter of food supply.** If all active galaxies indeed do have massive black holes in their centers and it turns out that many nonactive galaxies do as well, the difference may really come down to whether or not the black hole has a ready supply of gas to devour or not. If it does, the galaxy's nucleus will be brilliant and "active." If it doesn't, it won't. The fact that quasars are very distant and therefore are seen by us the way they appeared long ago suggests that there may have been more fuel in the centers of galaxies when these galaxies were young (and the universe was young) than there is today. Perhaps, long ago, our galaxy was host to a quasar.

852 **Black holes make time slow down and also act as time machines.** A fascinating prediction of Einstein's General Theory of Relativity states that in a gravitational force field time slows down and the stronger the gravitational force, the slower time ticks. It's important to point out that Einstein wasn't talking about *psychological* time, as in, "time

flies when you're having fun" or "time crawls when you're waiting in the dentist's office" but real *physical* time by whatever means or whatever type of clock you choose to use to measure it (hourglasses, wristwatches, atomic clocks, and so on). Since black holes have very intense gravitational fields, Einstein's theory says they also have a significant effect on the passage of time in their neighborhood. This is true for both stellar and galactic black holes, but galactic black holes could someday prove to be the ultimate time machines.

Consider the adventures of two identical twins. After waving good-bye, one blasts off in a rocket, flies out into space, does a U-turn around a big black hole, and returns to Earth. Upon returning home, the traveling twin discovers that he or she is much younger than the sibling that was left behind. Less time would have gone by for the traveling twin, so he or she would have aged less. Indeed, if the traveling twin flew close enough to the black hole or around and around it for a sufficiently long time, he or she could return to Earth in the distant future—long after the other twin was dead.

853 **Black holes would allow us to *last* into the future rather than *jump* into the future.** Time in the vicinity of a black hole flows more slowly than farther from it, so not only do mechanical and atomic clocks tick more slowly in their vicinity but so do biological clocks. It's important to note, therefore, that if you should travel to the neighborhood of a black hole and then return to Earth, you would not have magically jumped to some point in the future. Instead, because you would have aged more slowly than persons who were left back on Earth, you would simply have "lasted longer" than they would and hence you would have lasted into a period of time that you would not have otherwise lived to see.

854 **Black holes might allow you to go *into* the future but not *back to the future.*** When it comes to time travel, black holes are only "one-way streets," that is, they can only slow time down, not reverse its direction. Thus, black holes might allow some future space travelers to last farther into the future than they otherwise would have done, but they cannot be used to travel back in time as happens in many popular science fiction movies. This indeed may be one of nature's "safety valves," for if we could indeed travel into the past, we could change history and create paradoxes, such as seeing to it that our parents never met and hence eliminating our own existence.

---

855 **Hanging around black holes can be dangerous to your health.** There is, of course, the risk of falling in and getting crushed by the singularity, but with some black holes, things can get really ugly long before that. The reason has to do with a phenomenon known as *differential gravitational force.* It deals with how hard an object pulls on the *near* side of another object vs. how hard it pulls on the *far* side of that object with its gravitational force. The closer two things are to each other, the stronger will be the gravitational force between them. But since

things usually have a finite size, the gravity on the near side of an object is greater than on the far side. The side of the Earth that faces the Moon at any particular time is about 8,000 miles closer to the Moon than the opposite side of the Earth and the fact that the Moon is pulling harder on the Earth's near side than its far side creates the tides. The stronger the gravitational force field around something, the stronger is the differential gravitational force induced across something else nearby. The Moon exerts very little difference in force over different parts of your body. However, venturing too close to certain black holes can mean a mighty big difference between the gravitational grip the hole exerts on the near part of your body vs. the parts of you that are a little farther from the hole. Indeed, if you were heading for some black holes feetfirst, the gravitational pull on your feet would grow to be so much greater than the gravitational pull on your head that it would feel as though you were hanging off London Bridge by your fingers with the entire population of London holding on to your toes.

---

856 **The bigger the black hole, the safer it is for time travel.** The bigger a black hole, that is, the larger its event horizon, the less is the differential gravitational force near its event horizon. This, in turn, means that you could theoretically get close enough to a galactic black hole to do some decent time traveling without as much of a risk of getting torn to shreds. However, you would still need to take other precautions, including making sure that you didn't accidentally drift across the hole's event horizon. Given a large enough black hole, you wouldn't have your spaceship being torn apart to warn you and, once you drifted across this invisible boundary, you would be doomed because the velocity that you would then need to achieve to avoid being dragged relentlessly toward the singularity would be greater than the speed of light.

857 **Black holes probably all rotate.** Since virtually all stars rotate and the rate of rotation of a collapsing star increases as it collapses (just as that of a spinning skater who pulls in her arms), so all black holes might be expected to rotate.

858 **Galactic black holes may allow for more than travel through time.** Some of the equations that describe such *Alice in Wonderland* adventures as time travel via black holes also suggest another intriguing possibility. For a *rotating* black hole, certain equations suggest that another type of trip is possible. In such a black hole, two types of horizons would exist with the widest separation between them occurring around the hole's "equator." According to the equations, if a spaceship flew through the outer horizon but remained above the inner horizon, it could escape the hole without having to ex-

ceed the speed of light. Furthermore—and most intriguingly—the spaceship would not only emerge in another time but also "somewhere else," either at another point in our universe or in another universe. Such a route is known as an *Einstein-Rosen Bridge* or a *wormhole.* Such wormholes have popped up in popular science fiction in recent years.

859 **There is no physical proof that wormholes actually exist.** It must be stressed that the above scenario is simply based on one interpretation of a set of equations. Many mathematical models are based on equations, but just because a model exists or equations describe a particular situation, that doesn't mean that the situation in fact corresponds to physical reality. Only when theory is substantiated by several independent observations is it taken as possibly describing a real situation.

860 **There are more of some types of galaxies and less of others.** Out to several hundred million light-years from Earth, about 60 percent of all galaxies are ellipticals (with the vast majority of these being dwarf ellipticals), about 20 percent are spirals and barred spirals, and the remaining 20 percent are peculiars and irregulars. Farther out, we see a much higher percentage of irregular galaxies.

CHAPTER 14

COSMOLOGY: FROM THE END OF THE UNIVERSE TO THE DAWN OF TIME

861 **Cosmology is a branch of astronomy.** Cosmology is the branch of astronomy that seeks to determine the answers to some of the biggest and most profound questions that can be asked. Things like: "What is the nature of the universe as a whole?" . . . "What was the origin of the universe?" . . . "How old is it?" . . . "What will be its ultimate fate?" . . . "How do we fit into the big picture?" Astronomers who try to come to grips with such issues are known as *cosmologists.*

862 **Cosmology draws its talent from a rather diverse gene pool.** There are both observational and theoretical cosmologists. Theoretical cosmologists are pretty real, they just deal in theoretical mathematical models of the universe. In addition to "pure astronomers," however, cosmology also attracts people whose primary training is in such areas as mathematics and the branches of physics that deal with things that have high amounts of energy and things that are very small. These areas are known as *high-energy physics* and *particle physics.* This is because the early universe was a very high-energy kind of place and

when it comes to studying the universe in the very distant past, sometimes *atomic accelerators* are more useful than telescopes.

863 **Like stars, galaxies tend to congregate in clusters.** Like star clusters, some clusters of galaxies have many members, while others are relatively small.

864 **Our Milky Way galaxy is a member of a small group.** Our Milky Way galaxy and the Andromeda galaxy are members of a small group of about 35 local galaxies appropriately called the Local Group. The Milky Way is the largest galaxy in the group with Andromeda a close second. A third spiral galaxy named M33 also belongs to the ensemble and is about half as big as Andromeda. Most of the other members are dwarf ellipticals and irregulars. As a whole, the Local Group stretches across more than 3 million light-years of space.

865 **Once again, we find that we're not in the center of things.** After Copernicus, we had to abandon the belief that we were at the center of the universe. After Shapley studied the distribution of globular clusters, we had to accept the fact that we weren't at the center of our galaxy, but for a few years we could cling to the notion that the Milky Way was the only galaxy in the universe. After Hubble, we had to accept the fact that ours was not the only galaxy. And now that we have charted the Local Group, we find that, while we do seem to live in the largest galaxy in the Local Group, the Milky Way isn't in the center of the Local Group.

---

866 **Our galaxy has a zone of avoidance.** Near the plane of our galaxy, there appears to be a zone of avoidance where few galaxies can be seen. However, this apparent anomaly is due to the obscuring dust within the plane of the Milky Way and has nothing to do with the actual distribution of galaxies in the universe. At the farthest limits of the observable universe, galaxies seem to be distributed randomly around the sky. Nevertheless, the obscuring dust must be taken into account in our study of galaxies from Earth because it both dims and reddens the light of these galaxies, making them seem a different color and more distant than they really are.

867 **If you look far enough out into space away from the plane of our Milky Way, you will see far more galaxies than stars.** As you look along the plane of the Milky Way, the number of stars is virtually uncountable. And because of the number of

stars crowding our view as well as the obscuring dust, galaxies are very difficult to see. Thus, the stars we see in such places in the sky far outnumber the galaxies. But if we take a large enough telescope and look in areas of the sky far from the galactic plane, such as up around the Big Dipper, the galaxies in our field of view actually far outnumber the stars. If we were in deep intergalactic space, with sensitive enough eyes, the galaxies we would see shining all around us would far, far outnumber the stars we see on a dark starry night from Earth.

868 **In general, galaxy clusters are divided into *regular* and *irregular clusters.*** Regular clusters tend to be heavily concentrated toward their centers and have many member galaxies, most of which are elliptical and S0 galaxies. Such clusters are also characterized by one or two giant elliptical galaxies near their centers or a subgroup of what are known as *supergiant diffuse galaxies,* which are large but rather dim. Irregular galaxy clusters can have many or relatively few members, but there tends to be no central concentration as with regular clusters. Irregular clusters also usually contain more spiral galaxies. The Local Group is an example of an irregular group of galaxies.

869 **The Local Group is puny compared to its closest cluster neighbors.** The Virgo Cluster located in the constellation of Virgo is the closest irregular cluster to ours. It contains about 1,000 member galaxies. In spite of the fact that it's an irregular cluster, it

A section of the Virgo Cluster, the nearest example of an irregular cluster of galaxies to the Local Group. Near the center and along the right edge of the image can be seen M84 and M86, two of the cluster's giant ellipticals. (NOAO)

lays claim to three giant elliptical galaxies: M84, M86, and M87. The cluster as a whole stretches about 12° across the southern skies of spring, which means that it stretches about 13 million light-years across space at its distance of almost 60 million light-years from us. The closest regular galaxy cluster is the Coma Berenices Cluster located just above Virgo in the sky. It contains over 2,000 members in a space about 16 million light-years across and is located at a distance of 325 million light-years. Note that while the Coma Berenices Cluster is only about 5 times the diameter and 25 times the volume of the Local Group, it contains about 80 times as many galaxies. Thus, on average, the galaxies in the Coma Berenices Cluster are closer together.

870 **The giant elliptical galaxies frequently located near the centers of large galaxy clusters have a very sordid past.** Supercomputer simulations show that collisions between spiral galaxies in clusters tend to strip these galaxies of much of their interstellar gas, which then collects near the center of the cluster. There the gas is consumed by an elliptical galaxy that grows larger and larger as the collisions between spirals continue. Thus, we can not only explain why regular galaxy clusters frequently have giant elliptical galaxies located where they do but also why these galaxies are so large and massive. They're cannibals and they eat their neighbors.

871 **Galaxies or groups of galaxies can actually bend light so much they create phantom galaxies in space.** In recent years, astronomers have come across places in the sky where several galaxies appeared to be clustered very closely. When spectra were taken, it was found that not only were the spectra of these galaxies similar to each other—they were *identical*. This was no coincidence, for each of the galaxies in the group was really a *mirage* image of the *same* galaxy. The illusion was created because yet another galaxy or group of galaxies just happened to lie along the same line of sight between us and the more distant one. The gravitational force field of the closer galaxy or galaxies bent the light of the more distant galaxy around itself and on toward Earth and, in so doing, created ghost images of the more distant galaxy around itself.

Depending on the position of the foreground and background objects and the distribution of mass within them, the ghost images can take on a variety of appearances. Sometimes the ghost images look like pieces

Curved wisps of light are mirage images of a distant galaxy created by the bending of its light by the gravitational field of a closer galaxy in the same line of sight. The discovery of such "gravitational lensing" offers dramatic proof of one of the predictions of Einstein's General Theory of Relativity. (NASA/STScI)

of an arc, sometimes they form four arcs or spots of light surrounding the foreground galaxy (known as an *Einstein Cross),* and, on rare occasion, can even look like a complete ring or halo of light. The phenomenon is a consequence of Einstein's General Theory of Relativity, which states that mass bends light. On a larger scale, it is the same type of phenomenon seen during a total solar eclipse when the Sun's gravitational force is seen to bend the light of distant stars.

872 Edwin Hubble and his associates not only measured the distances to many galaxies, they also measured their motion. Just as we can use the shift in the spectral lines of a star to tell us if that star is moving toward us or away from us and how fast, the same technique can be applied to galaxies. (In this case, the spectrum of the galaxy is simply the combined spectra of all its stars

with the spectra of the brightest stars or brightest feature dominating.) Between 1912 and 1925, Hubble and his associates compiled the spectra of 41 galaxies. When the spectra were examined, it was found that a few galaxies that were in the Local Group were drawing closer to us (their spectra were blue-shifted), but all of the other galaxies in the sample had spectra that were red-shifted, meaning that the galaxies were moving away from us.

873 Edwin Hubble drew one of the most profound conclusions ever made about the universe. By the mid- to late 1920s, Hubble was using the new 100-inch telescope to not only see individual stars in nearby galaxies but to see individual Cepheid variables. By timing their periods of variability, Hubble was able to calculate the distances to the galaxies and, by 1929, he had enough distances and corresponding red shifts measured to see a definite trend. Simply put, on average, the farther away a galaxy was, the faster it was moving away from us. And that, simply put, could only mean one thing—the universe . . . was *expanding.*

874 If, in general, the other galaxies in the universe are moving away from us, does that mean that we are in the center of the universe? No. To understand why, we have to bake some raisin bread.

We're going to do a little "thought experiment" and to keep it simple we're only going to use four raisins as shown in the illustration on page 298. We are also going to use special transparent dough and enlist the aid of an intelligent heat-resistant bug who has agreed to help by riding on one of the raisins. We place the bug on one of the raisins and ask him to measure the distance from his raisin to each of the other raisins. He gets the distances shown in Illustration 1. Now we put the bread in the oven and turn on the heat. Soon the bread begins to rise and after an hour we pull the bread out of the oven. We again ask the bug to measure the distance from his raisin to each of the others. This time he gets the numbers in Illustration 2. Thus, in one hour Raisin A has moved from 1 inch away to 2 inches away, Raisin B has moved from 2 inches away to 4 inches away, and Raisin C has gone from 3 inches away to 6 inches away. This means that in that hour Raisin A moved with a speed of 1 inch per hour, Raisin B moved at the rate of 2 inches per hour, and Raisin C traveled at the speed of 3 inches per hour.

By now, I hope the analogy is clear. The raisin bread is the expanding universe, the raisins are galaxies in the universe, and the bug is us. The farther away a raisin was from the bug, the faster it moved away from him, even though the universe as a whole expanded at a constant rate. In an hour, the bug's "raisin bread universe" doubled in size and each galaxy got twice as far away from him as it was the hour before.

But now, imagine that our bug is on one of the other raisins and we are going to repeat the experiment. If you do so, by measuring the distances from, say, Raisin A to each of the other raisins at both the begin-

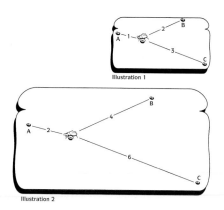

Baking some raisin bread offers a good illustration of the expanding universe. As the bread expands, the more distant a raisin is from the bug, the more rapidly that raisin moves away from the bug.

ning and the end of the experiment, you will find that you again reach the same conclusion. The farther a raisin is away from the bug, the faster it will be moving away from him. And it will work for any raisin. The point is that in an expanding universe, the other galaxies would be moving away from you and the farther away a galaxy was, the faster away it would be moving, regardless of which galaxy you were on when you made your observations.

875 This point is summed up in a very fundamental concept known as the *Cosmological Principle.* The Cosmological Principle essentially says that observations from any point in the universe would yield the same results, that is, that we do not live (nor does anyone else live) at a special place in the universe. In effect, this realization takes us the final step away from the concept of an Earth-centered universe.

876 **If the universe is expanding, why do some galaxies in the Local Group move toward us?** Throughout the universe, galaxies all have small individual motions. Since the expanding universe doesn't carry nearby galaxies away from us at a very high speed, nearby galaxies can actually wind up having a net speed in our direction. As we look farther and farther into space, the general expansion of the universe carries galaxies away from us at faster and faster speeds and these speeds soon prove far greater than the individual motions of the galaxies. So except for some of the closest galaxies, the galaxies have a net motion away from us.

877 **The motions of individual galaxies must be accounted for in trying to measure the general expansion of the universe.** Just as there are individual motions to the galaxies in the Local Group, the same is true for galaxies in other clusters. And clusters of galaxies have individual motions relative to other clusters. When astronomers talk about the expansion of the universe, they are referring to the expansion of the universe as a whole, or more correctly, the expansion of space in which the galaxies are embedded (known as the *Hubble Flow*). To arrive at this number, we must measure the red shifts of individual galaxies, but then, by carefully examining individual motions, remove these localized effects to get a true handle on the flow rate of the space in which the galaxies find themselves.

878 **Localized galaxy motions help plot how matter is distributed in the universe.** Our Local Group as a whole appears to have a localized motion in the direction of the galaxies of the Virgo Cluster (and indeed forms a kind of supercluster with it). Thus, our Local Group may be thought of as "falling" in the direction of Virgo because there is somewhat more mass in that region of space. At the same time, the Local Group *and* the Virgo Cluster may be "falling" in the general direction of the constellations Centaurus and Hydra toward an even larger concentration of matter known simply as the Great Attractor. Nevertheless, while the Local Group may be attracted *toward* the Virgo Cluster at about 150 miles per second, the general expansion of the universe is actually carrying us *away* from the Virgo Cluster at about 750 miles per second. So the net result is still an expanding universe. The challenge is to figure out the actual Hubble Flow rate after taking out all the individual motions of galaxies and galaxy clusters. It's a little like watching a bunch of flies swarming around some garbage on the back of a truck and trying to figure out how fast the truck is moving away from you by only being able to measure the speeds of the individual flies.

879 **When astronomers speak in terms of the expansion of the universe, they speak in terms of a thing called the *Hubble Constant*.** In sampling a group of relatively nearby galaxies, Hubble discovered that, on average, the farther away a galaxy was from us, the faster it was moving away from us. A galaxy twice as far away was, on average, moving away from us twice as fast. Thus, the rate at which the universe was expanding, at least in our own local part of the universe, could be described by a number that was *a certain rate of speed . . . for a particular distance from us.*

Specifically, Hubble found that galaxies that appeared to be about 326 million light-years away were traveling away from us at a speed of a little over 300 miles per second, while galaxies that appeared to be about 650 million light-years away were traveling away from us at a speed of over 600 miles per second. (Twice as far—twice as fast.) Because astronomers typically use kilometers instead of miles and, for really large distances, they use megaparsecs instead of light-years (1 megaparsec equals 1 million parsecs, which equals 3.26 million light-years), the Hubble Constant is usually written in units of kilometers per second per megaparsec (or km/sec/mpc). Again, this is just a speed . . . that grows larger . . . as the distance away from us increases. Converting the above numbers that Hubble got into kilometers per second from miles per second and into megaparsecs from millions of light-years, we get a Hubble Constant, that is, a rate of expansion of the universe of about 500 kilometers per second per megaparsec.

880 **As it turns out, Hubble was wrong in his estimation of the speed of universe expansion.** Hubble wasn't wrong about the fact that the universe was expanding. It was. And he wasn't wrong about the fact that, on average, the farther a galaxy is away

from us, the faster it is moving away from us. He was right about that, too. He was just wrong in his estimate of the *distances* to the galaxies, which meant that his estimate of how *fast* the universe was expanding was wrong. The distance of about 2.2 million light-years that I gave to the Andromeda galaxy above is actually the modern accepted value. But the number Hubble came up with at the time was only a little over 850,000 light-years and his estimates of the distances to more distant galaxies was even farther off the mark. The problem was that the Period-Luminosity Relationship he used for the Cepheid variables in Andromeda turned out to be incorrect. Since then, astronomers have corrected the error and obtained the correct distances. This in no way, however, diminishes what Hubble did. The discovery that we live in an expanding universe was still one of the most profound discoveries ever made.

881 **Okay, so what's the real value of the Hubble Constant—the real rate of expansion of the universe?** The answer is: We're still not sure. There have been a number of problems in trying to nail down the Hubble Constant. Hubble began a terribly profound piece of work: figuring out that we live in an expanding universe. But his data were based on a small sample of galaxies that were all pretty close to us. Would astronomers get the same expansion rate for the universe as a whole if they looked farther and farther from home? As telescopes got bigger and bigger, film got more and more sensitive, and finally the CCD came in to replace film, astronomers were able to measure the red shifts of more and more distant galaxies (that is, fainter and fainter galaxies). That was *half* the battle. To measure the Hubble Constant to greater and greater distances from Earth, however, astronomers also had to determine the *distances* to these galaxies. Using Cepheid variable stars only worked out to a certain distance. Beyond that, even the brilliant Cepheids became too faint to see. So astronomers cleverly turned to other ways of determining the distances to galaxies.

882 **Other clever methods were dreamed up for measuring the Hubble Constant, all aimed at determining the distances to more and more distant galaxies.** In each case, the basic concept was the same. First, think of a type of celestial object that could be seen over greater distances than the Cepheids. Second, find such an object in a distant galaxy and measure how bright it is. Third, find the same type of object in a galaxy whose distance you already know. Fourth, calculate the distance to the distant galaxy based on how much fainter the celestial object looks in that galaxy than in the galaxy of known distance. Clever stuff.

So astronomers set about . . .

✧ Looking for brilliant O and B stars in other galaxies and comparing their apparent brightnesses in distant galaxies to their brightness in closer galaxies of known distance.

✧ Watching for nova explosions in other galaxies and comparing their apparent peak brightnesses to novae in our galaxy or other galaxies of known distance.

✧ Comparing the brightest globular star clusters in other galaxies with the brightness of the brightest globulars in galaxies of known distance.

✧ Watching for Type Ia supernovae in other galaxies and comparing their apparent peak brightnesses to the peak brightnesses of Type Ia supernovae in galaxies of known distance.

✧ Comparing the brightnesses of entire distant spiral galaxies with the same amount of spin as spiral galaxies of known distance with the same amount of spin. (It turns out that the faster stars are spinning around the inner parts of a spiral galaxy, the intrinsically brighter the galaxy is.)

✧ Comparing the apparent brightness of the largest giant elliptical galaxy in different clusters of galaxies (assuming that the largest such galaxy in each cluster was the same size and intrinsic brightness).

. . . and more. Again, the idea was to use different techniques to stretch the distance measurements to galaxies that were successively farther and farther away (like the rungs on a ladder). O and B stars could be brighter than Cepheids and so could be seen over greater distances than the Cepheids. Type Ia supernovae were really bright and so could be seen over even greater distances. Entire galaxies, if consistent in brightness and size, were giant conglomerates of stars that could be seen over even greater distances. And so on. Clever stuff, indeed.

883 Nevertheless, problems have prevented astronomers from setting what the exact Hubble Constant is. The problems have included:

⬥ Different celestial objects served as good distance indicators over different distance ranges, but, in some cases, these ranges did not overlap.

⬥ Where there was overlap, measurements using different celestial objects frequently gave different distance estimates to the same galaxies.

⬥ At times, astronomers weren't sure how much correction to make for any interstellar dust that might be along the line of sight. (Remember, dust dims the light of more distant objects and thus makes them seem to be farther away than they really are.)

⬥ As astronomers looked to greater and greater distances, the celestial objects under scrutiny got progressively fainter and fainter, which, in turn, made the data less reliable.

⬥ As astronomers looked farther and farther into space, they were also looking farther and farther back in time and the nature of galaxies that were farther away seemed to be somewhat different than galaxies that were closer. This, in turn, challenged the assumption that the celestial objects being compared for brightness at different distances were indeed the same.

884 Therefore, different astronomers using different techniques have calculated rather different values for the rate of expansion of the universe. Until recently, cal-culated values for the Hubble Constant have ranged from as low as about 50 kilometers per second per megaparsec to as much as about 100 kilometers per second per megaparsec. That's a factor of two, which didn't narrow things down too much, and that's not very good.

885 Within the last few years, astronomers have made some major strides at nailing down the expansion rate of the universe. Using the Hubble Space Telescope, a team of 20 international astronomers led by Wendy Freedman of the Carnegie Observatories in Pasadena recently made some major breakthroughs. Because of the HST's keener vision, the astronomers were able to see and monitor Cepheid variable stars in galaxies out to distances of 60 million light-years in the Virgo Cluster and another smaller cluster known as the Fornax Cluster.

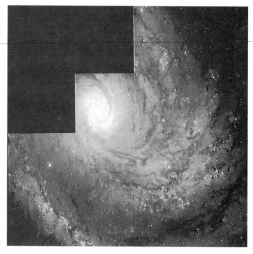

A portion of the spiral galaxy M100 as seen by the HST. (NASA/STScI)

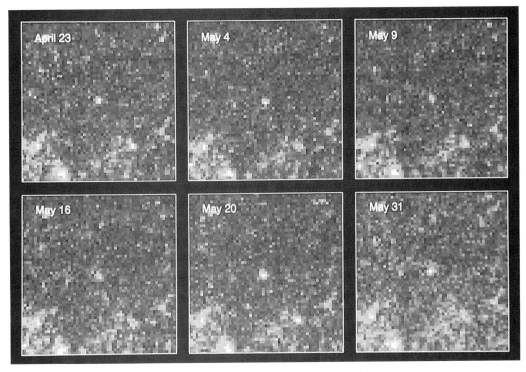

A sequence of images of a Cepheid variable inside M100. The HST has enabled astronomers to see such variables out to greater distances than ever before and thus extend the "cosmic yardstick" such stars provide to help determine the rate of expansion, age, and ultimate fate of the universe. (NASA/STScI)

This allowed for better and more refined data on the Cepheid technique itself and also allowed for more overlap with some of the other methods of determining distances. The team reported results for the expansion rate of the universe that consistently were in the range of 68 to 78 kilometers per second per megaparsec. An expansion rate of 75 kilometers per second per megaparsec, for example, would mean that galaxies would appear to be moving away from us at a rate of 162,000 miles per hour for every 3.26 million light-years farther out we look. This study is continuing, so further refinements may be expected.

886 **Another group of astronomers led by a former assistant to Edwin Hubble has come forward with a somewhat different expansion rate.** In 1996, another research team led by Allan Sandage, also of the Carnegie Observatories, used the HST to measure the distance to a particular barred spiral galaxy in the Virgo Cluster (known as NGC 4639) using Cepheids. The results indicated that this galaxy is 78 million light-years from Earth, farther than any other galaxy in which Cepheids have been measured. NGC 4639 was of particular interest because, in 1990, a Type Ia supernova exploded within it. This, therefore, gave Sandage and his team the chance to tie down the distance to NGC 4639 with two different measuring sticks, namely Cepheids and a Type Ia supernova, and to extend the distance scale thus established to much farther galaxies, since Type Ia supernovae, at peak brightness, can be seen over distances 1,000 times greater than Cepheids. In so doing, however, Sandage and his team got a value of only 57 kilometers per second per megaparsec for the expansion rate of the universe—well below the range reported by Freedman's team. The research, the debate, and the discrepancies continue and serve to point out both the difficult and exacting nature of the work involved and the fact that science is an evolving and human process.

887 **Plotting the distances to more and more distant galaxies has also given us a three-dimensional map of the universe.** Beginning in the 1980s, several groups of astronomers, including Margaret Geller, John Huchra, and their colleagues at the Center for Astrophysics at Cambridge, Massachusetts, began the massive undertaking of mapping the universe. To date, the red shifts of tens of thousands of galaxies have been obtained and the distances to these galaxies have been estimated. Most of the work covers several thin "pie wedge"-shaped slices of the sky out to over 600 million light-years in the northern sky and almost three times as far in a survey done from the southern hemisphere in Chile. In spite of the great distances and large numbers of galaxies surveyed, the maps thus far constructed only cover a portion of the universe smaller in relative terms

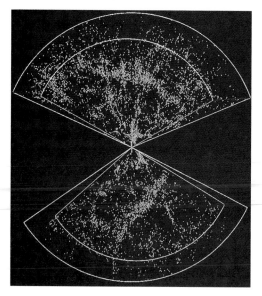

Observations made in "pie wedge"-shaped regions of the sky begin to reveal the large-scale structure of the universe. They show that galaxies are not distributed uniformly through space but are distributed as if on the surface of gigantic soap bubbles surrounding great regions quite devoid of luminous matter. (Margaret Geller and John Huchra, Smithsonian Astrophysical Observatory)

than the area of Vermont compared to the area of the rest of the Earth. Over the coming years more "pie wedge" pieces will be put into place as, gradually, large parts of the sky come under surveillance.

888 **Today's technology makes such large-scale surveys of the universe possible.** Thanks to incredibly sensitive imaging devices called charge-coupled devices, or CCDs, and other clever applications of technology, today's astronomers can study far more objects in far less time. In his day, Edwin Hubble frequently had to take all night or even a succession of nights to record the spectrum of a single faint galaxy. Today, astronomers can record the spectrum of a galaxy or even measure the spectra of dozens of galaxies simultaneously in only minutes. And it's a good thing because it's a big universe and as Margaret Geller is fond of saying, "So many galaxies . . . so little time."

---

889 **If you could construct a giant jigsaw puzzle of the universe, what would it look like?** This is one of those really big questions. And while the above surveys have only put a few of the puzzle pieces into place so far, some striking large-scale detail can already be seen. In short, throughout the universe, clusters and superclusters of galaxies seem to align themselves along great curved structures that surround huge, roughly spherical regions (called *voids*) where few galaxies exist. Simply put, the universe looks like a giant clump of soap bubbles with the galaxies distributed on the surface of the bubbles. Some of the voids so far discovered are more than 300 million light-years across, yet each amounts to no more than 1 percent of the size of the volume of observable universe.

---

890 **China may lay claim to a Great Wall, but there is a Great Wall of Galaxies that has it beat by more than a few miles.** The most prominent feature within the Center for Astrophysics galaxy map is a colossal "structure" dubbed the Great Wall of Galaxies. Stretching some 200 million light-years high and 500 million light-years long, it forms the boundary between two voids and con-

tains thousands of galaxies, including the great Coma Berenices Cluster.

891 **Other recent studies extend these surveys to even greater distances.** Other recent studies have pushed the galaxy mapping surveys out to over 4 billion light-years in some directions. They confirm the "bubbles and voids" structure of the universe out to a billion light-years or so. But beyond that point, the distribution of galaxies seems to become more uniform.

892 **The concept of an expanding universe suggests that, at some point in the past, the universe had a beginning.** If the universe is expanding and, on average, the galaxies are getting farther and farther apart, then this suggests that in the past the galaxies were closer to each other than they are now and that, if we could trace the universe far enough back in time, we would reach a point when the universe originated from a very dense state.

---

893 **The birth of the universe is referred to as the *Big Bang*.** Once scientists realized that the universe was expanding, it was a simple matter to "run the movie backward in their minds" and envision a moment when the universe came into being as a very dense, very compact entity. They refer to this moment as the birth of the universe, or the Big Bang. The name may not be elegant, but no one has come up with a better one or at least a catchier one—yet.

894 **It is beyond modern science to say what happened before the Big Bang.** Our universe, including space, matter, energy, and time, all came into being at the moment of the Big Bang. Since time did not exist before the Big Bang, the notion of "before the Big Bang" is meaningless and tantamount to asking, "What's north of the North Pole?"

895 **In spite of its name, the Big Bang should not be thought of as an explosion.** The Big Bang was *not an explosion* in which matter and ultimately galaxies began to fly apart. Instead, it was *a sudden expansion of space* that *carried* matter and energy along with it.

896 **The Big Bang should also not be thought of as having taken place at a particular point in the universe.** This is also an incorrect concept but one that's sometimes hard to lose. The point is that the Big Bang did not occur at a particular point *in* the universe. The entire universe was the point that began to expand at the moment of the Big Bang. Hence, there is no center to the universe today and there is no spot in the universe today where you can point and say, "The Big Bang happened *there.*" The Big Bang involved the *entire universe.* So the Big Bang happened *everywhere.*

897 **If the universe is expanding, what is it expanding into?** This is a frequently asked question, but the answer isn't always very satisfying to people. In point of fact, the universe isn't expanding into *anything.* When astronomers say that the universe is expanding, they mean that *space itself* is expanding—all the space that defines the universe is expanding—the space that you and I and all the galaxies are in. In two dimensions, it's as if all the galaxies are pins stuck in a rubber sheet and the sheet is being stretched larger and larger. In getting farther and farther apart, the galaxies aren't expanding into a vast emptiness beyond. The universe itself is expanding and the galaxies are just going along for the ride. Put another way, the galaxies aren't expanding *into* space but rather *with* space. When seen in this way, red shifts more correctly apply to the expansion of the universe itself rather than the individual recessional velocities of individual galaxies.

898 **Does this mean that the galaxies are expanding, too?** Essentially, no. All the stars, gas, and dust in individual galaxies are gravitationally bound together and this keeps the galaxies themselves from really expanding as the space around them does.

899 **Following World War II, a group of scientists suggested that the Big Bang should have "left a mark" detectable to this day.** By 1948, theoretical physicists George Gamow, Ralph Alpher, and Robert Herman developed a rather formal theory. In part, it suggested that if a Big Bang had occurred at the birth of the universe, a very hot fireball would have resulted and the radiation from that fireball would have expanded, as the universe has expanded, ever since. Furthermore, they predicted that in the expansion, the radiation from the primordial fireball would have continued to cool until, today, it would be barely 3° above absolute zero. The radiation would be coming from all over the sky because it would have filled the universe and would thus be a kind of *background radiation*, a kind of cosmic wallpaper, on which and in which we and all the galaxies dwell.

900 **Scientists (and most of the rest of the world) use different temperature scales than the one commonly used in the United States.** In the United States, the Fahrenheit temperature scale is in common use. Indeed, because most people are used to it, I have used it extensively throughout this book. But in much of the rest of the world and in scientific circles, other scales are used. On the Fahrenheit scale, water freezes at 32°F and boils at 212°F. In common use in much of the rest of the world is the Centigrade or Celsius scale. On this scale, water freezes at 0°C and boils at 100°C. Since it takes 180 Fahrenheit degrees to go from freezing to boiling and only 100 Centigrade degrees to do the same, we see that each Centigrade degree is equal to 1.8 Fahrenheit degrees.

901 **The introduction of the Kelvin temperature scale, or How low can you go?** While there seems to be little limit to how high temperatures can get in the universe, there is a definite limit to how low they can go. In experiments that continue to this day in a branch of science known as *cryogenics,* scientists have performed experiments taking substances to very low temperatures. Both theory and experiment confirm the fact that there is, in effect, an absolute lowest temperature in the universe—a temperature at which all motion at the atomic level within an object would cease. This temperature happens to correspond to about −457° on the Fahrenheit scale or −273° on the Celsius scale. Since this corresponds to the absolutely coldest temperature in the universe, scientists decided to establish a new temperature scale whose degrees would each be the same size as a Celsius degree but where "zero" would correspond to the absolutely coldest temperature possible—a temperature that was aptly named *absolute zero* or zero degrees Kelvin.

902 **Scientists realized that actually detecting the 3° Kelvin background radiation coming from space would serve as proof that the universe was born in a Big Bang (and just might be worth a Nobel prize).** In 1965, Arno Penzias and Robert Wilson of the Bell Telephone Laboratories built a horn-shaped antenna to see if they could find a sign of the predicted background radiation. From all over the sky, their antenna indeed picked up a low uniform level of microwaves that could not be accounted for by any previously known object in space, the antenna itself, or anything in the laboratory environment. The microwaves were coming from everywhere, filling all of space, and had precisely the strength at different wavelengths (that is, they had the spectrum) that would be expected to be coming from something that had a temperature barely 3° (actually, about 2.73° Kelvin) above absolute zero. The existence of Gamow's, Alpher's, and Herman's 3° Kelvin cosmic background radiation was dramatically confirmed, the Big Bang was confirmed, and Penzias and Wilson were eventually awarded the Nobel prize.

---

903 **The cosmic background radiation contains more energy than the entire rest of the universe.** Although it is very weak at only 2.73° above absolute zero and requires very sensitive instruments to detect, the cosmic background radiation still contains more energy than all the stars and galaxies in the universe. This is because this radiation is all-pervasive and covers the entire sky, in effect, all of space. In contrast, the stars and galaxies only cover tiny points on the sky and the fact that the stars (and their galaxies) are very hot by comparison doesn't make up for their small apparent sizes.

---

904 **In 1990, a satellite named *COBE* took our most detailed look yet at the cosmic background radiation.** In its first nine minutes of operation, the *COBE (Cosmic Background Explorer)* satellite gathered more data about the cosmic background radiation from the Big Bang than Penzias, Wilson, and a host of other astronomers had been able to do in all the years since 1965. (Such is sometimes the power of using a new satellite for the first time, especially when it's looking

with great sensitivity at an area of the spectrum not studied in great detail before.) *COBE* confirmed in stunning detail that the universe was radiating all over at 2.735°K and, more significantly, that this radiation was the same in every direction to an accuracy of better than 0.0004 percent. Since this radiation was the very remnant of the Big Bang itself, its smoothness across the sky meant that the Big Bang and the radiation that was present in the very early universe was very smooth.

905 **The universe today is far from smooth, which leads to a mystery.** The universe we see around us today is very *lumpy*. Planets are lumps with very little matter in between. Stars are lumps separated by tenuous gas and dust. And the same is true for galaxies and clusters of galaxies. Yet the cosmic background radiation is incredibly smooth. How, then, did a universe that was incredibly smooth in its youth evolve into the very lumpy universe we have today? That's one of the great $64,000 questions of modern astronomy and we'll have more to say about it later in this chapter.

906 **How long ago did the Big Bang occur?** This, of course, is another way of asking, "How old is the universe?"—a pretty profound question. To get the answer, however, we need to know the rate of expansion of the universe but we also need to have a handle on how much matter there is in the universe. In effect, if there is a lot of matter in the universe, then the total gravitational force generated by all that mass would mean that the universe wouldn't have to be very old to have the galaxies be as far apart from each other as they are now. Hence, the universe could be relatively young. On the other hand, if there is less total mass in the universe, the overall gravitational force would be less and so more time would have had to elapse since the Big Bang for the galaxies to have reached their current distances.

---

907 **So how much mass is there in the universe?** At first, getting the answer to this question might seem like a laborious but straightforward task. We know the mass of individual stars. We have a good estimate of the number of stars in different types of galaxies as well as the amount of interstellar gas and dust. So, if we count up all the galaxies in the universe and add up the mass of all their stars, gas, and dust, we should have an answer. Astronomers have done precisely this, but the answer they get is far too low. In fact, when we add up all the matter that we can account for in all the searches and surveys done with all the telescopes in the world, we still only wind up with no more than 10 percent of all the matter that must be there. An embarrassing and disconcerting situation, to say the least.

---

908 **How do we know there must be more matter in the universe than we see?** In the 1920s, the Dutch astronomer Jan Oort realized that stars moving above and below the disk of the Milky Way behaved as if there was at least 50 percent more matter in the disk of our galaxy than could be accounted for. In the 1970s and 1980s, American astronomer Vera Rubin and her associates also did careful studies of how fast stars were orbiting around the centers of the Milky Way and other spiral galaxies. It was assumed that such stars should orbit the centers of their galaxies like the planets orbit our Sun. That is, stars closest to the center of the galaxy should orbit the fastest and those progressively farther out should orbit slower and slower. In reality, however, most of the stars beyond a few thousand light-years of the center all orbit at almost the same speed. This curious finding could also be accounted for if there was about 10 times more matter present in these galaxies than could be seen. Some of this matter would be in the disks of these galaxies, but most would have to be distributed in vast halos surrounding the galaxies and extending well beyond the halos occupied by the globular star clusters.

---

909 **The motions of galaxies in clusters also reveal that there is more to the universe than meets the eye.** Studies of clusters of galaxies only add to the mystery and make matters worse. If astronomers tally up all the mass they can see in the form of stars, gas, and dust in various clusters of galaxies from the Local Group to those in Virgo, Coma Berenices, and elsewhere, they are faced with the embarrassing conclusion that these clusters should not exist. That is, there is so little total mass and, therefore, so little gravitational force within these clusters that they should have broken up, losing their members to intergalactic space, long ago. Yet, obviously, the clusters do still exist, meaning there must be more matter present in these clusters than we can see. How much more? The results of the motion studies suggest that all the matter we actually see in the clusters is no more than 5 percent—and possibly as little as 1 percent—of all that must be there! The additional matter that we don't see has been dubbed *dark matter* by astronomers and is assumed to account for our universe's apparent "missing mass."

910 **Could more dark matter be located in the great voids between galaxies mapped out by Geller and others?** Yes. The voids may turn out to be regions that are devoid of luminous matter but still filled with dark matter. The places where we see galaxies may just be places where more dark matter happened to accumulate in the early history of the universe and provided enough local gravity for the galaxies to form. Should copious quantities of dark matter exist in the voids, it could add greatly to the amount of total mass we might expect to find in the universe and thus account for a significant fraction of the universe's missing mass.

911 **Dark matter represents one of the most profound puzzles in astronomy today.** It is important to note that the dark matter, which must exist in the universe, is *not* synonymous with the unilluminated in-

terstellar dust found in the Milky Way and other galaxies but rather some entirely different material. Whatever the dark matter is, it makes most of what *is* the universe, even though its nature remains unknown. In short, not only do we not know exactly *where* most of the matter in the universe is, we don't even know *what* it is.

912 **Searching for the dark matter: In this corner, we have the MACHOs.** Lots of ideas have been suggested for what might constitute the dark matter. In general, the current candidates for dark matter fall into two categories known officially as the WIMPs and the MACHOs. The MACHOs, or MAssive Compact Halo Objects, are all objects of significant mass. Suggestions for possible MACHOs include:

⬦   billions upon billions of brown dwarfs that are simply too faint to be seen over great distances in either the visible part of the spectrum or the infrared

⬦   numerous "unattached" black holes that have no nearby matter to feed upon and heat up to the point of visibility, and

⬦   numerous black dwarfs (white dwarfs that have cooled off and turned cold).

913 **And in this corner, we have the WIMPs.** WIMPs, or Weakly Interacting Massive Particles, in spite of that word *massive* in the title are assumed to be either:

⬦   one or more types of subatomic particles that are yet to be discovered, or

⬦   particles that are already known but

actually have more mass than we think they do.

The second group refers in particular to a family of tiny particles known as *neutrinos.* Neutrinos have long been assumed to have virtually no mass, but if they should prove to have even a little, it could be significant because there are so many of them flying all over the universe.

While most astronomers feel that there are probably a goodly number of various types of these MACHOs roaming through space, most feel that the vast majority of the missing mass must be in the form of a type or a variety of WIMPs that would not only occupy space within galaxies but also around them and, quite possibly, also significantly fill the great voids between the galaxies.

---

914 **The universe will have one of three ultimate fates.** One, the universe will continue to expand forever. Astronomers call a universe that behaves this way an *open universe.* Two, the universe will someday stop expanding and then collapse. Astronomers call such a universe a *closed universe.* Or three, the universe will continue to expand, but the rate of expansion will slow more and more over time such that, at an infinite time in the future, the universe would stop expanding—in other words, the expansion of the universe is coming to a stop, but it will take forever to actually get there. This third scenario is what astronomers call a *marginally open universe.*

915 **The universe will either end in *fire* or in *ice*.** If our universe is an open or marginally open universe, it will continue to expand well after the last star in the last galaxy has burned itself out—until the universe is a bunch of cold burned-out cinders separated by increasingly greater amounts of space. Such a universe will end in the ultimate *Big Chill.* If, however, our universe is closed, then someday it will begin to collapse, contracting to a denser and denser, hotter and hotter state—and, ultimately, a *Big* (and very hot) *Crunch.* In open and marginally open universes, time and space go on forever. In a closed universe, both time and space come to an end.

916 **Could a Big Crunch lead to another Big Bang?** Possibly. Another interesting possibility is that a collapsing universe might grow sufficiently hot and dense to overcome gravity at the last instant and, in effect, undergo a *Big Bounce.* This, therefore, leads to yet another possibility, namely an *oscillating universe* with, perhaps, an endless cycle of beginnings. In such a universe, technically, there would be no beginning of time. Time and space would have always existed.

917 **The kind of universe we live in depends on how much mass there is in the universe.** To help visualize this, imagine a "thought experiment" where a bunch of people all over the Earth are going to be tossing balls up into the air at the same time with the same amount of force. We're going to watch from a point out in space. At the count of three, everyone throws a ball into the air. As expected, each ball rises to a certain height, stops, and then falls back to Earth. But now let's make believe that we can magically change the mass of the Earth and run the experiment again. We decrease the mass of the Earth and again count to three. This time the balls rise higher before they stop and return to Earth because there is less mass and, therefore, less gravitational force pulling them back. If we could reduce the mass of the Earth enough, our players would finally be able to toss the balls so high they would just barely make it back to Earth. And reducing the Earth's mass even more would result in the balls flying off into space and never falling back to Earth. A parallel can be drawn between this situation and the expanding universe. The galaxies are the equivalent of the balls and their combined mass (including the mass of the dark matter) generates a certain total gravitational force. The more total mass there is in the universe (and the *higher the density* of the universe), the more likely we are to live in a *closed universe.* The *lower the density,* the more likely it is that the galaxies will keep flying away from each other forever and give us an *open universe.* And a certain *critical density* that's neither too great nor too small will be just the right amount to give us a *marginally open universe.*

918 **The density of matter in the universe also affects its *shape.*** The mass of an object actually warps space (and time) in its vicinity. This is a confirmed prediction of Einstein's General Theory of Relativity. With this in mind, Einstein and others took this notion to the next step and applied it to the whole universe. In effect, the total mass in the universe—or, more correctly, the average density of the universe—will determine the amount of warping of space that occurs throughout the universe and hence the geometry of the universe as a whole. If the average density of the universe is *greater* than the *critical density,* gravity is strong enough—that is, space is warped or curved enough—to actually *curve back on itself* and *close* the universe off. Hence, the term *closed universe.* Such a universe is also said to be bound and finite, but you still can't get to its edge because there isn't one. In three dimensions that's hard to imagine, but in two dimensions we can find a reasonable analogy. Imagine the universe is a balloon and the galaxies are little spots on the balloon. If we were restricted to only traveling on the surface of the balloon, we could travel from galaxy to galaxy all over the universe and never reach an end. There would also be no center to such a universe, since, living in two dimensions, we would not be able to see the center of the balloon—just its surface. (We would, however, be surprised to find that after traveling far enough in what we thought was a straight line, we would be back where we started from.) In contrast, if the universe has less than the critical density, it is an open universe. Thus, it will not curve back on itself but instead will be open and infinite.

919 **Whether we live in an open, a closed, or a marginally open universe de-**

pends on two things: the amount of mass in the universe and the rate of expansion of the universe. If we assume that the Hubble Constant (the rate of expansion of the universe) is about 75 kilometers per second per megaparsec and add up all the matter we can see in the universe, we wind up with only about 1 percent of the mass needed to close the universe. Under these conditions the universe is open, will expand forever, and end in the Big Chill. But as we have seen, studies of the motions of stars and galaxies tell us that we only see a small percentage of all the matter that must be there. If we add in all the dark matter that we know must be there and again assume that the Hubble Constant is 75 kilometers per second per megaparsec, we still don't have enough matter to close the universe, but we get somewhat close, by some estimates, to the amount of matter we would need to have a universe poised between open and closed. Nevertheless, the best data at present still suggest that we live in an open universe.

920 **Based on the uncertainties in the Hubble Constant and the amount of total mass in the universe, the best current estimate for the age of the universe is between 10 and 14 billion years.** That is, between about 10 and 14 billion years have elapsed since the Big Bang. But if you asked a dozen different astronomers, you would probably get twelve different answers.

921 **There is another way of estimating the age of the universe.** Another way to try to get a handle on the age of the uni-

verse conveniently ignores all the red shifts and galaxy counts and galactic distance measurements. Instead, it approaches the problem from a very different direction and tries to answer the question "How old are the oldest stars in the universe?" After all, the universe has to be at least a little older than the oldest stars in the universe. Globular star clusters contain some of the oldest stars, so if we plot their H-R diagrams and apply what we know about the evolution of such stars, we can determine their ages and hence put a lower limit on the age of the universe itself.

922 **The answer we get for the ages of the oldest globular clusters is a bit distressing.** There is some disagreement on the answer, but, in general, most astronomers estimate the ages of the very oldest of these clusters at between 13 and 17 billion years. As you can see, there is some overlap between the answer obtained by this approach and that obtained from the galaxy data—but not much. If the Hubble Constant turns out to be much over 75 kilometers per second per megaparsec and the universe is open, the age of the universe will drop to less than 13 billion years. Since you can't have stars in the universe older than the universe itself, some rethinking is going to be necessary in our understanding of stellar evolution. Conversely, if the universe is marginally open and we want to believe that some globular clusters have stars that are 16 to 17 billion years old, the Hubble Constant is going to have to come in around 50 kilometers per second per megaparsec—and that is out of the range of the best currently accepted values. Stay tuned.

923 **Many theoreticians today favor a marginally open universe.** Any value for the density of the universe below the critical density will lead to an open universe that will expand forever. Any value for the density of the universe that is above the critical density will result in a closed universe and a Big Crunch. And only one value for the density of the universe would lead to a marginally open universe—a universe perfectly poised between being closed and being truly open. Normally, one would not expect nature to come up with precisely the right density to do this, but some scientists believe that nature just may have done precisely that.

924 **The rate of expansion of the universe is not constant.** Whether the universe is open or closed, the galaxies, like balls thrown in the air, are not moving apart today as rapidly as they were in the past. Therefore, we should be able to get a handle on the age of the universe by measuring red shifts of progressively more and more distant galaxies. On average, are the more distant galaxies moving away from us faster than the closer galaxies? The answer seems to be yes, but so far the data is not nearly good enough to tell us exactly how much faster at exactly what distance. Problems again arise from the fact that very distant galaxies are very faint and therefore don't yield good spectra in short

periods of time and from the fact that methods of determining distances to such distant galaxies are still plagued with significant uncertainties. As telescopes and detectors get more and more efficient, this situation will change, but, for now, we have to try to get a handle on the age and destiny of the universe by using other means.

925 **The HST has looked at the most distant galaxies ever seen.** In 1996, astronomers used the HST to look for galaxies farther out into space than were ever seen before. The study, called the Hubble Deep Field study, covered an area of the sky near the Big Dipper only about one-thirtieth the diameter of the full Moon but revealed galaxies never seen in any telescope before. Some are almost as faint as magnitude 30 (that's 4 billion times fainter than the faintest stars that can be seen with the naked eye and 10

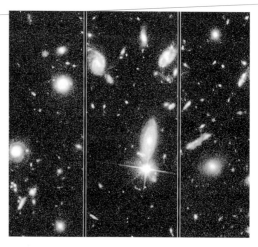

Images of distant galaxies from the Hubble Deep Field. (NASA/STScI)

times fainter than any galaxies ever seen with ground-based telescopes).

926 **The Hubble Deep Field images also allow us to see galaxies farther back in time than ever before.** Since as we look out into space, we are also looking back into time (in effect, seeing the successive "layers" of the universe as it looked at successive epochs in the past), the Deep Field study is allowing astronomers to compare what galaxies were like in the near past to what they were like at successively earlier times in the history of the universe, as much as 10 billion years back in time—to perhaps an era less than a billion years after the Big Bang. By comparing galaxies at different distances within the region imaged by the HST, astronomers are learning how galaxies evolved and, in particular, how the colors, sizes, and general types of galaxies changed from the time they first started forming until today. In short, they are getting important clues to how the universe itself has evolved.

927 **The Hubble Deep Field study is expected to tell us much, even though it covers a tiny part of the sky.** While three-dimensional maps of galaxies out to several hundred million light-years show great variations in the distribution of galaxies ("the bubbles and voids"), surveys that have looked farther seem to see a smoothing out of this distribution. For this and other reasons, including the great smoothness of the cosmic background radiation, astronomers are quite confident that, at the limits of the Deep

Field, the universe would look very similar in all directions. Thus, they feel that they will probably be able to generalize their findings about the tiny slice of the universe in the Deep Field study (kind of a "keyhole" view) to the early universe as a whole.

928 **The HST has revealed a bewildering menagerie of galaxies and galaxy types in the early universe.** While analyses will continue on the Hubble Deep Field study, preliminary results suggest that, in the early universe 1 billion or so years after the Big Bang, there were more galaxies than there are today (perhaps ten times as many) and that many more of these were irregular galaxies than we see today. The fact that there were more irregular galaxies in the past may make sense. In effect, if there were more

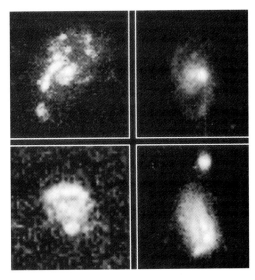

A gallery of images of peculiar, very distant galaxies as taken by the HST. Such galaxies seem to have been far more numerous in the early universe than they are today. (NASA/STScI)

galaxies in the past and, in addition, the universe was smaller than it is today, that means that the galaxies back then were much closer together. If they were closer together, more collisions between galaxies would have taken place, which would have led to more irregularly shaped galaxies. Furthermore, those frequent early collisions would have also resulted in some galaxies cannibalizing others, which, in time, would decrease the total number of galaxies.

929 **The Hubble Deep Field study should also help improve the cosmic distance scale and hence our estimates of the Hubble Constant.** As we have seen, one of the challenges of finding the expansion rate of the universe is accurately knowing the distances to more and more distant galaxies. And while astronomers have used a number of clever techniques, some of these techniques rely on comparing galaxies of similar apparent sizes and colors. For example, they might assume that if one giant elliptical galaxy appears half the size of another, the smaller one is twice as far away. But if galaxies gradually change their appearances and hence their intrinsic brightnesses as time goes by, we must understand how and when these changes occur and take them into account in determining distances. The Deep Field study will go a long way toward helping astronomers to do this.

930 **Does the Hubble Deep Field study allow us to look as far back in time as we can?** The Hubble Deep Field study allows us to peer back perhaps to within 1 billion years of the time of the Big Bang—back to an epoch shortly after the first galaxies formed. But it is possible to look even farther back in time to an even earlier stage in the history of the universe. In fact, we already have. The radiation that brings us this picture comes from so early a time in the history of the universe that galaxies had not yet formed. The radiation is the 3°K cosmic background radiation from the Big Bang itself.

931 **Does seeing the 3°K cosmic background radiation mean that we are seeing back to the moment of the Big Bang itself?** No. The cosmic background radiation allows us to see hundreds of millions of years farther back in time than we see when we look at the most distant galaxies from the Hubble Deep Field study but it still only takes us back to within a couple of hundred thousand years of the Big Bang itself. We cannot see farther back in time than this because our telescopes cannot penetrate the leading "edge" of this radiation, which looks like a smooth dense "fog" in all directions. (We'll explain why this veil of "fog" is there and what it means later.)

932 **Using other equipment, scientists can penetrate this veil of radiation almost back to the moment of creation.** Ironically, although we cannot *see* closer to the Big Bang than a couple of hundred thousand years, scientists now believe they have a pretty good understanding of events that occurred as far back in time as a *tiny fraction of a second* after the instant of the Big Bang itself. The equipment used to perform this feat doesn't consist of some new breed of supertelescope but rather the current breed of supercolliders, also known as *high-speed atomic accelerators.* In such devices, atoms and subatomic particles can be smashed into each other at nearly the speed of light and, in such moments, scientists can mimic the enormous temperatures that existed in the very earliest moments of the universe. Before we seek to understand more, however, and take an incredible journey back in time to the very dawn of creation, we need to know a little more about forces and matter and energy.

933 **There are four forces in the universe.** Luke Skywalker and Darth Vader aside, there are four—and only four—forces that govern the entire universe. One of these forces is the *gravitational force,* or *gravity.* All objects that have mass generate a gravitational force around themselves that extends out to the limits of the universe. The gravitational force obeys what is called an *inverse square law.* Simply put, if you double your distance from something, its gravitational force on you becomes one-quarter as great as before. Triple your distance and the force becomes only one-ninth as great, and so on. Normally, a lot of mass is needed to generate much of a gravitational force, but if you squeeze enough mass into a small enough space, it can overpower every other force in the universe—as evidenced in the formation of a black hole.

A second force is known as the *electromagnetic force.* This is the force that causes clothes to cling or hair to "fly away" on a dry day. It's also the force that is exerted by magnets. Unlike gravity, electromagnetic forces can repel as well as attract. Like the gravitational force, the electromagnetic force obeys an inverse square law, but in everyday experiences the electromagnetic force is far more powerful than the gravitational force. (You can demonstrate this by the simple act of picking up a nail with a small refrigerator magnet. Even though the entire Earth is pulling *down* on the nail with its gravitational force, the electromagnetic force the tiny magnet can muster is more powerful, as can be seen when the nail defies the Earth's gravity by jumping *up* to and sticking to the magnet.)

The third force is known as the *weak force.* This force is present in the nucleus of the atom and is responsible for the process known as *radioactive decay* by which some atoms spontaneously change into others.

The fourth force is known as the *strong,* or *nuclear, force* and is the force responsible for holding the atomic nucleus together. It is a very strong attractive force over very short distances (as in the size of an atomic nucleus) but becomes a repulsive force over greater distances.

934 **In the 1800s, what had been thought of as two separate forces were shown to be one.** The English physicist James Clerk Maxwell showed that electric and magnetic forces were, in fact, one force. In the language of physics, these forces were said to be *coupled.* In time, the question arose as to whether or not the other forces in the universe might, under certain circumstances, also be coupled so that all might be looked upon as manifestations of a single force. Einstein spent a good deal of his later career trying to find such a Grand Unifying Theory. After years of work, he still failed, but the work goes on at a variety of universities around the world and enough pieces are in place to allow scientists, also armed with scientific data from collider experiments, to take a "mind journey" from the dawn of time.

935 **Within the first 0.00001 of a second after the Big Bang, the universe was pure energy.** Until about 0.00001 of a second after the Big Bang, the temperature of the universe was still over a trillion degrees—so hot that the universe was simply a blinding sea of radiation. No matter could yet exist. Then suddenly the universe cooled enough for the first subatomic particles to form—particles that would ultimately give rise to protons, neutrons, and electrons, from which atoms would form. At the same time, antimatter particles for each of these particles also formed and soon they annihilated each other in brilliant bursts of gamma rays. The number of particles of matter outnumbered the corresponding particles of antimatter and this remaining matter ultimately built the universe we have today.

936 **About three minutes after the Big Bang, the first atomic nuclei formed.** By this point in time, the universe had cooled to a couple of billion degrees—cool enough for protons and neutrons to begin to bond together to form the first nuclei of atoms. Nuclear isotopes of hydrogen and helium formed, as well as a small amount of the third-most-complex element, lithium. Within another twelve minutes, however, the universe had cooled below the threshold temperature at which nature can fuse elements and so the process stopped. Thus, the early universe consisted of just 3 elements. All the other 89 elements in nature, all the carbon and oxygen and iron and uranium and the rest, would not be forged until billions of years later—in the hearts of stars or the fiery infernos of supernovae.

937 **A few hundred thousand years or so after the Big Bang, the first stable atoms began to form.** By now, the temperature of the universe had cooled to between 3,000 and 4,000 degrees. Previously, electrons had periodically dropped into orbit around atomic nuclei, but the radiation all around them was still so fierce that they were immediately ripped from orbit and thrown back out into free space once again. Now such free electrons could permanently settle in stable orbits and the first atoms came to be. As this happened, something else equally significant began to happen. With more and more electrons coming out of free space and dropping into orbit around atomic nuclei, radiation would not be gobbled up and immediately spit out in a new direction, as had been the case when this radiation had been at the mercy of free electrons. Instead, radiation could begin to travel somewhat unimpeded from place to place in the universe. The universe, dominated until now by blinding radiation, began to become transparent. And that is why, today, as we look out into space and back into time, this is as far as we can see—back to the time when the radiation "fog" began to clear.

938 **By a billion years after the Big Bang, the large-scale structure of the universe was in place, quasars were shining brightly, and the first galaxies had formed.** These are probably among the most distant galaxies now observed by the HST. The galaxies likely formed around places where there were more than average concentrations of dark matter whose gravitational force acted as a "catalyst" for the luminous matter to congregate. And that brings us into epochs of time already covered elsewhere.

939 **Can we go back still earlier in time, closer than 0.00001 of a second after the Big Bang?** Yes. But before we do, we need to understand a little about a thing called *scientific notation,* which is simply a way of more easily writing numbers that are very big or very small.

Since many of the quantities we encounter in the very early universe are either very big or very small, this way of writing numbers can really come in handy. In writing numbers in scientific notation, we express the number in terms of its "power of 10," which is just another way of talking about how many times you have to multiply the number 10 by itself to get the number in question. For example, the number 100 is the same as 10×10, or 10^2, that is, 10 to the "power of 2," or the second power. Similarly, the number 1,000 is just $10 \times 10 \times 10$, or 10^3, that is, 10 to the "power of 3," or the third power. Notice that the power of 10 is just equal to the number of zeros behind the 1 in the long form of the number. So far, it might seem easier to just write 100 or 1,000, but what about the number one quadrillion? We could write out the whole number, which looks like 1,000,000,000,000,000, or we could simply write 10^{15}, which is 1 with 15 zeros behind it. In the universe, there are even bigger numbers, so scientific notation saves a lot of time and ink. Other numbers can also be written in scientific notation. For example, 2,000 is simply 2×10^3.

Similarly, very small numbers can also be written in scientific notation. $^1/_{10}$, for example, is written 10^{-1}, that is, "10 to the minus 1"; $^1/_{100}$ becomes 10^{-2}; $^1/_{1,000}$ becomes 10^{-3}; and so on. Again, the number of the power equals the number of zeros. And, for example, $^2/_{1,000}$ becomes 2×10^{-3}. That's all there is to it. And now that you have mastered scientific notation, you're ready for a ride in the ultimate "way-back machine."

About 0.00001 of a second after the Big Bang, the first subatomic particles of matter formed and became stable. To venture back earlier in time, we enter a universe that was pure energy and where the four basic forces of nature begin to lose their identities. At one-one-hundred-billionth (that's

1/100,000,000,000 or 0.00000000001) of a second after the Big Bang the temperature of the universe was 20 quadrillion°F (that's 20,000,000,000,000,000). (Using scientific notation, this just becomes—at 10^{-11} seconds after the Big Bang, the temperature of the universe was 2×10^{16}°F.) At this temperature, the electromagnetic and weak forces became one. That is, they became indistinguishable. Physicists say they *coupled* and nature got simpler, more *unified.*

940 **At 10^{-35} seconds after the Big Bang, the strong, or nuclear, force couples.** At this point, the temperature of the universe has reached an incomprehensible 2×10^{28} degrees. At this moment, the strong force couples to the electromagnetic and weak forces. There are now only two forces in the universe.

941 **You have to have GUTs to understand the universe at this stage.** The theories that take us to this very early point in the universe are known as Grand Unifying Theories, or GUTs. We say GUTs with an *s* because, over the last few decades, there have been more than one. Over time, some of them have done reasonably well at accounting for details in the early universe, as well as explaining how it evolved into the universe we have today and some of them have not.

942 **Inching still closer to the Big Bang requires a Theory of Everything.** At 10^{-43} seconds after the Big Bang, gravity, the last remaining unique force, finally couples to

the others. There is now only one force governing the entire universe—a universe that is far smaller than the nucleus of a single atom and whose temperature is more than 10^{33} degrees. But for now, this is where our journey must end. To go any farther back in time, a new theory is necessary—one that explains how gravity works in a world of truly minute dimensions, a theory that requires an understanding of what is called *quantum gravity.* Scientists simply refer to this theory as the Theory of Everything and, so far, it doesn't exist. Someday, should we have such a theory in hand, we may be able to understand what the universe was like back at the instant of the Big Bang itself.

943 **Some GUTs also helped to solve a couple of ugly problems in cosmology that were uncovered back in the 1970s.** In the 1970s, as cosmologists were trying to explain the evolution of the very early stages of the universe, three significant problems continued to plague them that the earlier Big Bang theory couldn't explain. The first of these came to be called the *horizon problem;* the second, the *flatness problem;* the third, the *matter-antimatter problem.* The horizon problem (which might also be called the *smoothness problem*) relates to how such a *smooth* early universe could have evolved into the *lumpy* universe we have today. The flatness problem relates to the density of the early universe compared to today. Simply put, if the density of the early universe had departed even *very* slightly from the *critical* density, that is, one that leads to a marginally open universe, the universe would have either collapsed in on itself long ago or would have become so diluted so quickly that today's galaxies would still not have formed. The matter-antimatter problem relates to the fact that since matter and antimatter were both created from energy in the early universe, nature nevertheless created more matter than antimatter, leaving a matter universe today. In effect, why should nature have preferred one over the other?

944 **Inflation may be bad for the economy, but it apparently was necessary for the universe to properly evolve at one time.** In the 1980s, astronomer Alan Guth of Stanford University realized that many of the GUTs being developed at the time could actually contain solutions to the above problems if, between 10^{-35} seconds and 10^{-32} seconds after the Big Bang, the universe very briefly changed its rate of expansion. If, during this very brief *inflation period,* the universe had doubled in size every 10^{-34} seconds or so and grew from being far smaller than a proton to the size of a softball (That's about 10^{50} times in a tiny, tiny fraction of a second!), the smoothness and the flatness problems would both be eliminated.

The smoothness problem can in some ways be seen as a *communications problem.* In effect, in the beginning the universe was very smooth throughout, but the current universe seems to be too large to have allowed parts of the early universe to communicate this smoothness to other parts of the universe in later epochs. Through inflation, the early universe expanded so rapidly that this communication could not occur and thus the "message of smoothness" could not be carried over into later epochs.

The Inflation Theory also proves valuable because it virtually guarantees that a critical density before the inflation epoch would have carried over into the postinflation universe. Just as an ant would find a 10-foot balloon far flatter than a 1-inch balloon, so a postinflation universe would appear much flatter than a preinflation universe.

945 **Why was more matter than antimatter created in the early universe?** In many situations, nature seems to like balance and symmetry. Yet in the early universe this seems to not have been the case when it came to the creation of matter and antimatter. Today, we live in a universe that is virtually all matter. Yet, nature might have been expected to have created just as much matter as antimatter. Some GUTs neatly explain the situation in the following way. At the time that matter and antimatter were first being created, most of these particles were in the form of what scientists call *X-bosons* and *anti-X-bosons.* If, however, the X-boson is unstable, as some GUTs predict, these particles would have decayed into protons before all the X-bosons and anti-X-bosons had time to annihilate each other. The surviving particles of matter, in turn, led to the matter universe we have today.

946 **While the *COBE* data showed that the cosmic background radiation was very smooth, they showed that it wasn't per-**

fectly smooth and that left cosmologists very relieved. The fact that all the initial measurements of the cosmic background radiation showed that it was very smooth—that is, very much the same temperature in all directions—posed a problem because we know that we live in a very lumpy universe today. The *COBE* instruments were finally sensitive enough to show very, very slight variations in the temperature of this background radiation—temperature *differences* of no more than 17 millionths of a degree K. These weren't much, but they were welcomed by astronomers, for they also represented regions of very, very slight *differences in density* in the early universe—differences that ultimately, somehow, must have led to the lumpy grand-scale structure of the universe we see today. With such tiny density variations finally found, astronomers could hope to try to explain how "the bubbles and voids" came to be and how many of the galaxies and clusters of galaxies could have begun to form within 1 billion years of the Big Bang. It wouldn't be easy, but they could try.

947 **The little WIMPs may have played a significant role in creating the overall structure of the universe.** It is generally believed that the WIMPs, in one form or another, not only constitute the vast majority of the dark matter in the halos of galaxies and permeating galaxy clusters but that these WIMPs are also responsible for dictating "the bubbles and voids" grand-scale structure of the universe itself. Remember, the *WI*

in WIMPS stands for *weakly interacting.* While this property means that these particles have been extremely difficult to detect (because they don't interact much with either non-dark matter in space or with our scientific instruments) it also means that these dark particles would not have interacted with radiation in the early minutes of the universe like normal matter. Put another way, these particles were unencumbered by radiation as soon as they were created and would have immediately moved out through the universe, carrying variations in their distribution (variations in the density they brought to the universe) with them. In time, these variations in density became the framework around which normal matter accumulated and ultimately became luminous clusters of galaxies filled with stars.

948 **Scientists categorize dark matter as either "cold" or "hot" dark matter.** The *M* in WIMPs stands for *massive,* meaning that these particles are relatively massive in subatomic terms. Cosmologists also refer to this type of matter as *cold dark matter.* In addition to the WIMPs, other weakly interacting particles have been considered as candidates for some or all of dark matter. These particles, which could include neutrinos if neutrinos could be shown to have mass, would have much lower masses than the WIMPs and, as a group, are known as *hot dark matter. Cold* and *hot* simply refer to the average speeds of the particles in question, since speed is usually a function of temperature.

949 **GUTs have incorporated both cold and hot dark matter with different degrees of success.** Most GUTs using hot dark matter have been successful at reproducing the large-scale structure of the universe in supercomputer simulations but have difficulty getting galaxies or clusters of galaxies to form. In contrast, GUTs using cold dark matter have been more successful at producing both large-scale structure and individual galaxies. No model, however, is flawless and the greater general success of the cold dark matter models should not be taken as proof that they are really correct. More work lies ahead for both observational astronomers and theoreticians.

---

950 **What does the future hold?** Progress will continue on many fronts in cosmology, from efforts to get better determinations of the distances to galaxies and better red shifts to more refined studies of where the dark matter may lie, how much of it there is, and what it is made of. The better the data, the more precise the calculations and conclusions that can be made. New breeds of telescopes and detectors will greatly speed the work and many astronomers are hopeful that within the next decade or two, we will actually know the age of the universe to within 5 percent and we will be more certain about the ultimate fate of the universe. And theoreticians will continue to work on a Theory of Everything in an effort to push our understanding back to the instant of creation and to create a grand unifying model of the entire evolution of the universe. It is an interesting time to be alive.

CHAPTER 15

THE SEARCH FOR EXTRATERRESTRIAL LIFE: ARE WE ALONE?

951 The search for extraterrestrial life and especially intelligent extraterrestrial life is a very natural human pursuit. As beings that can look up and wonder at the stars, perhaps the most human thing we can do is to wonder if we are the only creatures in the cosmos that do so. The force that drives us to find out is the same that drew our ancient ancestors out of the cave to pursue what was over the next hill or beyond the river or ocean. It is a force that *makes* us human. The answer will take us beyond where any and all human minds have gone before.

952 In trying to evaluate the likelihood of extraterrestrial life and seeking it out, the guiding principle is: Seek out conditions that are believed to have been important for life developing on Earth. Then search for those conditions elsewhere in the universe. If those conditions are common, life may be common. If those conditions are rare, life may be rare.

953 When life first developed on Earth, this planet was a very different place. Most scientist believe that life on Earth first developed in the seas, in shallow tropical tide pools or moist deposits of clay. At the time, however, the Earth's atmosphere was not rich in oxygen as it is today. Indeed, oxygen, a gas that we and most other forms of life on Earth today need to survive, was completely absent. Instead, the Earth's first atmosphere was probably a mixture of carbon dioxide, carbon monoxide, ammonia, methane, and other poisonous gases released from enormous volcanoes over millions of years.

954 The development of life on Earth from the first life-forms to the evolution of Homo sapiens has been a long and haphazard road. The first "life-forms" on Earth were barely more than long spiraling strings of molecules that had somehow learned the trick of replicating themselves. Searching in the surrounding sea waters, they at first found an abundant supply of food: new molecules to use for replication. In time, however, the food supply would have been exhausted and our planet's brief experiment with life would have ended in mass starvation. Before this occurred, however, a random, unexpected change in some of the life-forms—a mutation—took place and some life-forms acquired the ability to *synthesize* food from an abundant and quite inexhaustible source, namely sunlight. This gave rise to the first primitive plants. In the process of conducting *photosynthesis,* these plants released a waste product called oxygen. In time, it bubbled up from the oceans to change our planet's atmosphere and paved the way for more advanced life-forms to arise. Countless species evolved. Some underwent beneficial mutations and evolved on. Others suffered unfavorable mutations or could not adapt to changes in climate or food supply and became extinct. About 65 million years ago, after a reign of nearly 200 million years, a life-form known as the dinosaurs became extinct, perhaps as the result of a chance collision of the Earth with a small asteroid. And only then did nature provide the opportunity for a little creature that looked something like a rat to ultimately evolve into Homo sapiens—a species that would ultimately look up at the sky, ask questions, and build the tools to find the answers.

955 **Would nature likely repeat itself?** Given the millions upon millions of mutations and other chance occurrences that led from the first primitive life-forms to Homo sapiens, most biologists put the odds of nature repeating itself, even if another primitive Earth was a given, to be one in billions if not trillions.

956 **In our own solar system, only the Earth seems presently suitable for life.** Some planets are simply right for life and others are not. One essential ingredient for the *development* of life may well be liquid water—a fluid neutral medium in which life-forms can easily move about to find food or each other. Yes, life on Earth today is tenacious, existing in environments that range from polar ice to boiling hot springs to the sulfurous vents of volcanoes. But life *began* in water and only adapted through mutation later. As we look around our own solar system, it appears that the Earth may well be the only planet that presently has life. Indeed, with the possible exception of Mars, the other planets and satellites may have always been sterile.

957 **It seems logical that forms of life are also rare on planets around other stars.** Some planets might have too little mass (and therefore too weak a gravitational field) to retain an atmosphere. Others would be too close to their star to have liquid water or allow organic molecules to exist. Still others might be expected to be too far away from their suns and therefore so cold that liquid water would never exist on their surfaces.

958 **But one does not live on water alone . . .** There are other factors involved in a planet's ability to sustain life. These include the degree of tilt of a planet on its axis, the shape of its orbit, and its rate of spin. The greater the axial tilt of a planet, the more severe would be its seasons, while a very elliptical orbit could also create great extremes of temperature and climate as the planet's distance from its central star varied greatly during the course of its year. Finally, a planet that spins very slowly—or not at all—might experience high degrees of temperature change from day to night. All of these factors do not necessarily rule life out but certainly may be expected to "push the odds" against it.

959 **Can we see planets around other stars (extrasolar planets)?** For the time being, the answer is no. The problem is twofold. First, the stars are much, much brighter than any encircling planets. The stars simply give off millions of times more light than their smaller planets reflect. Trying to see planets around other stars is a bit like trying to see a firefly crawling around the rim of a searchlight that's pointed directly in your eyes. The second problem lies in the fact that planets around other stars, although separated from these stars by millions or hundreds of millions of miles, would still appear to lie virtually on top of the star as seen from the great distance of Earth.

960 **Nevertheless, in the not too distant future, we may actually be able to get pictures of planets around other stars.** Using techniques like *interferometry* and *adaptive optics,* astronomers soon hope to improve the "eyesight" of telescopes to the point where they can actually get images of planets around other stars. Another hope is to conduct such searches in the *infrared* (or heat)

part of the spectrum. While a star is typically millions of times *brighter* than its planets, it is only hundreds of times *warmer,* so the glare of the star in the infrared would not be nearly as overpowering.

961 **Astronomers have searched for planets around other stars in other ways.** Taking a cue from the fact that faint stars in binary pairs can betray their existence by causing their companions to do a "drunk walk" across the sky, astronomers went looking for even smaller wobbles that would betray the existence of objects with smaller mass, namely planets. From the 1960s through the 1980s, several observatories reported detecting such wobbles and excitedly announced the discovery of planets around several nearby stars, including Barnard's Star. In all such cases, however, astronomers at other observatories could not replicate the results (a fundamental procedure in science). It turned out the wobbles were due to such things as removing and replacing the lens or mirror in the telescope during servicing or changes in the temperature of the observatory or the photograph or the measuring device. All of these can and did contribute to slight apparent displacements of the stars in the pictures. After a while, astronomers announcing the discovery of planets around other stars were looked upon in the same way as the boy who cried "wolf."

962 **Finally, in the 1990s, yet another approach to searching for extrasolar planets paid off.** This technique also got its start in the study of binary stars. Astronomers can frequently tell if what looks like a single star in a telescope is really a close double star by examining the object's spectrum. The presence of more than one star is revealed when the combined spectra periodically split and come back together due to the *Doppler shift.* In systems where one star is much brighter than the other, only one overpowering spectrum may be seen. But the presence of the other star can still be detected as it pulls on its companion and causes its companion's spectral lines to oscillate back and forth. In 1995, using this technique, a pair of astronomers from the Geneva Observatory in Switzerland announced the discovery of a planet in orbit around a star called 51 Pegasi. The star is almost identical to our Sun, visible to the naked eye, and barely 40 light-years away. Confirmation from other astronomers at other observatories came within days. This time the results were real.

963 **51 Pegasi's planet proved to be a surprise.** From the magnitude and period of the oscillation in the spectral lines of 51 Pegasi, astronomers were able to determine the mass of the planet and the length of its year. The planet had a mass about equal to that of Jupiter but was orbiting the star once every 4.22 days at a distance of only a little over 4 million miles from 51 Pegasi—eight times closer than Mercury is to our Sun!

964 **Soon after 51 Pegasi's planet was discovered, other planets were found around other stars.** Within months, a team of American astronomers announced the discovery of a giant planet having about 6.4 times the mass of Jupiter in orbit around another star called 70 Virginis. The planet was orbiting the star at a distance of about 47 million miles, or about half the distance of the Earth from the Sun. Soon other announcements of other planets around other stars followed, including a planet almost 4 times the mass of Jupiter only about 4 million miles away from a yellow star named HR 5185.

965 **The discovery of planets that massive orbiting so close to their stars is requiring some rethinking of some old ideas.** Until these discoveries, most astronomers found it easy to explain why our solar system has small rocky planets near the Sun and giant gas planets farther out. The heat from the developing Sun, it was believed, had driven off the primordial atmospheres of the inner planets while allowing deep atmospheres to remain around planets that formed in the cooler outer reaches of the solar system. Here, however, were giant planets orbiting stars at distances much closer than the Earth is to our Sun; in some cases, closer than Mercury. Obviously, some rethinking is in order.

966 **What are these newly discovered planets like?** For now, all we really know about these planets is their approximate masses, the lengths of their years, and the distances they orbit from their central stars. We also know they must be *hot* because they are so close to the stars they orbit. Indeed, the surfaces of the planets around 51 Pegasi and HR 5185 are likely to have temperatures around 2,000°F—far too hot for life.

967 **Are giant massive planets the only kinds of planets other stars have?** All of the planets found around other stars so far have ranged from about half to more than 6 times the mass of Jupiter. Are there no smaller Earthlike planets out there? There probably are. It's just that the technique being used to search for planets around other stars favors the discovery of massive planets that orbit close to their stars. Simply put, the more massive a planet is and the closer in it is to its star, the greater will be the Doppler shift it causes in the spectral lines of its star. Furthermore, the closer the planet is to its star, the shorter will be its year or, to put it another way, the sooner the cycle of shifting in its star's spectral lines will repeat for astronomers to catch and verify. In time, as more data is gathered and instrumentation becomes even more sensitive, the discovery of Earth-sized planets will be within reach. It will eventually even become possible to scan the spectra of such planets for the all-important sign of oxygen, which, in turn, would likely signal the presence of life.

968 **Only certain kinds of stars are likely to have habitable planets in orbit around them.** Clearly, stars like our Sun are good candidates, since . . . we exist. But many other types of stars, while they may have planets, are probably not right for life. O and B stars, for example, leave the Main

Sequence after only 10 million to 100 million years and blow up as supernovae. Life on Earth took 1 billion years to get started and it has taken another 3.6 billion years to get to us. Red dwarf Main Sequence stars burn as stable stars for even longer than our Sun. But to get enough warmth for life, a planet would have to be very close. Tidal forces could then slow the planet's spin (like the Sun did to Mercury) so that one side would bake and the other freeze. So F, G, and K Main Sequence stars (essentially solar-type stars) really seem to be the best candidates for having planets with life orbiting around them.

969 **Certain solar-type stars would also have to be eliminated in a search for life.** Stars in binary or multiple stars systems produce complex gravitational fields around themselves, which, in turn, can make stable orbits for encircling planets difficult or impossible. There are two exceptions. If two or more stars are very widely separated, planets could have stable orbits around any of the stars and not be greatly affected by the gravity or radiation of the others. Also, if two stars in a binary pair were very close, planets might be able to maintain stable orbits around both—in effect, seeing them as "one star." Here, however, the strange evolutionary patterns of interactive binaries could lead to one of the stars going nova or supernova before very long.

970 **Still, this galaxy has many good candidate stars for harboring planets with life.** Eliminating all the wrong spectral types and "poorly designed" binary and multiple star systems, we are probably still left with between 10 and 15 billions stars in our galaxy alone that are right for life.

---

971 **The first extrasolar planets were actually discovered around a very unlikely candidate.** In 1992, radio astronomers reported the existence of two planets, each with masses around 3 times the mass of the Earth, in orbit around a millisecond pulsar named PSR 1257+12 using the Doppler shift of the pulsing in the pulsar itself. Since then, planets were reported to be orbiting two other pulsars as well. At first, finding planets in orbit around stars that have undergone supernova explosions seemed strange. It has been suggested that such planets may actually be the reformulated remains of one or more companion stars that was blown to pieces in the supernova explosion. Material from the destroyed companion(s) might have formed a disk around the pulsar that later accreted into the planets.

972 **Knowing which stars are right for life and finding planets around other stars is exciting but still a far cry from finding life.** The recent discovery of extrasolar planets suggests that solar systems in general might be very common and, therefore, increases some scientists' hopes that life too may be reasonably common. Finding planets, however, is one thing and finding life is quite another. To confirm extraterrestrial life, *contact* is necessary.

973 **Traveling to the stars in a search for life is not presently a practical pursuit.** The distances to the stars are enormous. Upon leaving the solar system, the *Voyager 2* spacecraft was traveling over 40,000 miles per hour. Yet even at this speed, it would take it about 67,000 years to cover a distance equal to that from Earth to the nearest star. In time, we may learn to travel faster, but round trips will still be long and largely boring ordeals. Alas, real interstellar spaceflight just isn't a *Star Trek* episode.

974 **There is a natural speed limit in the universe.** That limit is the speed of light—186,000 miles per second. Even using sophisticated matter-antimatter engines, whose development is still far in the future, a starship could not go as fast as, let alone faster than, the speed of light. Allowing time for speeding up and slowing down on both ends of a journey would still make the trip decades long—a situation that may not be impossible but clearly is impractical— and, probably *unnecessary.*

975 **In searching for certain types of extraterrestrial life, there is a faster, cheaper alternative: *communication.*** Searching for nonintelligent extraterrestrial life will require physically going to its location either in person or through the use of robots. However, if our goal is the search for *intelligent* life and, in particular, *technically advanced* intelligent life, physical contact would hardly be necessary. In this case, all that would be necessary would be *communication.* And for that, we can apply a technology that has already worked well on this planet for nearly a century. The technology is called *radio* and it's the way E.T. phoned home. In short, radio waves could be used to send messages across space in search of intelligent extraterrestrials as well as to carry messages from them to us.

976 **Radio waves have great advantages for communication with extraterrestrials.** First, while no spaceship can travel as fast as the speed of light, radio waves do. Second, it costs a great deal less to send a radio message to the stars than a spaceship. Third, radio waves can be redirected from one star system to another in seconds in a search for life, which is a great deal faster and cheaper than redirecting a spaceship.

977 **Radio waves also seem to have additional communications advantages over other forms of electromagnetic radiation.** Any form of electromagnetic radiation can be used to carry communications, including light (as the fiber optics used in modern telephone systems clearly demonstrate). We could use high-powered lasers to beam flashes in Morse code to the stars, but over interstellar distances, radio waves are more practical because they are far less absorbed by interstellar dust than light waves and also require much less energy to transmit. There are also fewer natural emissions from the universe in the radio part of the spectrum, so interference is minimized.

978 **Inadvertently, we have been broadcasting to the universe for years.** Shortly after the turn of the twentieth century, Guglielmo Marconi successfully sent and received the first radio signals across the Atlantic, creating the first radio link between North America and Europe. Part of those signals, however, leaked up through the Earth's atmosphere and out into space, where they have been traveling forever outward at the speed of light. By now, those radio signals are nearly 100 light-years from home and have passed dozens of stars. In the intervening years, all the other radio and non-cable TV broadcasts ever made have also, in part, escaped into space in an ever-expanding sphere that radiates out from our little blue planet—all the news broadcasts and commercials and symphonies and soap operas and *I Love Lucy* episodes. Like it or not, we have already announced our existence to the cosmos and the "dirty laundry" is on the line.

979 **Fortunately, the actual content of these domestic broadcasts would probably be too weak to decipher at interstellar distances.** Such signals become very weak over interstellar distances and also tend to get "smeared" into each other because of the Earth's motion through space. Given sophisticated enough technology, alien civilizations could be watching reruns of *The Gong Show,* but high-powered, highly directed transmissions over a narrow range of wavelengths could be detected more easily. Similarly, we could hope to receive such transmissions from space here on Earth.

980 We have the tools needed for interstellar communication via radio waves. The tools are *radio telescopes.* They were constructed to detect naturally emitted radio waves from stars, galaxies, and interstellar gas but are just as suited to the reception of radio waves used to carry information. In short, radio telescopes give us the technology to seek out and communicate with other technically advanced life-forms that have similar equipment anywhere in our galaxy. In short, radio telescopes are the keys to the Galactic Club.

981 The first intentional search for radio signals from advanced extraterrestrial life occurred in 1960. In a project he whimsically called OZMA, radio astronomer Frank Drake used a 140-foot radio telescope at the National Radio Astronomy Observatory in West Virginia to listen for signals that might be arriving from would-be planets around two nearby solar-type stars, tau Ceti and epsilon Eridani. His receiver was only set on one frequency and the listening only went on for several hours. The results were negative, but through this exercise, a respected scientist demonstrated the feasibility of interstellar communication using available means and lent an air of legitimacy to the pursuit. SETI (the Search for ExtraTerrestrial Intelligence) was born.

982 Frank Drake likened the search for extraterrestrial life to "looking for needles in a cosmic haystack." Given the millions upon millions of stars scattered all over the sky and the billions upon billions of different radio frequencies that aliens might conceivably choose, searching for needles in the cosmic haystack was a formidable challenge indeed. Since the 1960s, other radio telescopes have been enlisted in the search. Most significantly, the sophistication of radio receivers and computers has increased a millionfold. Scientists are still not using pitchforks on the cosmic haystack, but at least they have graduated from tweezers to dinner forks. In particular, Harvard University physicist Paul Horowitz heads up a project called META (Mega-channel ExtraTerrestrial Assay). Using one radio telescope in the northern hemisphere and a twin in the southern, he and his assistants systematically scan the skies and look for signals on over 8 million different radio frequencies. The results so far are still negative, but the search continues.

983 What sort of signal are scientists looking for? In short, any radio signal that *nature cannot produce on its own.* The initial discovery of pulsars proved intriguing, since, at the time, nothing was known in the natural universe that could produce repeated radio pulses with the accuracy of an atomic clock. The realization that pulsars were rapidly spinning stars provided an important lesson. But a series of pulses in a mathematical series, such as 1, 2, 4, 8, and so on, repeated over and over, for example, could not be created by a star or other natural phenomenon and so would likely be a clear sign of intelligent life, as would a series of pulses corresponding to the mathematical value of *pi* carried out to a certain number of decimal places or any of an infinite number of other nonnatural sequences.

---

984 **How do you talk to an extraterrestrial?** Chances are they aren't as fluid in English as they happen to be on most TV programs. Indeed, communication may at least initially have to rely on what may be the only universals we have in common, namely mathematics and science. From there, basic vocabulary might be established and true exchanges of new information and ideas take place.

---

985 **Interstellar conversations are going to be slow.** Even though we will be communicating at the speed of light, a round-trip radio call to the nearest star would take almost 9 years (the Alpha Centauri system is 4.3 light-years away). So communications may well take the form of them sending us their equivalent of the Encyclopaedia Brittanica and us doing the same for them. The vast distances between the stars make quick chats an impossible proposition.

986 **If contact is ever made, the aliens we encounter are likely to be far more advanced than we.** The argument is pretty straightforward. Here on Earth, we first developed radio telescopes and, therefore, the ability to communicate with extraterrestrials only about 65 years ago. In cosmic terms, that's about as long ago as a nanosecond. The universe has been around for something like 10 to 14 billion years and Earthlike planets have probably been part of it for the better part of the last 7 or 8 billion years. The Earth itself was formed only about 4.6 billion years ago. Therefore, it is conceivable that Earthlike planets developed intelligent forms of life that reached our level of development at least a few billion years ago. Thus, by now, they could be a few billion years ahead of us. That probably gives you an idea of about where we might stand and how much we can expect to teach them compared to what we might learn from them, if we have the good sense to pay attention.

987 **If incredibly advanced civilizations exist in the universe and we have now scanned hundreds of thousands of stars on millions upon millions of radio frequencies, how come we still have not made contact?** A very good question. Indeed, some scientists point to this very fact and use it to argue that life in the universe is a very rare entity. Do some extraterrestrials know we exist and have they decided for now to leave us alone until we demonstrate that we can first live peacefully with our own species? Would advanced aliens think we are interesting enough to contact? Are advanced civilizations sufficiently rare that space and time act as natural quarantines that work against contact? Are we indeed alone in our galaxy if not the universe? Perhaps advanced extraterrestrials simply gave up using anything as primitive as radio waves for communication eons ago. Perhaps they have no desire to communicate.

Perhaps, like the whales and dolphins, they are intelligent but lack the physical ability to fabricate tools and technology. Or perhaps lots of other civilizations also have radio telescopes, but, like us, they are all just *listening.* For now we can only look to the stars and wonder if other eyes elsewhere also look up and wonder.

988 **Do most astronomers believe that there is other intelligent life in the universe?** It is simply a matter of probabilities. Most scientists would indeed guess "yes" on the existence of intelligent alien life somewhere else in the universe. There are just too many stars in too many galaxies to play the odds any other way.

989 **Do most astronomers believe we are being visited by alien spacecraft?** No. As Carl Sagan has noted, the touchdown of a spacecraft from another world would be a truly extraordinary event, so claiming it has happened is an extraordinary claim. And that requires extraordinary evidence as proof. In the opinion of virtually all professional scientists, such evidence does not exist. If, for example, someone would come forward with part of an alien spacecraft made of materials or containing some technology that could be shown by independent testing to have not been fabricated on Earth or if someone came forward with a biological organism not based on DNA or carbon-based chemistry, then professional science would be truly chal-

lenged and a claim of extraterrestrial visitation might well be verified. There have been many stories and TV shows and magazine and newspaper articles of sightings and contact but no hard, irrefutable evidence. And good science must be based on strong, independently verified evidence.

990 **The first "attempt" to intentionally communicate with extraterrestrials took place in the 1970s.** In the 1970s, two pair of spacecraft, *Pioneer 10* and *11* and *Voyager 1* and *2,* were launched to study the outer planets of our solar system. Knowing that they would then leave the solar system to wander among the stars, Carl Sagan and Frank Drake could not resist the temptation to put a "note" into each of these "bottles tossed out on to the cosmic ocean"—a kind of "Greetings from Planet Earth." The *Pioneer* spacecraft each carry a small plaque that illus-

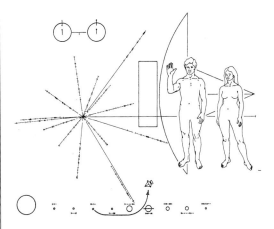

The plaque on the side of the *Pioneer 10* and *11* spacecraft depicts a man and a woman and other information designed to illustrate from where and when in the galaxy the spacecraft was launched.

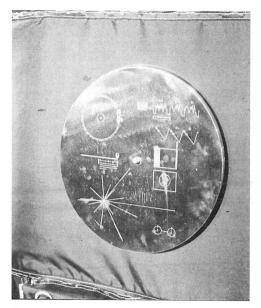

The covering over a record placed on the side of the *Voyager 1* and *2* spacecraft. The record contains sights and sounds from planet Earth. (NASA/JPL)

The Arecibo radio telescope is the largest radio telescope in the world. It measures 1,000 feet across and is constructed in a natural valley in the mountains of Puerto Rico. You may recognize it from the James Bond move *Golden Eye*, where it was supposed to be a Russian transmitter located in Cuba. (NAIC/Arecibo Observatory)

trates a man and a woman and symbols that aliens could conceivably decipher that would tell them from where and when the spacecraft was launched. The *Voyager* spacecraft each carry a more sophisticated "note"—a record that contains greetings in many languages, a variety of natural and man-made sounds (from whale songs to jackhammers), selected pieces of music (from Bach to Chuck Berry singing *Johnny B. Goode*), and 116 pictures of everything from a human birth to Monument Valley to the Taj Mahal. (See page 113 for a picture of the *Voyager* spacecraft.)

991 The likelihood of aliens ever finding any of these "notes" is nil, but . . . they still serve a purpose. Outer space is a very big place, so the chances that Klingons will someday bump into one of these little spacecraft is very close to zero. Nevertheless, the exercise of designing the plaques and records was worthwhile because it caused us to pause and look at ourselves through "alien eyes," that is, to think about what is truly representative of our world and our civilization and how to communicate it.

992 In 1974, we intentionally broadcast our first radio message to the stars. Using the 1,000-foot Arecibo radio telescope (the largest in the world) at the Arecibo Observatory* in Puerto Rico, Carl Sagan and Frank Drake beamed a message into space. At that moment, on the one frequency used for the transmission, the Earth "lit up"

---

*The Arecibo Observatory is part of the National Astronomy and Ionosphere Center, which is operated by Cornell University under a cooperative agreement with the National Science Foundation.

brighter than any other object in the galaxy. The message consisted of a sequence of tones that, if arranged in the correct order and converted to light and dark spots, creates a crude picture.

993 **The Arecibo message contains a lot of information.** A carefully crafted picture is indeed worth a thousand words, but I shall try to use fewer here. From top to bottom, the message depicts: the numbers from 1 to 10 in binary, the atomic numbers of the five elements that go into making DNA, the formula for the basic components of DNA, the number of pairs of bases in a human DNA molecule surrounded by the spiral structure of DNA itself, a humanoid form sandwiched between the number representing the current human population of Earth and the height of the average human in multiples of the wavelength of the radio wave carrying the message, a crude depiction of our solar system with the third planet displaced toward the human form, and finally, a depiction of the Arecibo radio telescope and the diameter of its dish, again in multiples of the wavelength of the message's carrier wave. The act of sending the message was simply to illustrate its technical feasibility and to attempt to craft a message in the "language of science."

994 **We are not likely to get a reply to the Arecibo message for some time.** For the sake of publicity, the Arecibo message was sent to a visually spectacular object, a giant globular cluster of several hundred

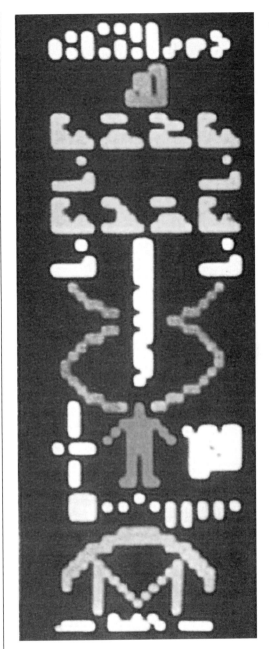

The Arecibo radio message to the stars consisted of a string of radio pulses (like Morse code) made up of only two distinct radio frequencies. If these frequencies are depicted as light and dark squares and arranged into a rectangle, they form a crude pictogram that is indeed a picture worth a thousand words.

thousand stars in the constellation Hercules known as M13. Granted, we got to beam the signal at several hundred thousand stars all at once (what a party line), but M13 is about 24,000 light-years from Earth, so we're not going to hear any return messages for 48,000 years. Furthermore, globular clusters contain old Population II stars, which are very poor in metals and other heavy elements; therefore, many scientists feel that Earthlike planets are unlikely in globular clusters. Again, the transmission of the message *was* only an academic exercise aimed at demonstrating the feasibility of using the technology at hand.

995 **There is an equation that predicts the number of technically advanced civilizations in the galaxy with whom we could hope to communicate.** This equation was devised many years ago by radio astronomer Frank Drake and has since become known as the Drake Equation. In one form, it looks like this.

$$N = N^* \times fs \times fp \times fr \times fl \times fi \times ft \times L$$

996 **N in the Drake Equation equals the total number of such civilizations in the galaxy.** N^* is the number of stars in the galaxy, fs is the fraction of those stars that burn stable and long enough to support life, fp is the fraction of those that have planets, fr is the fraction of those planets that are suitable for life, fl is the fraction of those on which life has actually developed, fi is the fraction of those on which *intelligent* life has

developed, and ft is the fraction of those on which the intelligent life has developed the technology to communicate over extraterrestrial distances. And finally there is L, which we will discuss next.

997 **L in the Drake Equation is the length of time a civilization can communicate over interstellar distances before it either chooses to no longer do so or becomes incapable of doing so.** Some civilizations may have searched for signals for a time and then given up (or had their congressional funding cut off). Other civilizations may retrogress in technological capability or cease to exist because of some large-scale natural disaster or war. This latter point is particularly interesting, since the ability to build and use radio telescopes requires a knowledge of the workings of the atom, which is also the knowledge one needs to create weapons of mass destruction. Indeed, like us, other civilizations would likely have come upon both of these capabilities at virtually the same moment in their evolution as a species and a civilization. It will be interesting to someday find out how others fared with the same challenge we face ourselves today.

998 **So what kind of a number does the Drake Equation predict for N?** To date, the only one of the factors in the entire Drake Equation that we can estimate with any degree of certainty is N^*. For other factors, we can try to make educated guesses and for L we have no idea. So, not surprisingly, different people come up with very different answers. It is noteworthy, however, that biologists, in general, come up with much smaller numbers for N than do astronomers because many biologists feel that the probability of intelligent life or even life, in general, arising on a planet is very, very low.

999 **Many radio searches have centered around a couple of special frequencies.** With billions of potential radio frequencies to choose from in searching for alien signals, how can radio astronomers hope to make an intelligent choice? In the late 1950s, MIT physicist Philip Morrison first suggested the use of radio waves for extraterrestrial communication and pointed to a particular frequency as a potentially good place to tune. The frequency he suggested was 1420 megahertz, which corresponds to a wavelength of 21 centimeters or about 8 inches in length. Why this frequency? Because it is the fundamental radio frequency on which hydrogen, the most abundant element in the universe, broadcasts. We know that, and aliens with radio telescopes would have to know it, too. Most actual searches have followed this notion and listened at the hydrogen frequency or multiples thereof.

1000 **Others have sought to gather around the "cosmic watering hole."** A fanciful idea was also put forward in a search for the "right frequency." In addition to the hydrogen transmission at 1420 MHz, there was another frequency in the radio spectrum, at about 1666 MHz, on which OH molecules (those having 1 atom of oxygen bonded to 1 atom of hydrogen) naturally transmit energy. Since $H + OH = H_2O$ or water, perhaps, alien life-forms would gather to communicate somewhere between these frequencies, just as animals of different species gather around watering holes.

1001 **What might be the consequences of contact with advanced extraterrestrial beings?** For better or worse, it will likely be the most profound event in the history of the human race. Perhaps, we will be the beneficiaries of a veritable intellectual treasure trove—the combined knowledge and wisdom of an advanced alien civilization or the accumulated knowledge and wisdom of numerous civilizations passed down from one great galactic civilization to the next over the last several billions of years. These are, indeed, interesting times in which to live.

A

Abell, Dr. George, 58
Adams, John Couch, 134
Air Pump (constellation), 166
Alcor (star), 217
Aldebaran (Alpha Tauri) (star), 173, 268
Aliens, search for, 333–36, 339
Alnilan (star), 173
Alnitak (star), 173
Alpha Centauri C (Proxima Centauri) (star), 179
Alpha Centauri star system, 167, 179, 225, 281, 282, 334
alpha Herculis (Ras Algethi) (star), 218, 219
Alpha Regio (Venus mountain), 94
Alpher, Ralph, 306, 308
Alrischa (star), 173
Altair (star), 179, 202
American Association of Variable Star Observers, 216
Amor (asteroid), 147, 148
Andromeda (Princess) (constellation), 277
Andromeda galaxy (M31), 277–78, 280, 281, 283, 294, 300
Aneghito meteorite, 159–60
Antares (star), 184, 263, 267
Antimatter, 33
    See also Matter-antimatter problem
Antiproton, 33
Anti-X-bosons, 321
Apennine Mountains (Moon), 46, 47, 48
Aphrodite Terra (Venus highland region), 93
Apollo asteroids, 147, 148
Apollo astronauts, 59, 60, 105

Apollo 13, 59
Aquarius (Water Carrier) (constellation), 166, 227
Arabs (ancient), 211, 217, 220
Arecibo radio telescope, 26, 336–38
Aries (constellation), 165
Aristotle, 36–38, 39, 55
Asterisms, 165
Asteroids, 144–48
    and Earth, danger to, 147–48
    numbered, 148
Astrologers, 166
Astronomers
    amateur vs. professional, 9–10
    as astrophysicists, 2
    and constellations, 165
    as cosmologists, 294
    creativity of, 3
    and eclipses of Sun, 82, 83
    naming stars, 173
    observational, 2–3
    and observatories, 13–14, 17–18
    and parallax concept, in measuring distances to stars, 178–79, 181–82
    and stars, spectra of, 185–91
    as theoreticians, 2–3
    tools of, 18–20, 213–14
    and twinkling stars, 12–13
    and universe, expansion of, 301–4
    and variable stars, 214–16
    and zodiac, 166
Astronomy
    creativity in, 3
    defined, 2
    as detective work, 3
    extragalactic neutrino, 244

gamma-ray, 251–52
progress of, 41
X-ray, 278
Atens (asteroid), 147, 148
Atoms, 29–33, 76, 237, 318
Aurora, 2, 71–73
Autumnal equinox, 53

B

Baade, Walter, 236, 238
Bailey's beads, 85
Baliunis, Dr. Sallie, 69
Barnard's Star, 264, 328
Barringer Meteor Crater (Arizona), 148, 159
Bell, Jocelyn, 236, 238
Bessel, F. W., 224
Beta Centauri (star), 167
beta Cygni (Albireo) (star), 218
Beta Pictoris (star), 197–98
Beta Regio (Venus mountain), 94
Betelgeuse (Alpha Orionis) (star), 173, 179, 183, 184, 189
Big Bang, 305–9, 311, 313, 315–22
Big Bounce, 311
Big Chill, 311, 313
Big Crunch, 311, 314
Big Dipper (Plough), 165, 167, 170, 217, 262, 264, 269, 295, 314
Binoculars, 9, 12, 184, 273
Black holes, 3, 247–51, 263, 278, 284
event horizon of, 248–49
and galaxies, 287–91
and light, 250–51
mini-, 251
singularity of, 248–49
and X rays, 249, 272–73
Black Widow pulsar, 245, 246
BL Lacertae objects, 285, 286, 287, 289
Bok, Bart, 204
Bok globules, 204
Brahe, Tycho, 39, 232, 234, 241

C

Caesar, Julius, 155
Callisto (Jupiter satellite), 116, 117–18, 121, 125, 138

Caloris Basin (Mercury), 89, 90
Camelopardalis, (Giraffe) (constellation), 279
Cancer (Crab) (constellation), 165, 268
Canis Major (Great Dog) (constellation), 179, 183
Capella star system, 16, 268
Capricornus (constellation), 166
Carbon dating, 31
Carina (Keel) (constellation), 215
Cartwheel Galaxy, 282, 283
Cassini (spacecraft), 128
Cassini Division (Saturn), 123
Cassiopeia (Queen) (constellation), 232, 241, 268
Castor (star), 218
Cat's Eye Nebula (NGC 6543), 230
CCD (charge-coupled device), 9–10, 18–19, 20, 304
Celestial equator, 11
Centaurus (constellation), 179, 299
Centaurus A (radio galaxy), 285
Cepheid variables (Cepheids), 212–14, 276, 297, 300–4
Cepheus (King) (constellation), 212, 221
Ceres (asteroid), 144–45
Cernan, Gene, 59
Cerro Tololo Inter-American Observatory, 13
CGRO (Compton Gamma Ray Observatory), 251–52
Charon (Pluto satellite), 140, 141
Chinese (ancient), 36, 53, 68, 71, 82, 83, 165, 167, 216, 235
Chryse Planitia (Plains of Gold) (Mars), 103
Clarke, Arthur C., 76
Cleopatra (Venus crater), 94
COBE (Cosmic Background Explorer), 308, 321–22
Color, defined, 22
Columbus, Christopher, 55, 179
Coma Berenices Cluster, 296, 305, 309
Comets, 149–55
coma of, 150
and Earth, collision with, 154
heart of, 149, 150–51
history of, 154–55
and Jupiter crash, 114–15
tail of, 150–53
Communication, and extraterrestrial life, 331–39
Communities, function of, 14
Computers, 12, 16–19
Conservation of Matter and Energy law, 78
Conservation of Momentum law, 166
Constellations, 165–67, 172, 180

shape of, 263–64
Coordinates, celestial, 11–12
Copernicus, Nicholas, 38, 39, 294–95
Copernicus (lunar crater), 45, 46, 104
Corvus (Crow) (constellation), 282
Cosmic rays, 27
Cosmological Principle, 298
Cosmology
  defined, 294
  problems in, 320–23
Crab Nebula (M1), 235–36, 238–40, 241, 245
Crawford, Dr. David, 14
Cryogenics, 307
Cygnus (Swan) (constellation), 173, 184, 202,
  218, 232, 241, 249, 250
Cygnus Loop, 241
Cygnus X-1, 249

D

Dactyl (satellite), 146, 147
Dark matter, 33
  cold or hot, 322–23
Declination (Dec.), 11–12
Degeneracy, 229
Deimos (Mars moon), 107
Delta Cephei (star), 212, 220, 221
Deneb (star), 173, 179, 202, 232
Deutsch, Armand, 30
"Diamond ring effect," 81, 84, 85
Dinosaurs, 155, 266, 327
Dione (Saturn satellite), 122, 123
Discovery Rupes (Mercury scarp), 90
DNA, 337
Doppler effect, 265
Doppler shift, 220, 329, 330, 331
Double Cluster (h & chi Persei), 268, 270
Draco (Dragon) (constellation), 168
Drake, Frank, 333, 335, 336, 338
Drake Equation, 338–39
Duhalde, Oscar, 242–43

E

Eagle Nebula (M16), 200, 201, 276
  EGGs, 201
Earth

age of, 95, 334
and asteroids, danger from, 147–48
atmosphere of, 12–13, 16, 25, 27, 55–56, 71,
  72, 75, 95, 127, 134, 156, 236, 326
axis of, 65, 66, 130, 190
birth of, 155
as black hole, 247–48
as center of universe, 36
changes in, 79
climate of, 69–70
collision with, and Moon, 58
and comets, collision with, 154
craters on, 46
day on, 73
diameter of, 91, 130
escape velocity of, 247–48
evolution of, 155
gravitational field of, 60, 111
and Halley's Comet, 151, 152
ice ages on, 104
vs. Jupiter, 110–12, 114
life on, 116, 119, 128, 155, 326–27, 330
magnetic field of, 70–71, 114
vs. Mars, 97–106
mass of, 312
vs. Mercury, 90
and meteorite impacts, 159–60
phases on, 61
poles of, 72
rocks on, 60, 105
rotation of, 167–68
roundness of, 54–55
size of, 74, 225
and Sun, passage around, 172
tidal forces on, 56–57, 120, 230, 290
Van Allen radiation belts, 114
vs. Venus, 91–95
Earthshine, 50
Eclipse
  of Moon, 50, 54–56, 80
  of Sun, 49, 50, 75, 80–85, 250
Egg Nebula, 228
Egyptians (ancient), 165, 166, 168
Einstein, Albert, 28, 40–41, 78, 244, 250, 289,
  297, 312, 318
Einstein Cross, 297
Einstein-Rosen Bridge, 291
Electromagnetic force, 317
Electromagnetic radiation, 22–25, 332
Electrons, 29, 30, 237

Elements, 29–30, 36, 191, 318
    in stars, 187
    types of, 30
Ellipses, described, 39
Enceladus (Saturn satellite), 126–27
Endeavour (Space Shuttle), 18
Epsilon Aurigae (supergiant), 221
epsilon Eridani (star), 333
epsilon Lyrae (quadruple sun), 218
Equinoxes, 65
Eskimo tribes, 71
Eta Carinae (star), 215
Europa (Jupiter satellite), 116–19, 121, 138
European Southern Observatory, 10
European Space Agency, 154
Extraterrestrial life, 326–39
and communication, 331–39

F

51 Pegasi's planet, 329, 330
4179 Toutatis (asteroid), 147
False dawn, 161
Fermi, Enrico, 233
Flatness problem, 320–21
Fluids, nature of, 31–32
Fornax Cluster, 302
Fox (constellation), 166
Freedman, Wendy, 302, 304
Friction, generation of, 32
Furnace (constellation), 166

G

Galaxy(ies)
    and black holes, 287–91
    clusters, 294–96, 309–10
    collisions of, 281–82, 316
    colors of, 280
    distances to, determining, 301–5, 314–16
    motion of, 297–300, 309–10
    Seyfert, 283, 287, 289
    shapes of, 282–83
    spectrum of, 297
    and stars, 281, 295
    types of, 277–89, 291, 295–97, 315
    voids between, 305, 310
    See also Milky Way
Galileo, 38–39, 68, 117, 121–22, 124, 136, 256–57
Galileo (spacecraft), 114, 115–16, 120, 146–47
Galle, Johann Gottfried, 134
Gamma ray(s), 22, 23, 24, 78, 250, 252
    astronomy, 251–52
    telescopes, 25
Gamow, George, 306, 308
Ganymede (Jupiter satellite), 116, 117, 118, 121, 125, 127, 138
Gases, 31–32
Gaspra (asteroid), 146, 147
Geller, Margaret, 304, 305, 310
Gemini (Twins) (constellation), 157, 165, 166, 218, 268
Geminid meteor shower (Geminids), 157
General Theory of Relativity (Einstein), 28, 41, 244, 250, 289, 297, 312
George III, King, 130
Georgium Sidus (Georgian Star), 130
Giotto (spacecraft), 154
GL 105A (star), 200
G1229 (star), 200
G1229B (Glise 229B) (brown dwarf), 199, 200
Global Oscillation Network Group (GONG), 70
Global Surveyor (spacecraft), 105
Globular clusters, 258, 313
Grand Unifying Theories. *See* GUTs
Gravity, 40, 290, 317, 320
    assists, 137
    and black holes, 247–49
    and light, 250
    and stars, 194
    wave detectors, 245
    waves, 28–29
Great Annihilator, 249–50
Great Attractor, 299
Great Nebula, 198
    in Orion, 195–97, 201, 203, 276
Great Square of Pegasus, 276–77
Great Wall of Galaxies, 305
Greeks (ancient), 36, 96, 107, 117, 138, 181, 216, 256, 269, 277
Grimaldi (lunar crater), 46
GROJ1744-28 (star), 252–53
Gula Mons (Venus volcano), 94
Guth, Alan, 321
GUTs (Grand Unifying Theories), 318, 320–21, 323

H

Hale-Bopp (comet), 154
Hall, Asaph, 107
Halley, Edmund, 149–50
Halley's Comet, 149–50, 151, 152, 154, 155
Hawking, Stephen, 251
HD 163296 (star), 198
Helioseismology, 70
Helix Nebula, in Aquarius, 227
Hellas (Mars basin), 102
Hen (constellation), 166
Hercules (constellation), 218, 266, 272, 284, 338
Hercules A (galaxy), 284
Herman, Robert, 306, 308
Herschel, William, 129–30, 258
Herschel (Saturn crater), 126
Hertzsprung, Ejnar, 206
Hindus (ancient), 36, 53
Hipparchus, 181
Hoba West (meteorite), 159
Hodges, Mrs. E. Hewlett, 160
Horizon problem (smoothness problem), 320, 321
Horowitz, Paul, 333
Horsehead Nebula, in Orion, 201, 202
Houtman, Frederik de, 166
HR 5185 (star), 329, 330
H-R (Hertzsprung-Russell) diagram, 206–10, 225, 228–29, 267, 270, 272
Hubble, Edwin, 276, 297, 299–300, 303, 304
Hubble Constant, 299–302, 313, 316
Hubble Deep Field study, 314–16
Hubble Flow, 299
Hubble Space Telescope (HST), 10, 18, 73, 97, 107, 114, 115, 120, 128–31, 135, 136, 138, 189, 196–97, 199–201, 215, 216, 227, 239, 251, 272, 276, 286, 288, 295, 302, 303, 314, 315
Huchra, John, 304
Huygens, Christiaan, 122
Huygens (probe), 128
Hyades Cluster, 269, 270
Hyakutake (comet), 153
Hyakutake, Yuji, 153
Hydra (constellation), 299
Hyperion, 127

I

Iapetus, 127
IC 2944 (nebula), 204
Icarus (asteroid), 148
Ida (asteroid), 146–47
Inflation Theory, 321
Infrared (IR), 22, 23, 24, 328
    telescopes, 25, 259–60
Interferometers, 26
Interferometry Principle, 16–17
Interplanetary dust, 161
Interstellar dust, 183, 202–4
Inverse square law, 182, 248, 317
Io (Jupiter satellite), 117, 119–21, 138
Ions, 30, 32
IRAS (Infrared Astronomy Satellite), 203, 259, 277
Ishtar Terra (Land of Ishtar) (Venus terrain), 93
Isotopes, 30–31
Ithaca Chasma (Tethys canyon), 126

J

Jews (ancient), 53
Juno (asteroid), 145
Jupiter (planet), 37, 38, 73, 110–21, 122, 129–32, 135, 137, 139, 144–50, 153, 189, 198–200, 221, 232, 236, 262, 329, 330
    comet crashes into, 114–15
    days in, 111
    vs. Earth, 110–12, 114
    exploration of, 115–17
    gravitational field on, 111
    Great Red Spot (GRS), 110, 111, 112, 125
    magnetic field on, 114, 120–21
    rings around, 113–14
    satellites of, 116–21, 125, 127, 136
    surface of, 112
    weather in, 110–13, 116

K

Keck Telescope, 15, 16, 17
Keck II Telescope, 16, 17

Kelvin temperature scale, 307–8
Kepler, Johannes, 39, 218, 232, 234
Keyser, Pieter, 166
Kitt Peak National Observatory (Arizona), 13, 14, 17, 66, 158
Kohoutek (comet), 151
Kuiper Belt (zone), 150

L

Lacaille, Nicolas de, 166
Lacerta (Lizard) (constellation), 285
Lagrangian points, 147
Laplace, Pierre, 250
Large Magellanic Cloud (LMC), 242, 243, 250, 279, 280
Lasers, 16
Law of Universal Gravitation, 40
Leavitt, Henrietta, 213
Leo (Lion) (constellation), 166
Leonid meteor shower (Leonids), 158
Lepus (Hare) (constellation), 200
Leverrier, Urbain, 134
Levy, David, 149
Libra (Scales) (constellation), 166
Lick Observatory, 4
Light, 6, 22
    and black holes, 250–51
    and gravity, 250
    pollution, 14
    and sound, 174
    speed of, 40–41
    waves, 22
Light-years, 179
Liquids. *See* Fluids
Little Dipper, 165, 170
Llama (constellation), 166–67
Local Group, 294–96, 299, 309
Lowell, Percival, 98, 102
Lyra (Harp) (constellation), 173, 214, 217, 226, 227, 266

M

M 13 (globular cluster), 272, 338
M32 (galaxy), 277, 280

M33 (galaxy), 294
M35 (star cluster), 268
M36 (star cluster), 268
M37 (star cluster), 268
M38 (star cluster), 268
M67 (star cluster), 270
M81 (galaxy), 260
M84 (galaxy), 296
M86 (galaxy), 296
M87 (galaxy), 288, 296
M92 (globular cluster), 272
M100 (galaxy), 302, 303
MACHOs (MAssive Compact Halo Objects), 310–11
McMath-Pierce Solar Telescope, 66, 67
Magellan, Ferdinand, 279
Magellan (spacecraft), 94
Marconi, Guglielmo, 332
Mariner 4 (spacecraft), 99
Mariner 9 (spacecraft), 99–100, 106
Mariner 10 (spacecraft), 88, 89, 92
Mars (Red Planet), 30, 32, 37, 58, 91, 96–107, 110, 129, 139, 144, 145, 147, 148, 221, 327
    canals on, 98, 102, 118
    climate on, 103, 104, 106–7
    craters on, 101, 104
    vs. Earth, 97–106
    "face" on, 104–5
    life on, 98–99, 102–3
    moons on, 107
    in opposition, 97–98
    rocks on, 104, 105–6, 107, 160
    seasons on, 96–97
    volcanoes on, 100
    water on, 103, 104
    winds on, 106
Mass-luminosity relationship, 209
Matter, 29–33
    antimatter, 33
    behavior of, 32
    dark, 33, 322–23
    forms of, 29–32
    states of, 31–32
Matter-antimatter problem, 320, 321
Maunder Minimum, 69
Maxwell, James Clerk, 317
Maxwell Montes mountains (Venus), 93, 94
Maya (ancient), 256
Medici family, 117
Mercury (planet), 79, 88–91, 92, 94, 95, 96, 110,

117, 118, 127, 128, 129, 139, 140, 144, 148, 219, 221, 329, 330
  craters on, 89, 91
  vs. Earth, 90
  vs. Moon, 89, 90
  scarps on, 90
Meridiani Sinus region (Mars), 101
Messier, Charles, 149, 235
META (Mega-channel ExtraTerrestrial Assay), 333
Meteorites, 155, 158–60
  categories of, 159
Meteoroids, 155, 158–59
Meteors (shooting stars), 155–59
  showers of, 156–58
Micrometeorites, 161
Microscope (constellation), 166
Microwaves, 22
Milky Way, 9, 166, 202, 242, 252, 256–73, 276, 277, 279–80, 281, 282, 283, 285, 286, 289, 309, 310
  age of, 266
  band of, 257–59, 262
  center of, 258, 262, 263
  general shape of, 257
  halo of, 271, 273
  and Local Group, 294–95
  magnetic field of, 263
  reclassifying, 278
  size and structure of, 260–61
  spiral arms of, 261–62, 266–68
  width of, 284
  zone of avoidance in, 295
  See also Galaxy(ies)
Mimas (Saturn moon), 126
Mintaka (star), 173
Mira (Algol) (star), 211, 214, 220
Miranda (Uranus moon), 133
Mizar (star), 217
Molecules, 76, 186, 187
  in interstellar space, 205
  types of, 29
Monocerous (Unicorn) (constellation), 250
Moon, 44–61
  at apogee, 81
  astronauts on, 59–60, 102
  "blue," 54
  craters on, 44–46, 47, 91, 101, 104, 148
  cycle of phases on, 48–50
  dark areas on, 47
  diameter of, 10, 117, 145

  eclipse of, 50, 54–56, 80
  formation of, 58
  full, 10, 44, 49, 52–54, 61, 157–58, 179, 182, 196, 235, 268, 314
  gravity of, 56–57, 60
  Harvest, 53
  history of, 58–59
  "illusion," 51
  imperfections on, 38
  influence of, on human beings, 57–58
  interior of, 59
  lunar features, 44–47
  "man in," 47–48
  vs. Mercury, 89, 90
  motion of, 12
  mountain ranges on, 46, 47
  new, 49, 80
  at perigee, 81
  phases on, 48–50, 61
  rising of, 51–52
  rocks from, 60
  "seas" on, 47–48
  setting of, 52
  sides of, 50–51
  size of, 44, 54, 58
  surface of, 51
  temperature on, 44
  tidal forces on, 56–57, 120, 230, 290
  volcanism on, 58–59
  waxing vs. waning, 49, 50
Moonquakes, 59
Morrison, Philip, 339
Motion, Newton's three laws of, 40
Mount Palomar observatory, 14, 15
Mount Wilson observatory, 14
Muslims, 53

N

NASA, 59, 251
NCG 1530 (galaxy), 279
Neap tides, 57
Nebulae, 194–98, 201–2
  See also Planetary nebulae
Neptune (planet), 73, 110, 113, 121, 134–38, 139, 140, 150
  Great Dark Spot, 134, 135
  satellites of, 136–37

Scooter, 134, 135
Small Dark Spot, 134
weather of, 135–36
Wizard's Eye, 134, 135
Neutrinos, 244, 311
solar, 27–28
Neutrons, 29, 30–31, 237
New Quebec Crater (Canada), 159
Newton, Sir Isaac, 5, 39–40, 41, 149, 218
Newtonian reflector, 5
NG 104 (globular cluster), 271
NGC 205 (galaxy), 277, 280
NGC 315 (galaxy), 284
NGC 4258 (galaxy), 288
NGC 4261 (galaxy), 288
NGC 4565 (galaxy), 261
NGC 4639 (galaxy), 303, 304
NGC 6251 (galaxy), 284
NGC 7027 (nebula), 228
North Celestial Pole, 167
Northern lights. *See* Aurora
North Star. *See* Polaris
Nova Cygni 1975 (V 1500 Cygni) (star), 232
Novae, 216
dwarf, 231
full–fledged, 231
recurrent, 231
Nova explosions, 231
Nuclear bombs, 70, 71
Nuclear fission, 77
Nuclear force, 317, 320
Nuclear fusion, 77

O

Objects in space, 23–25
Observatories
places for, 13–14
travel to, 17–18
Olympus Mons (Mars volcano), 100
Omega Centauri (globular cluster), 272
Oort, Jan, 309
Oort cloud, 150
Ophiuchus (Serpent Bearer) (constellation), 232
Optical telescopes. *See* Telescopes (optical)
Optics, adaptive, 16
Orion (Hunter) (constellation), 173, 179, 182,
183, 184, 189, 214, 267, 268

Great Nebula in, 195–97, 201, 203, 276
Horsehead Nebula in, 201, 202
Spur, 261

P

Pallas (asteroid), 145
Palomar Observatory Sky Survey, 10
Paraboloid, 4–5
Parallax concept, in measuring distances to stars,
178–79, 181–82, 190–91, 270
Parsecs, 179
Parsons, Thomas (Earl of Rosse; Lord Rosse),
235
Pathfinder (spacecraft), 105
Pauli Exclusion Principle, 229, 248
Peary, Admiral Robert E., 160
Penzias, Arno, 307, 308
Periodic Table of Elements, 30, 246
Period-Luminosity Relationships, 213, 300
Perseid meteor shower (Perseids), 157–58
Perseus (Hero) (constellation), 261, 268
Perseus Arm, 261
Persians (ancient), 96
Phobos (Mars moon), 107
Photography, in tool kit, 18
Photometers, 19–20
Photometry, 15
Photosynthesis, 327
Physics, branches of, 294
Piazzi, Giuseppe, 144, 145
Pioneer (spacecraft), 115
Pioneer 10 (spacecraft), 335
Pioneer 11 (spacecraft), 335
Pisces (Fish) (constellation), 165, 173
Pixels, 18–19
Planetary gap, 144–45
Planetary nebulae, 21, 79, 226–28
See also Nebulae
Planets
absorption spectra of, 21
auroras in, 72–73
birth of, 76, 155, 246–47
extrasolar, 328–29
life on, 327–28, 330–31
motion of, theories on, 37–38, 218
naming, 139–40
vs. stars, 199

without stars, 198–99

Plasma, 32

Plato (lunar crater), 47

Pleiades (M45; Seven Sisters) (star cluster), 268–70

Pluto (planet), 82, 110, 117, 118, 127, 138–41, 150, 166, 197, 198, 219
   satellite of, 140
   size of, 138, 139

Polaris (North Star), 167, 168, 170, 171, 172, 212

Pollution, light, 14

Population I stars, 273, 280

Population II stars, 273, 280, 338

Population III stars, 273

Positrons, 33, 250

Praesepe (Beehive) (star cluster), 268

Procyon (star), 224

Protons, 29, 30–31, 237

Protoplanetary disks (proplyds), 197

Protostars, 209–10

PSR 1257+12 (pulsar), 331

Ptolemy, 37–38, 39, 181

Pulsars, 3, 190, 238–42, 244–46, 252–53, 333

Q

Quantum gravity, 320

Quasars, 286–87, 289, 319

R

Radiation, 3–4
   cosmic background, in universe, 307, 308, 316, 318–19, 321–22
   electromagnetic, 22–25, 332
   from Sun, 23–24, 78–79, 83

Radioactive decay, 31, 317

Radio telescopes, 25–27, 259–60, 262–63, 284, 286, 333
   See also Arecibo radio telescope

Radio waves, 23–24, 236, 332–39
   vs. electromagnetic radiation, 332

Ray craters, 45

R Coronae Borealis (star), 214–15

Relativity. *See* General Theory of Relativity; Special Theory of Relativity

Rhea (Saturn satellite), 122

Rigel (Beta Orionis) (star), 173, 179, 182, 183, 184

Right ascension (R.A.), 11–12

Ring Nebula (M57), 226, 227

Ringtail Galaxies (Antennae), 282

R Lyrae (variable star), 214

Roche limit, of Saturn, 124

Romans (ancient), 92, 96, 122, 256

R Orionis (variable star), 214

ROSAT X-ray satellite, 278

RR Lyrae (variable stars), 173, 213, 214, 220, 258

Rubin, Vera, 309

Russell, Henry N., 206

S

70 Virginis (star), 329

Sagan, Carl, 102, 335, 336

Sagittarius (Archer) (constellation), 166, 250, 258, 261, 262

Sagittarius A* (object), 262–63

Sagittarius Arm, 261

Sakigake (spacecraft), 154

Sandage, Allan, 303–4

Saros cycle, 82

Saturn (planet), 37, 73, 110, 113, 114, 121–29, 130, 131, 135, 137, 139, 189, 221
   rings of, 121–24, 129, 132
   satellites of, 122, 123–29
   weather on, 125

Schiaparelli, Giovanni, 98

Schmidt-Cassegrain telescopes, 5–6

Schwabe, Samuel Heinrich, 68

Schwarzschild, Karl, 250

Scientific notation, 319–20

Scorpius (Scorpion) (constellation), 166, 167, 184, 262, 263, 267

Seasons, reason for, 65

Serpens (Serpent) (constellation), 200, 201, 276

SETI (Search for ExtraTerrestrial Intelligence), 333

Seyfert, Carl, 283

Seyfert galaxies, 283, 287, 289

Shapley, Harlow, 258, 272, 276, 295

Shelton, Ian, 242–43

Shoemaker, Caroline and Eugene, 149

Shoemaker-Levy 9 (S1–9) (comet), 9, 114, 115, 148, 149, 153, 154

Shooting stars. *See* Meteors

Sif Mons (Venus volcano), 94

sigma 2 Ursa Majoris (two stars), 219

Sirius (star), 11–12, 165, 179, 181–82, 183, 224, 225

Sirius B (white dwarf), 229

Sky
  constellations in, 165–67, 172, 180
  four-dimensional, 174
  by season, 180

Small Magellanic Cloud (SMC), 213, 279, 280

Smoothness problem. *See* Horizon problem

SN 1987A (supernova), 243–44

Soderblum, Larry, 136

Sojourner (robot), 105

Solar flares, 70–71

Solar neutrinos, 27–28

Solar storm, 70

Solar system
  in formation, 197–98
  life in, 327
  motion through space of, 266
  origin of, 90–91, 146

Solar wind, 76

Solids, 31–33

Solstices, 65

Sombrero Galaxy, 281

Sound, and light, 174

Southern Cross, 170, 202

Southern lights. *See* Aurora

Southern Sky Survey, 10

South Star, 168, 170

Space
  dust, 161
  objects in, 23–25
  and time, 41, 174–75, 311

Spaceships, 76

Space Shuttle, 18, 72, 251

Special Theory of Relativity (Einstein), 40–41

Spectra, 20–22
  absorption (dark line), 20–21
  continuous, 20, 21
  electromagnetic, 23
  emission (bright line), 20, 21–22

Spectrograph, 20

Spectroscope, 184–85, 219–20

Spectroscopic parallax, 191

Spring tides, 57

Star clusters, 194, 267–73
  globular, 271–73

open, 267–70, 272

Starlight, polarization of, 203

Stars, 164–75, 178–91, 194–221, 224–53
  age of, 313
  associations of, 267
  binary, 218, 219–21, 230–31, 244–46, 329, 330
  birth of, 194–96, 200–201, 205, 209, 217, 246, 273, 281
  and black holes, 248–50
  cannibalizing, 245
  circumpolar, 169, 172
  collapsing, 211–12
  colors of, 183–85
  composition of, 32
  "dark," 224
  death of, 246
  and disks, 213, 217
  distances to, determining, 178–79, 181–82, 190–91, 270
  double, 216–19, 225, 244
  dwarf, 189, 199–200, 206, 215, 225, 226, 228–29, 231, 240
  eclipsing, 220–21
  flare, 215
  and galaxies, 281, 295
  giant, 206, 226, 230–31
  gravity, importance of, for, 194, 225, 241
  halo, 213, 271
  heat of, 188
  and latitude, 170
  life on, 331
  lives of, 206–7
  magnetic fields of, 190
  magnitude (brightness) of, 181–83, 208–9
  and Main Sequence, 207–11, 225, 228–29, 230, 232, 267, 270, 272, 273, 330
  mass of, 200, 207–11, 214–15, 220, 225, 230–31, 248
  motions of, 218–19, 263–66
  multiple, 216–17, 225
  naming, 173–74, 214
  and neutrinos, 233–34
  neutron, 3, 27, 189, 236–38, 240–241, 244–46, 248, 252–53
  nova explosions of, 231–32, 33
  numbering of, 173–74
  O and B, 261, 267, 273, 280, 281, 301, 330
  orbiting of, 266
  parallax concept, in measuring distances to, 178–79, 181–82, 190–91, 270

vs. planets, 199
Population I, 273, 280
Population II, 273, 280, 338
Population III, 273
pulsars, 3, 190, 238–42, 244–46, 252–53, 333
radial velocity of, 265
seeing, 164, 167–68, 170–72
sizes of, 188–89
solar-type, 330
spectra of, 20–21, 185–91, 219–20, 265, 297
spots, 190
supergiant, 188–89, 206, 215, 221, 249, 261, 262
and thermonuclear fusion, 199, 207
trails, 168, 169
trigonometric parallax of, 190–91
twinkling, 12–13
types of, 205, 214
understanding, 205–6
variable, 211–16
wobbling, 224
  See also Constellations; Star clusters
Stickney, Angelina, 107
Stickney (Phobos crater), 107
Sublimation, 32
Suisei (spacecraft), 154
Summer solstice, 65
Summer Triangle, 179, 202
Sun, 64–85
  age of, 266
  appearance of, 79
  atmosphere of, 71, 73–76, 81
  and aurora, 71–73
  birth of, 76, 155
  blemishes on, 68
  brightness of, 79
  center of, 77
  as center of universe, 38
  changes in, 77, 79–80
  chemistry of, 66, 111
  chromosphere of, 73–75
  climate control of, 78–79
  color of, 67, 183, 184, 188
  and comets, 150–51, 153
  corona of, 74–76, 81, 84
  day on, 73
  diameter of, 75, 281, 282
  as dwarf star, 229
  Earth's passage around, 172
  eclipse of, 49, 50, 75, 80–85, 250

energy of, 77–78
explosions on surface of, 70
face of, 68
granules of, 68
gravitational field of, 250, 297
and gravity, 77
history of, 77
inside of, 78
interior of, 78–79
limb darkening of, 67
as magnet, 73
magnetic field of, 74, 75, 190
magnetic polarity of, 69
mass of, 197, 200, 208, 209, 225, 236, 237, 245–50, 263, 288
and naming planets, 139–40
and neutrinos, 233–34
paths across sky, 64–66, 167–68
prominences on, 74, 75
quadruple, 217–18
and radiation, 23–24, 78–79, 83
rising of, 50, 52, 64
rotation of, 73
setting of, 51, 64
shining of, 77–78, 182
size of, 66
stability of, 77, 79
as star, 64, 79, 128, 183, 184, 186, 187, 189–90, 204, 205, 209, 211, 216, 229, 232, 246, 271, 330
and sunspots, 68–70, 135–36, 190
surface of, 66–68, 70, 73–79, 81
temperature of, 3, 67, 74–79
and tides, 56, 57
ultraviolet rays of, 130
vibration of, 70
Superfluids, 32
Super Kamiokande solar neutrino detector, 28
Supernovae, 27, 241–44, 246–48
  See also Type I supernovae; Type II super-novae
Swift-Tuttle (comet), 157

T

3C 273 (quasar), 286
Tarantula Nebula, 242
tau Ceti (star), 333

Taurus (Bull) (constellation), 165, 173, 235, 268, 269

Tektites, 160

Telescopes
  buried underground, 27
  gamma-ray, 25
  infrared, 25, 259–60
  and Milky Way, 256–57
  and quasars, 28
  See also Hubble Space Telescope; Keck Telescope; Keck II Telescope; Radio telescopes; Schmidt-Cassegrain telescopes; Telescopes (optical); X-ray telescopes

Telescopes (optical)
  alternative, 5–6
  area of main lens or mirror, 7
  arrays, 17
  basic types of, 4
  and coordinate system, 12
  described, 3–4
  designs, 16
  diameter of main lens or mirror, 6–7
  eyepiece, for magnifying image, 7–9
  and focal length, 7–8
  and galaxy, 258–59
  "magical power" of, 6
  main lens or mirror of, 6–8, 10
  power, in types of, 7–10
  reflecting, 4–5, 15, 25
  refracting, 4–5, 9, 15
  resolving power of, 10
  size of, 15–16

Temperature scales
  Fahrenheit (F.) vs. Centigrade or Celsius (C.), 307
  Kelvin, 307–8

Terminator, on Moon, 50

Tethys (Saturn satellite), 122, 123, 126

Tharsis Bulge (Mars), 100, 102

Theory of Everything, 320, 323

Thuban (star), 168

Tides
  neap, 57
  spring, 57

Time
  forms of, 289–91
  historical approach, 316
  and space, 41, 174–75, 311

Titan (Saturn satellite), 127–28, 138

Titius, Johann, 144

Tombaugh, Clyde, 138

Total eclipses, types of, 80–82, 84–85

Trapezium (stars), 195, 196

Triton (Neptune satellite), 136–37, 138, 139

Trojan asteroids, 147

Tucana (Tucan) (constellation), 271

Tycho (lunar crater), 45

Type I supernovae, varieties of, 232, 234, 241, 301, 303, 304

Type II supernovae, 232–33, 234–35, 241

U

Ultraviolet rays (UV), 22–25, 226

Universe
  age of, 206, 308–9, 313–14, 323, 334
  Aristotle's view of, 36–38
  birth of, 305–7
  broadcasting to, 332
  chemistry of, 191
  closed, 311–13
  dark matter in, 310
  density of, 312, 314, 322
  early views of, 36–40
  evolution of, 315–23
  expansion of, 297–307, 311–14
  forces in, 317–18, 320
  and gamma rays, 252
  geocentric view of, 36
  heliocentric model of, 38
  laws of, 78, 166
  map of, 10–11
  marginally open, 311–13, 314
  mass of, 309–13, 311–13
  mysteries of, 252
  open, 311–13, 314
  oscillating, 311
  radiation (cosmic background) in, 307, 308, 316, 318–19, 321–22
  speed limit in, 331
  studying, 3, 14–15, 304–5
  supernovae as alchemists in, 246
  violence of, 252
  yardstick to, 213–14
  See also Big Bang; General Theory of Relativity; Special Theory of Relativity

Uranus (planet), 73, 110, 113, 121, 129–33, 134, 137, 138, 139, 140, 144, 258
  axis of, 130, 132

color of, 131
magnetic field of, 132
rings of, 131–32
satellites of, 131, 132–33
Ursa Major (Great Bear) (constellation), 165, 260
Ursa Minor (Little Bear) (constellation), 165

V

Valhalla (formation on Callisto), 118
Valles Marineris (Mariner Valley) (Mars), 100–101, 102
Vega (Alpha Lyrae) (star), 12, 168, 179, 184, 202, 217–18, 227, 266
Vega 1 (spacecraft), 154
Vega 2 (spacecraft), 154
Veil Nebula, 241
Vela (Sail) (constellation), 241, 242
Venus (planet), 39, 79, 88, 91–96, 100, 110, 128, 129, 139, 147, 182, 221, 232
   craters on, 94
   diameter of, 91
   vs. Earth, 91–95
   greenhouse effect on, 93, 95
   as "hell in space," 92–95
   and minieclipses, 95–96
   transits of, 96
Vernal equinox, 11, 53
Verona Rupes (Miranda ice cliff), 133
Vesta (asteroid), 145
Viking (spacecraft), 71, 100, 103–7, 256
Viking 1 (spacecraft), 102, 103
Viking 2 (spacecraft), 102, 103
Virgo (Virgin) (constellation), 166, 246, 284, 295, 296, 299, 309
Virgo Cluster, 295–96, 299, 302, 303
VLA (Very Large Array), 26–27, 263
VLBI (Very Long Baseline Interferometer), 27
Voyager (spacecraft), 73, 112, 115, 116, 119–20, 123, 125, 127, 139
Voyager 1 (spacecraft), 113, 123, 124, 335, 336
Voyager 2 (spacecraft), 113, 130, 131–32, 134, 135, 136, 137, 331, 335, 336
VV Cephei (variable), 221

W

War of the Worlds, The (Wells), 99
Wavelengths, 22
   of electromagnetic radiation, 24
Weak force, 317
Weaver, Harold, 250
Weber, Joseph, 28, 29
Welles, Orson, 99
Wells, H. G., 99
Willamette meteorite, 160
Wilson, Robert, 307, 308
WIMPs (Weakly Interacting Massive Particles), 310–11, 322
"Wind from the Sun, The" (Clarke), 76
Winter solstice, 65
Wolf-Rayet stars, 226
Wormholes, 291
W. Virginia stars, 213

X

X-bosons, 321
X rays, 22, 23, 24, 153, 252–53
   astronomy, 278
   and black holes, 249, 272–73
   bursters, 244
X-ray telescopes, 25

Y

Yerkes Observatory, 15
Yokhoh (satellite), 24

Z

Zeeman effect, 190
Zeta Lyrae (star), 218
Zodiac, 165–66
Zodiacal light, 161

Photographs in this book are courtesy of: UCO/Lick Observatory; National Optical Astronomy Observatories; Keck Observatories, University of California; National Aeronautics and Space Administration/Jet Propulsion Laboratory; European Southern Observatory; Yohkoh Soft X-Ray Telescope, Lockheed Missiles and Space Company, Inc.; National Radio Astronomy Observatory/Associated Universities, Inc.; Brookhaven National Laboratory; Institute for Cosmic Ray Research, University of Tokyo; University of Maryland; National Science Observatory; Anglo-Australian Observatory; Royal Observatory, Edinburgh, photos made from U.K. Schmidt plates by David Malin; University of Michigan; University of Virginia; Alan Dyer/Calgary Centennial Planetarium; Gart Westerhout; Palomar Observatory/Caltech; Lund Observatory; Max-Planck-Institut fur extraterrestrische Physik, Garching, Germany; Margaret Geller and John Huchra, Smithsonian Astrophysical Observatory; the Observatories of the Carnegie Institute of Washington; the Arecibo Observatory.

The Arecibo Observatory is part of the National Astronomy and Ionosphere Center, which is operated by Cornell University under a cooperative agreement with the National Science Foundation.